SPECIAL VOLUME IN MEMORY OF ILYA PRIGOGINE

ADVANCES IN CHEMICAL PHYSICS

VOLUME 135

THE WILEY BICENTENNIAL–KNOWLEDGE FOR GENERATIONS

$\mathcal{E}$ach generation has its unique needs and aspirations. When Charles Wiley first opened his small printing shop in lower Manhattan in 1807, it was a generation of boundless potential searching for an identity. And we were there, helping to define a new American literary tradition. Over half a century later, in the midst of the Second Industrial Revolution, it was a generation focused on building the future. Once again, we were there, supplying the critical scientific, technical, and engineering knowledge that helped frame the world. Throughout the 20th Century, and into the new millennium, nations began to reach out beyond their own borders and a new international community was born. Wiley was there, expanding its operations around the world to enable a global exchange of ideas, opinions, and know-how.

For 200 years, Wiley has been an integral part of each generation's journey, enabling the flow of information and understanding necessary to meet their needs and fulfill their aspirations. Today, bold new technologies are changing the way we live and learn. Wiley will be there, providing you the must-have knowledge you need to imagine new worlds, new possibilities, and new opportunities.

Generations come and go, but you can always count on Wiley to provide you the knowledge you need, when and where you need it!

WILLIAM J. PESCE
PRESIDENT AND CHIEF EXECUTIVE OFFICER

PETER BOOTH WILEY
CHAIRMAN OF THE BOARD

SPECIAL VOLUME IN MEMORY OF ILYA PRIGOGINE

ADVANCES IN CHEMICAL PHYSICS
VOLUME 135

Edited by

STUART A. RICE

Department of Chemistry
and
The James Franck Institute
The University of Chicago
Chicago, Illinois

AN INTERSCIENCE PUBLICATION
JOHN WILEY & SONS, INC.

For general information on our other products and services or for technical support, please contact
our Customer Care Department within the United States at (800) 762-2974, outside the United States
at (317) 572-3993 or fax (317) 572-4002.

Wiley also publishes its books in a variety of electronic formats. Some content that appears in print
may not be available in electronic formats. For more information about Wiley products, visit our
web site at www.wiley.com.

Wiley Bicentennial Logo: Richard J. Pacifico

Library of Congress Catalog Number: 58-9935

ISBN 978-0-471-68233-2

Printed in the United States of America

10   9   8   7   6   5   4   3   2   1

CONTRIBUTORS TO VOLUME 135

A. ARNEODO, Laboratoire Joliot-Curie et Laboratoire de Physiques, Ecole Normale Supérieure de Lyon, 69364 Lyon Cedex 07, France

B. AUDIT, Laboratoire Joliet-Curie et Laboratoire de Physique, CNRS, Ecole Normale Supérieure de Lyon, 69364 Lyon Cedex 07, France

RADU BALESCU*, Association Euratom—Belgian State for Fusion, Université Libre de Bruxelles, B-1050 Brussels, Belgium

E. B. BRODIE OF BRODIE, Laboratoire Joliet-Curie et Laboratoire de Physique, CNRS, Ecole Normale Supérieure de Lyon, 6934 Lyon Cedex 07, France

Y. D'AUBENTON-CARAFA, Centre de Génétique Moléculaire, CNRS, 91198 Gif-sur-Yvette, France

PIERRE GASPARD, Center for Nonlinear Phenomena and Complex Systems, Université Libre de Bruxelles, B-1050 Brussels, Belgium

ALBERT GOLDBETER, Service de Chimie, Physique, et Biologie Théorique, Faculté des Sciences, Université Libre de Bruxelles, B-1050 Bruxelles, Belgium

S. NICOLAY, Laboratoire Joliet-Curie et Laboratoire de Physique, CNRS, Ecole Normale Supérieure de Lyon, 6934 Lyon Cedex 07, France

GONZOLO ORDONEZ, Center for Studies in Statistical Mechanics and Complex Systems, The University of Texas, Austin, Texas 78712, USA; and International Solvay Institutes for Physics and Chemistry, CP231, B-1050 Brussels, Belgium. *Present Address*: Butler University, Indianapolis, Indiana 46208, USA

TOMIO PETROSKY, Center for Studies in Statistical Mechanics and Complex Systems, The University of Texas, Austin, Texas 78712 USA; and International Solvay Institutes for Physics and Chemistry, CP231, B-1050 Brussels, Belgium

YVES POMEAU, Laboratoire de Physique Statistique de L'Ecole Normale Supérieure, 75231 Cedex 05, France. *Present Address*: Department of Mathematics, The University of Arizona, Tucson, Arizona 85721, USA

ILYA PRIGOGINE[†], Center for Studies in Statistical Mechanics and Complex Systems, The University of Texas, Austin, Texas 78712, USA; and

*Deceased.
[†]Deceased.

International solvay Institutes for Physics and Chemistry, CP231, B-1050 Brussels, Belgium

JOHN ROSS, Stanford University, Department of Chemistry, Stanford, California 94305, USA

P. ST. JEAN, Laboratoire Joliet-Curie et Laboratoire de Physique, CNRS, Ecole Normale Supérieure de Lyon, 6934 Lyon Cedex 07, France

C. THERMES, Centre de Génétique Moleculaire, CNRS, 91198 Gif-sur-Yvette, France

M. TOUCHON, Centre de Génétique Moleculaire, CNRS, 91198 Gif-sur-Yvette, France

C. VAILLANT, Laboratoire Statistique et Génome, 91000 Evry, France

CHRISTIAN VAN DEN BROECK, Hasselt University, Dept. WN1, B-3590 Diepenbeek, Belgium

MARCEL VLAD, Department of Chemistry, Stanford University, Stanford, California 94305, USA; and Institute of Mathematical Statistics and Applied Mathematics, Casa Academiei Romane, 76100

INTRODUCTION

Few of us can any longer keep up with the flood of scientific literature, even in specialized subfields. Any attempt to do more and be broadly educated with respect to a large domain of science has the appearance of tilting at windmills. Yet the synthesis of ideas drawn from different subjects into new, powerful, general concepts is as valuable as ever, and the desire to remain educated persists in all scientists. This series, *Advances in Chemical Physics*, is devoted to helping the reader obtain general information about a wide variety of topics in chemical physics, a field that we interpret very broadly. Our intent is to have experts present comprehensive analyses of subjects of interest and to encourage the expression of individual points of view. We hope that this approach to the presentation of an overview of a subject will both stimulate new research and serve as a personalized learning text for beginners in a field.

STUART A. RICE

ACKNOWLEDGMENTS

The Symposium *In Memoriam Ilya Prigogine* on"Time, Irreversibility and Self-Organization" was held by the International Solvay Institutes for Physics and Chemistry at the University of Brussels on December 2–3, 2004.

This Symposium was made possible by the sponsorships of the European Commission, of the Fonds National de la Recherche Scientifique (FNRS), of the Nationale Fonds voor Wetenschappelijk Onderzoek (NFWO), of the Belgian National Lottery, of the "Communauté Franaise de Belgique," and of the "Hotel Métropole," which are gratefully acknowledged.

PROF. RENE LEFEVER

Université Libre de Bruxelles Belgium

CONTENTS

ILYA PRIGOGINE: HIS LIFE, HIS WORK[*]

RADU BALESCU[†]

*Association Euratom—Belgian State for Fusion,
Université Libre de Bruxelles, B-1050 Brussels, Belgium*

CONTENTS

[*]This chapter was originally published in the French language as "*Ilya Prigogine, sa vie, son oeuvre*, Mémoire in-8°, Classe des Sciences, Académie Royale de Belgique, 2006."
[†]Deceased, June 1, 2006

Special Volume in Memory of Ilya Prigogine: Advances in Chemical Physics, Volume 135,
edited by Stuart A. Rice

I. ILYA PRIGOGINE'S LIFE AND WORK

A. Introduction

In the history of science, there are few examples of such a flashing and immense ascent as that of Ilya Prigogine (Fig. 1). The little Russian Jewish immigrant arriving in Brussels at the age of 12 would end his life at age 87 with all the honors anyone—what is more, an intellectual—could dream of earning! Of course, the Nobel Prize in chemistry opened all the doors for him: He was able to use this opportunity for promoting a new vision of science. This extraordinary success is due, in the first place, to the importance of his works, but also to their novelty, to the introduction of the physicist to biology and the humanities, to his willingness to encourage dialogue, and to his leadership qualities in several international teams of researchers.

Figure 1.   Photograph of Ilya Prigogine.

I had the privilege of being one of his oldest disciples and, subsequently, a researcher who remained close to Ilya Prigogine. In the present chapter, I wish first to relate the main events of his life (Section I.B). This presentation is based on my own actual experience and on my perception of Prigogine's personality. One will find another source (sometimes a little different in the interpretations!) in his autobiography, available on the internet:

www.nobel.se/chemistry/laureates/1977/prigogine-autobio.html

The second part of this chapter (Sections I.C–I.F) is devoted to an analysis of his work. Let me say immediately that, within the reasonable limits of this chapter, I could not cover the totality of his very diverse oeuvre. I thus made a selection, somewhat arbitrary, of course, but I tried to discuss its most characteristic aspects. I thus gave up the analysis of his works in quantum statistical mechanics (which are essentially a transposition and generalization of the central ideas of the classical theory), his work on vehicle traffic, his contributions to European science policy, and a few isolated papers. I do not present his works in chronological order, because of the strong entanglement in time of Prigogine's activities. Most often he was working on several problems at the same time. During some periods, one of the subjects would dominate, then another came to the surface, after which his interest would focus again on a problem that was shelved. I shall therefore divide my exposition into four groups of subjects, respectively, macroscopic physics and chemistry (Section I.C), microscopic physics (Section I.D), cosmology (Section I.E) and the philosophical aspects (Section I.F). The last section is an appendix, where I develop some more technical comments intended for the readers having a certain mathematical background. Finally, in Section II, I establish a systematic list of his numerous publications, classified according to their subjects.

B. Short Biography

Ilya Romanovich Prigogine was born on January 25, 1917 in Moscow. His father was a chemical engineer who owned a little soap factory, whose success was modest. His mother, Julia Wichman, had studied the piano at the conservatory. Young Ilya would inherit her love of music. It is said in his autobiography that he was able to read a music score before knowing how to read a book!

Ilya had an older brother, Alexander, who would become a chemist and make a career in the mining industry in the former Belgian Congo. He had a hobby that he developed very seriously: ornithology. He even discovered a novel species of bird, to which his name was attributed. After his return to Belgium, he was elected member of the "Académie Royale des Sciences d'Outremer" (the Academy of African sciences).

The reader surely noted that Ilya Prigogine was born just a few months before the Russian revolution. Of course, the dramatic events of the time influenced unfavorably the business of his father. The family decided to leave Russia in 1921, spending a year in Lithuania, then going to Berlin, where young Ilya received a very good German education, which made him familiar with the classics of that nation. Ilya's father did not succeed in starting a new business in the disastrous economic circumstances of Germany of that time. Moreover, the threat of Nazism was visible on the horizon. The Prigogine family emigrated again and arrived in Belgium in 1929, where they settled for good. Young Ilya registered at the Athénée Royal d'Ixelles, an excellent Brussels high school. He had not yet made a final choice of a future career. On one hand, he was dreaming of becoming a professional pianist, but his teacher made him understand that he was not destined for that career.[1] Ilya also had a passion for history and archaeology (he realized later his dream by acquiring progressively an extraordinary collection of very rare pre-columbian objects). But finally, following his parents' and his brother's advice, in 1935 he started, in parallel, his studies of chemistry and of physics at the Université Libre de Bruxelles (ULB). He did not leave his Alma Mater until his death (even though he later committed some "infidelity" by accepting a part-time chair at the University of Texas at Austin).

During his studentship he continued to read the works of his favorite philosophers. The one who left a deep imprint on him and whom he would incessantly quote up to the end of his life was Henri Bergson.

The young student, aged 20, published in the *Cahiers du Libre Examen* (a local student journal) two papers: "*Essay on physical philosophy*" and "*The problem of determinism*," followed by a third one, in collaboration with Hélène Bolle (who would become his first wife), "*The evolution*." Remarkably, the roots of his future interests were already present in these works of his youth: determinism, the interpretation of quantum mechanics, biological evolution, and, above all, the concept of time.

Prigogine (who at that time signed his name as "Prigoshin," a transliteration from Russian, doubtless inherited from his sojourn in Germany) then makes a fundamental decision, which will "put him in orbit": he chooses as director of his Master's thesis, and later his Ph.D. thesis, Professor Théophile De Donder.

The latter was a rather extraordinary person. Born in 1872, he started his career as an elementary school teacher, finally obtaining, as a self-taught man, the title of Doctor in physical and mathematical sciences in 1899. Very soon he became interested in relativity and started in 1916 a long correspondence with Albert Einstein. He also had contacts with many other great scientists of that

---

[1]When the king of Belgium organized a reception at the Royal Palace on the occasion of the Nobel Prize, Prigogine asked the king to invite his old piano teacher: He performed some Chopin pieces. It was a very moving moment.

time: Henri Poincaré, Henri Lebesgue, Arthur Eddington, and more. As soon as he was appointed professor at the ULB, he started forming a remarkable generation of disciples, who would work at the ULB as well as abroad. Very soon he introduced the university to the most modern trends of physics of the time: relativity, quantum mechanics, statistical mechanics, thermodynamics. In the latter field, applied to chemistry, De Donder made an important advance by introducing the precise concept of affinity, which allowed him to calculate explicitly the entropy production in a chemical reaction. The door was half-opened toward nonequilibrium: Prigogine would open it widely.

Appointed as an assistant to De Donder in 1940, Prigogine had to give up this job when the University closed its doors in 1941, as a resistance to German occupation. The period of war was, of course, very hard. He was able to escape from persecution in the beginning, due to the organization of his fellow Russians, who provided him with documents certifying "officially" that he was a White Russian and was baptized. An unfortunate circumstance occurred in 1943: He was living with his companion Hélène Boll in an appartment previously occupied (without his knowledge) by a group of resistants, and thus he was in the focus of the Gestapo. The couple was arrested. During that same evening a dangerous expedition was organized by his friend Victor Mathot in order to recover in the apartment the manuscript of his future treatise on thermodynamics. Fortunately, the imprisonment was not very long: Due to numerous interventions, including one of Queen Elizabeth of Belgium in person, the couple was set free after a few weeks.

The war period was not one of scientific inactivity. The publication record of Ilya Prigogine contains 13 papers on thermodynamics published between 1940 and 1944 in the Bulletin of the Royal Academy of Belgium, in the Bulletin of the Chemical Society of Belgium, and in the Journal de Physique et le Radium (France). One learns from the acknowledgments of these papers that the young researcher was subsidized by the Solvay Institutes.

But Prigogine devoted these years mainly to the elaboration and the systematization of the ideas of his master De Donder. The result was his first *"magnus opus"* written in collaboration with Raymond Defay: *Treatise of Thermodynamics, in Conformity with the Methods of Gibbs and of De Donder*, whose first two volumes appeared in 1944 and 1946 (LS.3).

1944: ULB reopens its doors, and new opportunities are present. On september 13, De Donder writes a letter (whose original is preserved) to Prigogine. I am translating its second part[2]:

"3°) Chair of math. Chemistry: **applied** thermodynamics, **applied** stat. mech., **applied** wave mech., **applied energetics**.

---

[2]The words in italics are underlined once, the words in boldface are underlined twice, the words in boldface capitals are underlined three times in the original handwriting.

In order to be appointed to this chair, it is *indispensable* to possess the title of **agrégé** *de l'Ens. sup.*[3] You must thus start immediately on the job. I believe that within *two months* you could, with the results already obtained (see **BOOK**, t. I, II, and III) and with the *new notes, in preparation, write a thesis of great originality and rich in important results.*

4°) Your memoir on the liquids would become your **ANNEX thesis**.

Your faithful, (s) Th. De Donder."

Ilya started working hard and obtained the title of *agrégé* in 1945. In 1947 he was appointed *Chargé de Cours* ($\approx$ assistant professor), in 1950 *Professeur extraordinaire* ($\approx$ associate professor), and in 1951 *Professeur ordinaire* ($\approx$ full professor) at the Université Libre de Bruxelles, where he succeeded Jean Timmermans and was in charge of the course of Theoretical Physical Chemistry for the students in Chemistry.

In 1953 he was elected as Corresponding Member of the Royal Belgian Academy and becomes a full Member in 1960. He was at that time the youngest member of that Institution, where he developed an intense activity.

In 1959 Prigogine was appointed Director of the *International Institutes of Physics and Chemistry, founded by E. Solvay*. The main mission of these institutes, created in 1911 by the famous industrialist was the organization of the *Solvay Councils*, which gathered the greatest scientists of the time for discussions about the new major problems of science. Some of these Councils had a considerable historical importance. Thus, the Councils of 1911 and of 1927 were crucial moments in the birth of quantum mechanics; the latter witnessed the famous confrontation of Niels Bohr and Albert Einstein regarding the interpretation of the new mechanics. Prigogine gave new momentum to these institutes. While continuing the organization of the Solvay Councils, he widened the activities of the institutes by transforming them into high-level research institutes. Prigogine remained in this position until the end of his life.

Prigogine maintained strong international relationships with his foreign—in particular, American—scientists. Thus, he sojourned several times in the 1960s to the *University of Chicago* as a Visiting Professor, and he established long-living links with his colleagues. In 1967 he was appointed Professor at the *University of Texas at Austin*. His chair was later transformed into the "*Ilya Prigogine Center for Studies in Statistical Mechanics and Complex Systems*," to which he was appointed as the Director. Every year he would spend a significant fraction of his time in Austin. He would form there a team of scientists and teachers who would accompany him until the end of his life.

The supreme crowning of this intense activity was, in 1977, the award of the Nobel Prize in Chemistry to Ilya Prigogine, for his works in thermodynamics,

---

[3]A post-doctoral title required in Belgium for an appointment as a University professor.

leading to the discovery of dissipative structures. As will be shown below, the existence of dissipative structures and their formation through bifurcations, made possible by the nonlinearity of the evolution laws, leads to the concept of creation of order by amplification of fluctuations. This concept of self-organization is central in biology, but also in sociology, economics, geography, and so on. The Swedish academicians were rightly inspired in awarding the Nobel Prize for this discovery, which plays an important role in chemistry and in physics, but also opens so many other doors. The dialogue, so much hoped for by Prigogine, between Science and Humanities clearly became possible in this framework. He effectively stimulated this dialogue through his interventions in the debates, colloquia, and other intercultural symposia which he organized.

From this moment on, Prigogine became a public person. He would make good use of this sudden popularity. He created a great tribune for the promotion of science in general, and of his own scientific and philosophical ideas in particular. Up to the end of his life, the signs of international recognition were accumulating. He was granted numerous international prizes rewarding his activity, both in science and in philosophy; he was awarded 54 honorary doctorates and was an elected member of numerous academies. He developed an intense activity as special counsellor at the European Union: Through his reports he contributed significantly toward a new orientation and momentum to the scientific policy of Europe. He also contributed to the opening toward the Eastern countries, particularly toward Russia. Institutes and high-level schools bearing his name have been created in Brussels, Austin, Moscow, Italy, and Argentina. He became a tireless traveler, transmitting his message around the world. One may also note during this period the numerous interviews given to the nonspecialized press and to television, both in Belgium and abroad.

I cannot finish this presentation without underlining his activity as a teacher. Up to his retirement in 1987, he influenced a large number of students at the ULB, mainly in chemistry. In the beginning he taught thermodynamics and the beginnings of quantum and statistical mechanics in third-year chemistry, and he delivered a specialized course on solutions in the fourth year. Later, his teaching acquired a much broader extension, covering "Theoretical Chemistry" over the whole undergraduate program. He was sharing his teaching duties with a strong team of associate professors (V. Mathot, F. Henin, C. George, G. Nicolis, R. Lefever, A. Goldbeter). In the physics curriculum, his course of theoretical physics was shared with P. Résibois and R. Balescu. The team was completed with a group of assistants who took care of the practical exercises. The style of *ex cathedra* teaching of Prigogine was quite singular. He loathed entering into details of calculations or into minutely detailed demonstrations ("you will see this at the exercises!"), thus transferring the burden to his assistants. On the other hand, he was great in providing an admirable overview of the subjects he treated. This would often lead him toward unexpected associations with music, philosophy, history, or neolithic art. This characteristic gave him a unique charisma, to which all his students responded.

But his teaching was not limited to his *ex cathedra* courses. His graduate students, researchers, and visitors had the privilege to participate in the discussions taking place in his office. Actually, Prigogine did not like working alone; he felt a strong need to share his ideas with his colleagues. (This explains why most of his scientific papers are published in collaboration with one or several authors.) He would then explain at length his latest ideas (again, without many mathematical details) and stimulate the reaction of his audience.

I would now like to evoke some more personal aspects of Prigogine's life. His first marriage with the poet Hélène Bolle (with whom he had a son, Yves) ended with a separation. After several years, during a visit in Warsaw in 1961, he met *Maryna Prokopowycz*. She was a chemical engineer and worked at the Warsaw University; it was love at the first sight.[4] Soon Ilya and Maryna married in Poland. But many months of strained waiting and numerous high-level interventions were necessary before the Polish authorities of that time authorized Maryna to join her husband. They became a very happy couple and had a son, Pascal. In the difficult moments, and also in the happy ones, Ilya could always count on the moral support—and especially on the love—his marvelous wife brought to him and which he brought to her.

All those who had the chance of knowing him remember in Ilya Prigogine a man of great generosity. He strongly supported and helped to form the careers of numerous researchers and teachers, not only in Brussels, but throughout the world: One finds disciples and admirers in Western and Eastern Europe, in Russia, in the United States of America, in Latin America, in Japan, in China, in India, and so on.

At the University his attitude was rather formal; for instance, he did not like to call his co-workers (even the old ones) by their first name (whereas he did this, necessarily, in America!). But privately, he would let down the barriers: I witnessed roars of laughter when he was feeling well. I also witnessed periods of anxiety, related to the lack of acceptance of his ideas (he mentions this also in his autobiography). But they were quickly overcome by his unfailing enthusiasm. He would sometimes tell me: "*Ah, Balescu, now we solved all the problems!*" And two years later: "*Ah, Balescu, now we **really** solved all the problems!*," and two years later, ...

---

[4]In the following months we, the young co-workers in Brussels, started to consider his repeated travels to Warsaw rather "unusual," as were also the repeated invitations to Brussels of a Polish professor who delivered several seminars on the "powder electrodes," a subject of no interest to anyone here. His last seminar finished with an apotheose: He told us about the "powder electrodes without powder!" It was only later that we understood that this professor was Maryna's boss, and moreover that he would not let her leave Warsaw.

The last 10 years of his life were very painful. He was progressively sapped by an illness that made him suffer and that handicapped him physically. But up to the last moment he kept his mental readiness and continued following the work of his co-workers and suggesting new ideas.

He died in Brussels on May 28, 2003.

C. Macroscopic Chemistry and Physics

In Section I.A I mentioned the first book by Prigogine and Defay: *Traité de thermodynamique conformément aux méthodes de Gibbs et de De Donder* (LS.3).[5] At that time it was a quite original presentation of equilibrium thermodynamics, addressed mainly to chemists. The most striking feature is the absence of any reference to heat engines (Carnot cycles, etc.), which forms the starting point of all classical textbooks. The authors start from a concise, but simple and well-illustrated, exposition of the two principles. After that, there appears the basic notion of thermodynamic potential, from which the whole "treasure" of equilibrium thermodynamics follows logically. It is interesting to note (in order to understand the forthcoming subject) that practically the whole second half of the book is devoted to the properties of mixtures and solutions.

This treatise served as a basis for many teachers and was translated into numerous languages. Toward the end of his life, Prigogine felt the necessity of updating this treatise. He put his former co-worker Dilip Kondepudi in charge of this work, which he closely supervised. The result is a treaty: *Modern Thermodynamics, from Heat Engines to Dissipative Structures* (LS.15) (1998), which incorporates, in an attractive pedagogical form, all the progress achieved during the 54 years that separate it from the first version.

During the period 1945–1960, Prigogine worked on an intensive research program on **Mixtures and Solutions**. It can be framed into what can be called "Classical physical chemistry." It is clearly inspired by the professor he succeeded at the ULB and to whom many references are made: Jean Timmermans, a remarkable experimental physico-chemist. The results of these research efforts were published in a monograph written by Ilya Prigogine, Victor Mathot, and André Bellemans: *The Molecular Theory of Solutions* (LS.7), published in 1957; today this is still considered to be an important reference.

After this publication, Prigogine suddenly quit this research line (this was, however, continued by a group of his co-workers). One may wonder *a posteriori* what motivated the choice of this field of research, so singular in Prigogine's work. (Note, however, that his Master's thesis (1939) was already devoted to solutions of strong electrolytes, and, as noted above, half of his treatise on thermodynamics treats the same subject.) The "problem of time" that would

---

[5]These symbols refer to the list of publications given at the end of this paper.

become Prigogine's obsession later is notoriously absent here. Was he, maybe, trying to get closer to the experimentalists? Whatever the answer, in a recent interview Prigogine declared about this period that "he did not regret anything!"

During the period 1945–1954 Prigogine continues to develop the project closest to his heart: **nonequilibrium thermodynamics**. His "thèse d'agrégation" (mentioned in Section I.B), *Etude Thermodynamique des Phénomènes Irréversibles* (1945) (LS.4), was the first book devoted exclusively to this subject. Whereas De Donder's works were devoted solely to the chemical reactions, Prigogine extended the formalism to all irreversible macroscopic processes, including transport phenomena in hydrodynamics and electromagnetism (diffusion, viscosity, thermal conduction, electrical conduction, and cross-effects, such as thermodiffusion). He derived the general expression (today, a classic!) of the entropy production, appearing as a bilinear form:

$$P = \sum_i J_i X_i \equiv \mathbf{J} \cdot \mathbf{X} \tag{1}$$

where J_i denotes the set of dissipative *fluxes* (e.g., matter fluxes, heat flux, chemical reaction rates), and X_i denotes the corresponding thermodynamic *forces* (e.g., density or concentration gradients, temperature gradient, chemical affinities). [The set of fluxes (J_i) and of forces (X_i) can be grouped into "*vectors*" $\mathbf{J}$ and $\mathbf{X}$, which lead to more compact formulae.] This formula is very general, being valid in the whole domain where the macroscopic equations of evolution are valid. The fluxes and the forces are interrelated by phenomenological transport equations. During this first period, Prigogine limited himself to the simplest case, where these equations are *linear*. He then derived his celebrated theorem of *minimum entropy production in the nonequilibrium stationary states*, which is valid precisely in this linear domain.[6] The latter is even applied in biology, in a paper by Prigogine and Wiame (THL.8), where the authors conjecture that the living systems (necessarily open systems, exchanging matter and energy—and entropy!—with the external world) evolve toward a state of minimum entropy production. This evolution goes together with a global decrease of entropy, thus toward a complexification, creating structures. This paper contains the first attempt of application of nonequilibrium thermodynamics to biology.

The "thèse d'agrégation" ends with a very brief chapter "*Time and entropy*," which contains the root of Prigogine's future preoccupations. He defines a "*thermodynamic time*" related to the entropy production. It is interesting to point out one of the last conclusions of this chapter: "*Originating from the second principle, the thermodynamic time necessarily appears as a statistical concept. It loses its meaning at the scale of elementary processes.*" This

---

[6]One may note that Prigogine inherited from his master, De Donder, his love of variational principles!

conclusion will be vigorously repudiated 25 years later (when Prigogine will insist on the universality of the irreversible time, on all scales; see Appendix).

In the following years, Prigogine developed various additional aspects of the new thermodynamics. He published in 1955 the little treatise "*Introduction to Thermodynamics of Irreversible Processes*" (LS.6), which was very successful and was translated into many languages.

The year 1954 is a landmark in Prigogine's research in thermodynamics: for the first time he ventures to break down the "barrier of linearity." As in all sciences, the simplest problems occur when one studies the phenomena that happen in the neighborhood of a known reference state. The tools necessary for this study have been handed down to us by the mathematicians of the nineteenth century: linear analysis, a complete, simple and elegant formalism, offering the solution of all problems in this realm. Unfortunately, when applied to physical problems, one must take into account that its validity range is very limited.

Beyond, one enters a "*terra incognita*," where all surprises are possible. Prigogine entered it resolutely, accompanied in the beginning by his old friend Paul Glansdorff. The latter, also a disciple and admirer of De Donder and a few years older than Ilya, was a man with a very warm character; he was also extremely refined and was erudite in French history of science. He developed first a career as an engineer, exploiting his mastery of thermodynamics (applied to the industry of refrigeration); he was at that time a professor at the Polytechnic Faculty in Mons, Belgium. Prigogine obtained his appointment at the ULB; from that time on, a very fruitful collaboration started between the two men. The problem that was open can be formulated as follows: *How can one describe and study the irreversible phenomena occurring in a system in which the relations between fluxes and forces are no longer linear, or even no longer exist (as univocal functions)?* Glansdorff and Prigogine tackled the problem using the variational methods, dearest to the heart of their former master De Donder.

In an important paper (TNC.1), they offered for the first time an extension of nonequilibrium thermodynamics to nonlinear transport laws. As could be expected, the situation was by no means as simple as in the linear domain. The authors were hoping to find a variational principle generalizing the principle of minimum entropy production. It soon became obvious that such a principle cannot exist in the nonlinear domain. They succeeded, however, to derive a "half-principle!" They decomposed the differential of the entropy production (1) as follows:

$$dP \equiv d_J P + d_X P = \mathbf{X} \cdot d\mathbf{J} + \mathbf{J} \cdot d\mathbf{X}$$

and then proved that

$$d_X P \leq 0 \qquad\qquad (2)$$

This principle is very general, relating neither to the linearity nor to the symmetry of the transport laws. On the other hand, it is difficult to attribute a physical meaning to $d_X P$. The authors later attempted to derive a "local potential" from this property, and they applied this concept to the study of the chemical and hydrodynamical stability (e.g., the Bénard convection). The results of this approach were published in Glansdorff and Prigogine's book: *Thermodynamic Theory of Structure, Stability and Fluctuations* (LS.10, 10a), published in 1971.

In two papers of 1954 and 1956 (TNC.3, 5) Prigogine and Balescu showed that the property (2) opened the possibility of existence, far from equilibrium, of *oscillating chemical reactions*. It was a statement that was highly "politically incorrect" at that time. In a subsequent paper (1959), Thor Bak concluded: "*It is pointed out that none of the chemical reactions alleged to show oscillatory behavior have been thoroughly investigated experimentally!*" But in fact, in 1958 in an obscure Siberian journal, there appeared a paper by Belousov where the author announces the discovery of a true chemical clock; this work was taken over in 1964 by Zhabotinsky, and the news became known in the West. It was a marvelously simple reaction: One put together in a test tube some appropriate reactants, and one witnessed a change of color of the liquid, turning from red to blue and back within a period of a few minutes, thus easily observable and without any sophisticated equipment. The theoretical predictions of Prigogine and his co-workers were thus admirably confirmed by experiment! Later, many oscillating chemical and biochemical reactions were discovered and studied.

The year 1967 appears as a crucial year: In an important paper by Prigogine and Nicolis, "*On symmetry-breaking instabilities in dissipative systems*" (TNC.16), there appears for the first time the term "dissipative structures." The filiation of this concept with the "half-principle" of Glansdorff and Prigogine can be clearly perceived in the works of that period (particularly in the paper TNC.17). However, the new approach required a radical change of the theoretical methods.

In their subsequent works, the authors treated directly the nonlinear equations of evolution (e.g., the equations of chemical kinetics). Even though these equations cannot be solved explicitly, some powerful mathematical methods can be used to determine the *nature* of their solutions (rather than their analytical form). In these equations, one can generally identify a certain parameter κ, which measures the strength of the external constraints that prevent the system from reaching thermodynamic equilibrium. The system then tends to a nonequilibrium stationary state. Near equilibrium, the latter state is unique and close to the former; its characteristics, plotted against κ, lie on a continuous curve (the thermodynamic branch). It may happen, however, that on increasing κ, one reaches a critical **bifurcation** value κ_c, beyond which the appearance of the curve suddenly changes. For $\kappa > \kappa_c$ a branching of trajectories occurs, and mutiple stationary states appear, some of them stable, others unstable. The system then has the possibility of *choice* of proceeding along one or the other of these curves. Another possibility (Hopf bifurcation) is the appearance, beyond κ_c, of a *limit cycle*. One then witnesses an oscillating behavior (like the one produced in the Belousov–Zhabotinsky reaction). In all these cases a **temporal symmetry breaking** occurs: The character of the evolution is radically different from the one present in the neighborhood of equilibrium. Last but not least, when inequalities of concentration exist in various parts of space, diffusion enters the game. In that case, the bifurcation may lead to stationary states that are spatially structured (for instance, spatially periodic): Here the **symmetry breaking is spatial** as well as temporal. One finds in all these cases the appearance of states whose properties are totally different from those of equilibrium, and which can only live at finite distance from equilibrium. Prigogine and Nicolis call these states **dissipative structures**. The latter must necessarily "feed" on fluxes of matter and/or energy (thus, on external constraints) that permanently maintain the system far from equilibrium. They can therefore only exist in open systems. All these "bizarre" phenomena (and many others) are consequences of the *nonlinearity of the evolution laws*, and possibly of the competition between nonlinearity and (linear) spatial diffusion. When one goes even farther from equilibrium, new secondary, tertiary,..., and so on, bifurcations may occur, leading to new structures and possibly to a transition to chaos.

In parallel with the studies described above, which concern perfectly deterministic equations of evolution, it appeared necessary to complete the theory by studying the spontaneous **fluctuations**. Near equilibrium, any deviation is rapidly damped; but near a bifurcation point, a fluctuation may may lead the system "across the barrier." The fluctuation is then stabilized, or even amplified: this is the origin of the phenomenon which Prigogine liked calling "*creation of order through fluctuations.*" More specifically, one witnesses in this way a step toward *self-organization.*

Once the door was opened to these new perspectives, the works multiplied rapidly. In 1968 an important paper by Prigogine and René Lefever was published: "*On symmetry-breaking instabilities in dissipative systems*" (TNC.19). Clearly, not any nolinear mechanism can produce the phenomena described above. In the case of chemical reactions, it can be shown that an *autocatalytic step* must be present in the reaction scheme in order to produce the necessary instability. Prigogine and Lefever invented a very simple model of reactions which contains all the necessary ingerdients for a detailed study of the bifurcations. This model, later called the "*Brusselator,*" provided the basis of many subsequent studies.

In 1969 a paper by I. Prigogine, R. Lefever, A. Goldbveter, and M. Hershkowitz-Kaufman was published: "*Symmetry-breaking instabilities in biological systems"* (TNC.21) started a new direction of research that later proved very fruitful. In living systems one finds many oscillating chemical reactions that determine the fundamental rhythms of the organisms. The spatial symmetry-breakings might explain the formation of biological structures. In fact, the first forerunning work on dissipative structures is due to the great British mathematician Alan Turing, who, in 1952, established a model of morphogenesis involving such a symmetry breaking. The same spatial symmetry-breaking may (perhaps?) provide a possible explanation to the origin of life.

The subsequent 10 years were filled with an intense activity developing the new ideas and seeking new applications in progressively wider fields, including problems of economy, sociology, and geography. (Let us quote, for example, the very original works of Prigogine, Peter Allen, Françoise Boon, and Michèle Sanglier in the problem of urban development). These results were collected and synthesized in a remarkable book by Nicolis and Prigogine: *Self-Organization in Non-equilibrium Systems* (1977, LS.12); this matter was completed and updated in 1989 in *Exploring Complexity* (L.S.14) by the same authors. All these works were rapidly recognized and further developed by the international community of physico-chemists.

D. Microscopic Physics

In parallel with his work in thermodynamics, as soon as 1950, Prigogine took on the problem of the *microscopic foundation of irreversible phenomena.* The latter path, full of pitfalls, involving long fruitful periods, interrupted by periods of stagnation, or even of reversals, would preoccupy him for the next 50 years, until the end of his life (200 papers and several books). A communication to the Belgian Society of Logic and Philosophy of Science, delivered in 1951 under the title *"Probabilities and Irreversibility"* (GEN.5), is particularly illuminating. After an exposition of the paradox of irreversibility (irreversibility of macroscopic phenomena, but reversibility of the microscopic dynamical laws), Prigogine criticized the approaches of Ehrenfest and of Kirkwood, based on an operation of "smoothing" (*coarse graining*) in phase space. He concluded by establishing a program that would be his own for the remainder of his life:

"(One asks) three questions:

(1) What mechanism explains the independence of the final distribution with respect to the initial distribution . . . ?

(2) What is the relaxation time of the distribution, i.e., the time necessary for the establishment of the latter independence?

(3) How do the externally imposed constraints (temperature gradient, bulk velocity,. . .) modify the asymptotic distribution?

We may only raise these questions. The development of the main mathematical tools allowing their study (theories of Markoff chains and of linear operators in Hilbert space) may, however, allow us to hope for progress in the coming years."

A posteriori, one may insert "50" before "years"!

The first attempts (G. Klein and I. Prigogine, 1953, MSN.5,6,7) were very timid and not very conclusive. They were devoted to a chain of harmonic oscillators. In spite of a tendency to homogenization of the phases, there was no intrinsic irreversibility here, because an essential ingredient is lacking in this model: the interaction among normal modes. The latter were introduced as a small perturbation in the fourth paper of the series (MSN.8).

In 1955 a fundamental paper by *Léon Van Hove* (a great physicist, Prigogine's friend, former student at the ULB, and, at that time, professor at the University of Utrecht, Netherlands) was published. The paper was devoted to weakly coupled many-body quantum systems. The author realized the first sharing of the recently developed mathematical methods of quantum field theory (renormalization) with those of statistical mechanics. This sole aspect (beyond the importance of the results obtained) would suffice to establish the importance of this work: It was going to "refresh" statistical mechanics, by introducing a new "toolbox," which would become indispensable. In his work, Van Hove underlined the fact that even very weak interactions (measured by a parameter $\lambda \ll 1$) lead to contributions that grow limitlessly for long times ("secular terms"). It is thus necessary to perform a perturbation calculation *to all orders in* λ, select in each order the most divergent terms, and resume the resulting partial series in order to obtain a globally finite result.[7] Van Hove succeeded [by retaining the contributions of order $(\lambda^2 t)^n$] in deriving in this way Pauli's (irreversible) equation of evolution, thus avoiding the *ad hoc* assumptions introduced by earlier investigators.

This work (actually very difficult to read, and using a very heavy formalism) had the effect of a bomb in Brussels. Prigogine associated himself with Robert Brout (who was at that time a postdoc in Brussels) in order to understand, deepen, and develop Van Hove's ideas. The first result of this collaboration was a basic paper (1956, MSN.12) on the general theory of weakly coupled classical many-body systems.[8] Although still influenced by Van Hove's paper, this work by Brout and Prigogine is a generalization of the latter, as well as a simpler and more transparent presentation.

[7]As a trivial example, think of the exponential function $\exp(-t)$ for positive times: Its series expansion contains all positive powers of time, $t, t^2, t^3, \ldots$, which grow indefinitely for a long time; the sum of their series, however, is finite and decreases to zero.

[8]This paper was published as Part VII (and last!) of the series "Statistical Mechanics..." initiated by Klein and Prigogine. It represents, however, a radical change with respect to the preceding papers of the series!

Without discussing details (some of which are discussed in the Appendix), I should like to underline the fact that one can identify in this paper the starting point of an idea that Prigogine would pursue in all subsequent years. The classical problem of mechanics is based on the representation of the instantaneous state of the system by specifying the coordinates and the momenta of all its particles. This state is thus represented by a *point* in the many-dimensional "*phase space*." By specifying an initial condition—that is, the position of the representative point at a given time, as well as the forces that operate in and on the system—Hamilton's equations of motion determine the position of the point at any other earlier or later time. In other words, these equations determine a unique curve in phase space: the *trajectory* of the system. But, in quantum mechanics (because of Heisenberg's uncertainty principle), and also in classical mechanics of "nonintegrable" systems, it is shown that such a specification is impossible (or, at least, illusory). On the other hand, the study of an *ensemble* of systems (obeying the same equation of motion, but with different initial conditions), described by a *probability density* in phase space, briefly called a *distribution function* ρ, is perfectly univocal and leads, through Liouville's equation of evolution, to a statistical description of classical systems.[9] This concept would become the basis of the young science of nonequilibrium statistical mechanics. The idea was far from being new: It was introduced in the beginning of the twentieth century by Gibbs. Prigogine, however, had already in 1950 considered the "*death of trajectories*" to be an intrinsic property of unstable dynamical systems which, alone, would lead to an understanding of irreversibility. It will become the main axis of his theory and, more generally, of his vision of the world. The development of these ideas is briefly sketched in the Appendix.

Robert Brout would soon go back to the United States, but already a group of young enthusiastic students and future Belgian scientists was growing around Ilya Prigogine; they would develop and amplify the new nonequilibrium statistical mechanics. Among the earliest, in order of arrival, let me list them: Radu Balescu, Françoise Henin, Pierre Résibois, Claude George. It is interesting to note that we were all chemists, converted to physics by the charisma of Prigogine. The period 1956–1970 was certainly one of the most fertile for the Brussels group.

The research was greatly facilitated by two important elements. The (formal, perturbative) solution of the Liouville equation is greatly simplified by a Fourier representation (see Appendix). The latter allows one to easily identify the various types of *statistical correlations* between the particles. The traditional dynamics thus becomes a *dynamics of correlations*. The latter is completed by

---

[9]In the present paper I shall only discuss explicitly classical systems. The extension of these methods to quantum systems is possible, and it has been done by Prigogine and his co-workers.

the elaboration of a *diagram technique* (MSN.25) that establishes a correspondence between the numerous terms of the perturbation expansion and some well-adapted and efficient diagrams. It thus becomes possible to estimate visually, before any calculation, the order of magnitude of any term in the perturbation expansion of the distribution function. [The idea of a diagram technique was previously introduced in quite different fields: Mayer in the 1930s, for equilibrium statistical mechanics, and Feynman in the 1950s, for quantum field theory]. From here on, the machine could start moving. In the 1960s, Prigogine and/or his collaborators obtain new kinetic equations for systems as varied as: anharmonic solids (Henin), plasmas (Balescu), liquids (Nicolis, Misguich), quantum gases (Résibois), ferromagnets (Résibois), gravitational systems (Severne), relativistic systems (Balescu), scattering theory (Mayné), and so on. At the same time, the method provided a sound theoretical basis for macroscopic physics.

The vision of irreversibility that appeared in this first group of works, which formed the object of the first monograph on nonequilibrium statistical mechanics by Prigogine (1962, LS.9), was the following. The necessary conditions for an irreversible evolution were:

1. The large size of the system (i.e., a large number N of particles) enclosed in a large volume V. The calculations are greatly simplified in the thermodynamic limit: $N \to \infty$, $V \to \infty$, $N/V = n : finite$.

2. A (somewhat technical) condition on the resonances.

3. Finite range of the interactions and of the correlations.

4. The presence of a "small parameter."

Under these conditions, the distribution of the action variables (e.g., the momenta) [the *vacuum*, ρ_0] tends irreversibly toward the thermodynamic equilibrium after a sufficiently long time. Under the same conditions, the correlations are determined by the vacuum (technically, they become functionals of the vacuum distribution ρ_0) (see Appendix).

It should be stressed that in this first group of works (1956–1970) *there is nowhere any deviation from the Hamiltonian laws of dynamics*. This feature was taken as a preliminary postulate of our work. Irreversibility appeared as an **asymptotic** property of the evolution of certain classes of systems. The term "asymptotic" refers to the large size of the system, as well as to the long time scale of observation.

After 1969, the "Prigoginian" statistical mechanics started to change its aspect. I shall try to outline here the chronology and the significance of these changes. The more technical aspects will be discussed in the Appendix.

A first purpose consisted "only" of generalizing the domain of validity of the theory developed during the years 1956–1970. Prigogine's ambition was to show that the irreversible behavior found in the simple dynamical models was not limited by the approximations related to the presence of a small parameter (weak coupling, weak density,...). In all those cases, the existence of widely separated characteristic time scales leads to the representation of irreversibility as an *asymptotic* property, manifesting itself after a sufficiently long time, compared to the shortest characteristic time scale (e.g., the duration of a collision). In what way could one get rid of the restriction to small values of the characteristic parameter and generalize this behavior to systems with arbitrarily strong coupling?

A first answer to this difficult question was elaborated from 1966 on. A new methodology was introduced by C. George in 1967. It consisted of identifying a "piece" of of the distribution function that would evolve irreversibly to equilibrium, following a "**subdynamics**," *independently of its complementary part*. This identification is made operational by introducing a set of **projectors**, leading to a new, very elegant formulation of statistical mechanics.

By that time, a new member, of sizable stature, appeared in the Brussels group. Near the end of his life, *Léon Rosenfeld* left Copenhagen (where he had been for many years the right-hand man of Niels Bohr) and came back to his native Belgium. He became converted to Prigogine's ideas, which he actively supported and of which he became an enthusiastic promotor.[10] This collaboration became concrete in 1973 in a long review paper by I. Prigogine, C. George, F. Henin, and L. Rosenfeld (**PGHR**): "*A unified formulation of dynamics and thermodynamics*" (MSN.75).

The first motivation of Prigogine (which guided him actually since the beginning of his researches in this field) was to obtain a *general microscopic definition of entropy*, the universal indicator of irreversibility. This is a crucial question for establishing the molecular basis of thermodynamics. It was stated, and solved in the special case of dilute gases, by Ludwig Boltzmann in 1872. But this solution required a partial abandonment of the laws of mechanics, along with entering into the game of **probabilities**. The result was an avalanche of criticisms from the supporters of a purely deterministic evolution (i.e., the majority of the physicists of that time). In spite of this, "it worked!": Boltzmann's theory led later to a determination of the macroscopic transport coefficients of gases with a great precision, and it was fully supported by experiment.

The generalization of Boltzmann's solution turned out to be especially difficult. In their 1973 paper, PGHR performed a synthesis of the *projector* method of C. George and the idea of a **transformation** of ρ. The PGHR paper was considered for several years as the "bible" of Prigogine's group. The

---

[10]A significant anecdote: During a summer school in Sitges, Spain, in 1972, an anonymous student replaced the lettering of the announcement of a lecture by Rosenfeld about Prigoginian statistical mechanics by substituting the anagram: "EL DEFENSOR"!

authors developed a theory leading to a "causal irreversible dynamics" and to a definition of entropy. When the Indian physicist *Bandyanath Misra* arrived in Brussels, the final touch was achieved for this construct. For this reason, in order to avoid repetitions, I prefer to discuss here only this final form of the theory.

The new formulation, which approached the problem from a totally different point of view, appears in a series of papers, from which we extract mainly: *B. Misra, I. Prigogine, and M. Courbage* (**MPC**) (1979, MSN.98) and *S. Goldstein, B. Misra, and M. Courbage* (**GMC**) (1981, *J. Stat. Phys.*). The set of all Prigogine's works on macroscopic dissipative structures and on microscopic dynamical systems are the object of a review in the book *"From Being to Becoming,"* published in 1981.[11] The presentation is "semitechnical," intended for readers having a basic mathematical and physical background, without, however, entering the details of the derivations. The book was conceived as a complement to *La Nouvelle Alliance*, and therefore it inscribes the scientific results into a more general philosophical framework.

Given the importance of the work of Misra, Prigogine, and co-workers, I am providing in the Appendix a detailed, but simplified, analysis. The main result of these works can be concisely formulated as a unique *theorem*:

> There exists a class of dynamical systems, whose distribution function ρ obeys the (deterministic) Liouville equation, for which one can prove that, as a result of a transformation Λ of ρ, their evolution toward the future is "similar" to an irreversible stochastic evolution towards equilibrium, obeying a probabilistic evolution law. The transformed distribution function leads to a simple definition of an entropy. The members of this class will be called **intrinsically stochastic dynamical systems**.

A few remarks will help us in understanding the importance of this result. One should first stress the fact that this proposition (when precisely formulated) has the status of a *theorem*, proven with the full rigor required by mathematics. Next, one should note that we have here an "existence theorem": The class of systems to which it applies is defined in a univocal, but abstract, non-constructive way. In the paper GMC, the authors show that the necessary and sufficient condition of intrinsic stochasticity is that the systems belong to the class of *"Kolmogorov flows (K-flows)."* Although this class is mathematically well-defined, it is extremely difficult to prove that a given *physical* system belong to this class. On simplifying to the extreme, it may be said that these are systems exhibiting a high degree of *dynamical instability*. For these systems, a very small variation of the initial condition may lead to enormous deviations after a finite time.

---

[11]The Italian version, *"Dall'essere al divenire,"* of 1986, is considerably updated and includes in particular the MPC theory.

Note also that the formulation of the theorem contains the following restriction: "evolution toward the future." This implies that the evolution "toward the past" is described by a different law (splitting of the evolution Group into two irreversible Subgroups). Prigogine calls this feature a *temporal symmetry-breaking*," related to the *"arrow of time*." (Unfortunately, he fails to state that this "symmetry-breaking" has in no way the same meaning as the one found in the macroscopic dissipative structures: Here there is no critical threshold, no bifurcation, no multiple stationary states, and so on.)

It is important to stress the fact that in the proof of the MPC theorem, *the laws of classical dynamics are never violated.* One could summarize the significance of the MPC theorem by saying that, *for a well-defined class of dynamical systems, the new formulation "lays bare the arrow of time" that is hidden in the illusorily deterministic formulation of these unstable systems.*

The intrinsically stochastic systems are defined in an abstract way: It could be said that they live in the world of Platonic ideals. In order to transform this *mathematical* theory into a *physical* theory, one should be able to prove that there exist material systems satisfying the instability criteria required by MPC. At present, however, such a demonstration (satisfying the same criteria of mathematical rigor) does not exist. Therefore, in order to illustrate the consequences of their theory, MPC made do with a dynamical system reduced to its simplest expression: the *"baker's transformation."*[12] In this purely mathematical model, all desired quantities can be exactly and explicitly calculated; this was done by Prigogine and his co-workers for many coming years. The baker's model thus became the paradigm of the intrinsically stochastic systems. It is discussed in detail in all the books published by Prigogine in the following years.

But the major physical problem remained open: Could one prove *rigorously* that the systems studied before 1979—that is, typically, systems of N interacting particles (with N very large)—are intrinsically stochastic systems? In order to go around the major difficulty, Prigogine will take as a starting point another property of dynamical systems: **integrability**. A dynamical system defined as the solution of a system of differential equations (such as the Hamilton equations of classical dynamics) is said to be *integrable* if the initial value problem of these equations admits a unique analytical solution, weekly sensitive to the initial condition. Such systems are mechanically stable. In order to

---

[12]In this model the continuous evolution in time is replaced by successive transformations of the space on itself ("mappings"). The baker mapping consists of starting from an $L : L$ square and stretching it in one dimension while shrinking it in the other dimension, thus forming a rectangle $2L : L/2$; one then folds the latter on itself, in order to make again a square $L : L$. It is easily conceivable that after sveral transformations a finite region in the interior of the initial square is deformed, fragmented, and dispersed in the final square: This is the property called *"mixing."*

describe these systems, it is always possible to introduce a representation (actions and angles) in terms of degrees of freedom evolving independently of each other. In this way the interactions are "suppressed" (says Prigogine)—or, more exactly, "hidden." In the contrary case of **nonintegrable systems**, there exist in every arbitrarily small neighborhood of any point P, taken as initial condition, an infinite number of points generating solution that deviate exponentially from the solution passing through P. Given that it is not physically possible to determine an initial condition with infinite precision, it follows that for these systems it is impossible to make a precise long-time prediction. The initial small domain will be violently deformed after a sufficiently long time (without fragmentation, however, as in the baker's map!) and will end up covering a considerable fraction of space. Nonintegrability thus leads to a behavior of *K-flow "type,"* but it is generally impossible to prove *rigorously* the identity of the two concepts.

Starting in 1987, Prigogine, in collaboration mainly with *Tomio Petrosky* (University of Texas at Austin), attacked a new paradigm, which he calls **Large Poincaré Systems (LPS)**, classical or quantum mechanical (for simplicity, I only discuss here the classical version). This is, in fact, just a new name for the systems usually studied in statistical mechanics—that is, systems of N interacting particles, enclosed in a volume V, at the thermodynamic limit of *infinite systems*. (The reference to Poincaré is due to one of his theorems about nonexistence of analytical solutions for nonintegrable systems.) In order to treat this problem, Prigogine and his co-workers use a combination of the "old" ideas, synthesized in the PGHR paper, and the concepts of MPC. The main difficulty facing Prigogine and Petrosky (MSN.155), as well as their "ancestors," was related to the *resonances*—that is, apparently infinite contributions in the perturbative solutions of these problems. Following the old methods of C. George, they treat this problem by a "regularization" of the infinite terms. They also use a more precise definition of the functional spaces framing the distribution functions (1993, MSN.140), necessary for attributing a univocal meaning to the mathematical expressions of the theory. The difficulties showing up in this formulation are discussed in the Appendix.

In his final book, *La Fin des Certitudes* (LG.7), Prigogine summarized the achievements realized by his vision of dynamics in three statements:

"1. The irreversible processes (associated with the arrow of time) are as real as the reversible processes described by the traditinal laws of physics; they cannot be interpreted as approximations of the fundamental laws.

2. The irreversible processes play a constructive role in nature.

3. Irreversibility requires an extension of dynamics."

This very compact formulation requires some further precision. In my view, statement 2 has the firmest foundation. Whenever the evolution is effectively perceived as irreversible, as at the macroscopic level, the constructive role of irreversibility is manifest in the formation of the dissipative structures discovered by Prigogine.

Concerning statement 1, I believe that one should first define what exactly is meant by "approximation." In *La Fin des Certitudes* (p. 29), Prigogine rightly attacks the rather widely present view according to which statistical mechanics requires a (brute) coarse graining (i.e., a grouping of the microscopic states into cells, considered as the basic units of the theory). This process is, indeed, an arbitrary approximation that cannot be accepted as a basis of the fundamental explanation of the very real macroscopic irreversible processes.

In the MPC theory, the problem is not even posed. One starts defining the purely mathematical concept of "dynamical system" without any reference to a representation of reality. (The baker's transformation or the Bernoulli shift are obvious examples.) From here on, one proves mathematically the existence of a class of abstract dynamical systems (K-flows) that are "intrinsically stochastic"—that is, that possess precise mathematical properties (including a temporal symmetry breaking that can be revealed by a change of representation).

The problem appears when one tries to find a model representing the reality. In *La Nouvelle Alliance* (LG.1, p. 216), one finds the following statement, which clearly poses the problem: "In fact, we think, although we were not yet able to prove, that all chemical systems (and, a fortiori, all biological systems) possess a dynamical instability comparable to the baker's." Prigogine expresses here (very optimistically!) a fact that faces every theoretical physicist. The latter never describes a *real system*, but rather only a "*model of a real system*." He accepts this simplifying representation in order to construct his theory, hoping that it keeps the essential elements for the understanding of the phenomenon that is studied. Experiment will validate (or invalidate) *in fine* the chosen model. In *La Fin des Certitudes* (LG.7), Prigogine quoted a very significative statement of Léon Rosenfeld, expressing the same idea: "Every theory is based on physical concepts associated with idealizations that make possible the mathematical formulation of these theories."

Thus, Prigogine and Petrosky (PP) introduced the model of a "*Large Poincaré system*" (LPS). As stated above, the latter is, in fact, a large system, to which the operation of "*Thermodynamic limit*" is applied. Clearly, there exists no *real* system satisfying *strictly* the definition of a LPS: This infinite system is an "*idealization*," on which, by the way, all of statistical mechanics is based. One should thus be more specific about the statement: "The irreversible processes . . . cannot be interpreted as approximations of the fundamental laws" (statement 1). Quite explicitly, the approximations that are avoided in the PP theory are (a) the arbitrary coarse-graining and (b) the restriction to small parameters.

I now consider statement 3: How should "*an extension of dynamics*" be understood? In the MPC theory the problem does not exist: For the intrinsically stochastic systems there is no need for modifying the laws of dynamics. As for the LPS theory, one notes the presence of two essentially new concepts. The introduction of non-Hilbert functional spaces only concerns the definition of the *states* of the dynamical system, and not at all the law governing their evolution. It is an important precision introduced in statistical mechanics. The "extension of dynamics" thus only appears in the operation of *regularization of the resonances*. This step is also the one that is most difficult to justify rigorously; it is related to the (practical) necessity to use perturbation calculus (see Appendix).

After this long discussion of the successive formulations of statistical mechanics that appeared during 50 years of intense work on the microscopic theory of irreversibility, I should like to conclude by taking a distant view of the global result. Nobody could question the fact that an impressive edifice has been constructed under Prigogine's direction. Upon rereading the works on non-equilibrium statistical mechanics prior to 1950, one cannot avoid being impressed by the immense progress represented by the "Brussels school" (a name that I do not personally like very much, given the diversity of its members!). It provided, in the first place, a very general basic formalism, which turned out to be a very fruitful working tool. As stressed above, it gave rise to a flowering of results of enormous diversity. It is true that other groups (such as the Russian school initiated by N. N. Bogoliubov, and the Dutch school directed by G. Uhlenbeck) worked independently in the same direction; the results always turned out to be equivalent. Prigogine was always in the forefront: A proof is the large number of foreign visitors, co-workers, and students he attracted. His associated philosophical preoccupations mark an interesting difference between his "style" and his "competitors."

E.   Cosmology

Between 1986 and 1993 a limited but very interesting period in Prigogine's scientific activity took place: After the "infinitely small," he approaches the "infinitely large." He begins a collaboration with his colleague (and also former disciple of De Donder) Jules Géhéniau and with two younger physicists (Edgard Gunzig and Pasquale Nardone) in the field of **cosmology**. Einstein's equations of general relativity led to the celebrated model (initially due to the Belgian abbot-physicist Georges Lemaître) of the "*Big bang*," known today even by children. Most of the predictions of this model (completed with the notion of initial inflation) have been verified at this time by astronomical observations. But the phenomena taking place during the first tiny period after the primordial explosion still remain mysterious (their understanding requires a quantum theory of gravitation, at present one of the major open problems of science). The primordial event itself implies the existence of a universe infinitely concentrated at the "origin of time." The formation (*ex nihilo*) of such a singularity as well as the existence of an absolute lower limit of time are concepts that are difficult to accept in physics.

Basing their ideas on an important prior work of R. Brout, F. Englert, and E. Gunzig (1978), Prigogine and his co-workers introduced the idea of a *quantum vacuum* as an "eternal" framework. It is well known that this notion does not imply the absence of everything (as in the classical picture), but, on the contrary, the *potential* existence of all possible particles (Prigogine nicely says: "*The vacuum is a potential world, full of suspended possibilities*") Like all quantum systems, the vacuum is subject to fluctuations. Prigogine et al. then assume that the occurrence of a fluctuation, whose mass exceeds a critical value, destabilizes the vacuum (bifurcation!), and triggers off a massive production of matter at the expense of gravitational energy. This process is irreversible and is accompanied by entropy production. A slight modification of Einstein's equations allows such a scenario which, after a certain short time rejoins the evolution scheme predicted by the standard model. The theory thus avoids the initial singularity and the lower limit of time ("*time preceeds existence*" says Prigogine). We have here a very appealing idea that must, of course, be confirmed in the future by quantum gravity.

F. Philosophical Vision

After 1977, Prigogine's scientific activity was parallelled by an intense *literary and philosophical* activity. Its purpose was, clearly, the propagation of his ideas outside of the limited world of chemists, physicists, and biologists. In the beginning he was backed up in this activity by *Isabelle Stengers*.[13] The first fruit of their collaboration was the book *La Nouvelle Alliance—Métamorphoses de la Science* (1979, LG.1). [The english translation bears the title *Order Out of Chaos*.] It became an extraordinary best-seller, translated into numerous languages. The book is, however, not easy to read. It is directed, in principle, to a wide, "enlightened," but not specialized, audience. But, a reader of "literary type" would not grasp the more technical aspects, often suggested rather than explained, whereas a "scientific-type" reader would probably be hindered by the philosophical erudition. Prigogine exposes here his vision of the world, which he presents as a new dialogue between Science and Philosophy, made possible by his conception of time.

This book was followed by a series of other general books, which had comparable commercial success: IP & IS, **Entre le Temps et l'Éternité** 1988,

---

[13]Isabelle Stengers first studied chemistry, then philosophy, which later became her main field of activity (she is presently professor of philosophy of science at ULB). She did her PhD thesis under the direction of Prigogine, on the problem of time.

LG.3); IP, **Les Lois du Chaos** (1994, LG.6); IP, **La Fin des Certitudes—** *Temps, chaos, et les Lois de la Nature* (1996, LG.7); IP (ed.); **L'Homme devant l'Incertain** (2001, LG.8). Given that these books contain the progressive development of the same ideas, enriched by successive updates, I shall discuss them together.

The somewhat bizarre but appealing title of *Nouvelle Alliance* is explained in the introduction as a response to a celebrated sentence of the well-known book of the great biologist Jacques Monod, *Chance and Necessity*: "The old alliance is broken: Man finally knows that he is alone in the indifferent immensity of the Universe wherefrom he emerged by chance." IP and IS would like to demonstrate that this tragic view is related to "classical science" and that present-day science underwent a metamorphosis leading to a "new alliance" between Man and Nature. Prigogine's world vision originates in his scientific work, by a flashing extrapolation, for which he was quite gifted. It should, however, be underlined reciprocally that his research program was, from the very beginning, inspired and guided by his philosophical ideas.

According to Prigogine, the metamorphosis of science is accompanied by a new vision of time. On one hand, he notes the mechanical definition formulated by Newton in his *Principia*: "Absolute time, true and mathematical, in itself and by its own nature, flows uniformly without relation to anything external." Prigogine qualifies this situation as *static*, quoting his favorite philosopher Henri Bergson: "Everything is given!"

But during the nineteenth century new phenomena that do not enter this framework are discovered: transport of heat, diffusion of matter obey irreversible laws, governed by the second principle of thermodynamics. The evolution of living species discovered by Darwin also show evidence of a directional time. In all these cases it is no longer possible to describe temporal evolution as the motion along a trajectory: In a diffusive process the trajectory disintegrates, and the concept itself loses meaning. It is not possible, according to IP and IS, to inscribe irreversibility in the Newtonian framework. The concept of time that emerges from irreversibility is closer to Bergson's formulation, stated in *L'Évolution Créatrice*: "The Universe lasts. The deeper we look into the nature of time, the better we understand that duration means invention, creation of shapes, continuous elaboration of the absolutely New."

The strictly deterministic world vision of classical science led the philosophers (notably I. Kant) to introduce the notion of the "two cultures," opposing the deterministic, cold, and rigid science to the humanistic domain of intuitions and unpredictable sentiments. IP and IS strongly state that classical science is dead and that the new science, "metamorphosed" by new Prigoginian vision of time, would allow, in the long term, a reunification of the two cultures. Indeed, Bergson's "creative" time is now inscribed into science. It manifests itself at the level of macroscopic thermodynamics through the bifurcations, creating dissipative structures. The latter could lead in the future to a possible path toward the physical explanation of the vital (and human!) phenomena. This creative process can only be produced in the presence of external constraints, thus of fluxes of matter and of energy, that maintain the systems far from equilibrium.

Using a big extrapolation, IP and IS climb directly to the uppermost level: "One of the most promising perspectives opened by this metamorphosis is the end of the cultural split that made science a foreign body. . . . The mathematical sciences of nature, when they discover the problems of complexity and of becoming, also become able to understand something about some questions expressed by myths, religions and philosophies. . . ." The jump is enormous; but after this purely rhetorical statement, Prigogine calls to prudence! He expressed it very clearly (rather much later!) during a remarkable television interview[14]: ". . . nowadays the humanistic sciences can adopt new models: instability, chaos, etc. One must, however, remain prudent because the mechanism of decision, essential element in the description of sociology and economy, is obviously quite different in the case of molecules and in the case of human beings."

Prigogine often uses this rhetorical procedure: a big, shocking statement, followed (at a distance) by precise explanations that put back the problem on its natural level. Another example is the title of his final book: *The End of Certitudes* (1996, LG.7); this is a mind-catching phrase that may produce a feeling of uneasiness, or may be completely misunderstood. One could ask oneself: "What is going on here? Are we witnessing a surrender of Science? Do we live in a world without law and without order, where everything is possible and nothing can be predicted or controlled?" But if one carefully reads the book, he will quickly understand that a quite rational reality is hidden behind this rhetorical statement. The probabilities, which are henceforth governing nature, obey well-defined mathematical laws ("the laws of chaos," in Prigogine's words!). The predictions are, indeed, possible, but they address a different kind of object (probabilities, rather than trajectories). In the interview quoted above, Prigogine provides us with this magnificent key: "Scientific truth is not certitude, and uncertainty is not ignorance."

The opposition of the two concepts of time is particularly striking in the famous fundamental problem of statistical mechanics: How can we understand the emergence of an *irreversible evolution* at the level of macroscopic physics (in particular, of thermodynamics) from the *deterministic and reversible* Newtonian laws of mechanics?

In order to prepare for the answer, Prigogine started by distinguishing integrable from nonintegrable dynamical systems (see above, Section I.D). For

---

[14]E. Blattchen, *Ilya Prigogine: De l'Être au Devenir*, Brussels, 1998, Belgian television series "*Noms de Dieux.*"

the former, all the results of reversible Newtonian mechanics remain valid. But they form a quite small exceptional set, compared to the set of nonintegrable systems (whose degrees of freedom are coupled by interactions). The key to Prigogine's answer starts with the following sentence: "Insofar as we are unable, not only in practice, but as a matter of principle, to describe the system by a trajectory, and are forced to use a distribution function corresponding to a finite (even arbitrarily small) region, we can only predict the statistical fate of the system."

The important word in this sentence is *"predict."* It is important, in my opinion, to make a distinction between *"existence"* and *"predictability."* Prigogine himself said (much later, in *La Fin des Certitudes*, LG.7): "Every dynamical system must, of course, follow a trajectory, solution of its equations, independently of the fact that we may or may not construct it." Thus, a trajectory "exists" but cannot be "predicted." The impossibility of prediction is therefore related to the impossibility of defining an instantaneous state (in the framework of classical mechanics) as a limit of a finite region of phase space (thus a limit of a result of a set of measurements). For an unstable system, such a region will be deformed and will end up covering almost all of phase space. The necessity of introducing statistical methods appears to me to be due to the *practical* (rather than *theoretical*) impossibility of determining a mathematical point as an initial condition.

Prigogine thus arrived at the idea that the description of unstable dynamical systems requires an "extension of dynamics." As a result, the temporal symmetry of the theory is broken. Irreversibility appears as a result of the scission of the unitary group describing the evolution into two dissipative, nonunitary semigroups.

We discussed in Section I.D the meaning to be given to the word "extension." It is notewothy that in *La Fin des Certitudes* (1996, LG.7) Prigogine states quite rightly: "An extension of dynamics? This appears as a very strong statement, that can easily be misunderstood. There is no question here about adding new terms to the equations of dynamics...." "...There is no question of adding new terms that would break the symmetry of the equations, like Epicurus' clinamen. What we shall show is that the situations where we expect a symmetry breaking are also those that require a new formulation of dynamics."

At the time of publication of *La Nouvelle Alliance* the Prigoginian theory was still at the MPC stage. It is thus significant that all general statements be illustrated only by the baker's transformation. It is only in *Les Lois du Chaos* (1994) and in *La Fin des Certitudes* (1996) that the "Large Poincaré Systems" (LPS) show up. As stated in Section I.D 4, this concept results from the quest of "real" physical systems satisfying the criteria of intrinsic stochasticity. In this case, however, Prigogine and Petrosky were led to introduce a true modification ("extension"?) of the laws of mechanics. The reason is the appearance of the Poincaré resonances.

In conclusion, one notes that the major part of Prigogine's body of work is constructed around a personal, almost obsessional, view of **time**. This vision guided him in the choice of problems he studied, and also in the methodology of their approach. What is impressive is the great extent of the themes he treated, going from atomic physics to cosmology. In every domain he approached, be it chemistry or physics, or even biology or humanities, he produced jewels. His influence was literally universal. I could cite as a proof of this statement not only the numerous institutes bearing his name throughout the world, but especially the existence of at least one center (in Italy) explicitly devoted to the study of Prigogine's ideas.[15]

Independently of the details of his works, through his extraordinary charisma, Prigogine marked with an indelible imprint all persons who had the privilege of knowing him and working with him. A great personality entered into eternal rest in 2003.

Acknowledgment

I sincerely thank Dr. Jacques Misguich and Prof. Irina Veretennicoff for very fruitful discussions during the writing of this text.

G.   Appendix

1.   Statistical Mechanics: First Version

The starting point of classical statistical mechanics is the exact equation of evolution of the distribution function ρ in phase space: the *Liouville equation*, which Prigogine always wrote in the form

$$i \frac{\partial \rho}{\partial t} = L\rho \tag{3}$$

where L is a hermitian operator (technically: $i\times$ the Poisson bracket with the Hamiltonian).

The variables characterizing the phase space are, usually, the positions q and the momenta p of the material points constituting the system. One is free, however, to choose other couples of variables, related to the former by a canonical transformation, without changing the Hamiltonian form of the equations. In their paper of 1956 (MSN.12), Brout and Prigogine brought back to life the angles α and the actions J, basic building blocks of Hamiltonian mechanics, but more or less relegated to the "corner" of celestial mechanics.

---

[15]"Istituto di documentazione e ricerca sull'opera di Ilya Prigogine" in Brugine (Pd, Italy). Website: www.crs4.it/CISST/Istituto-Prigogine.html

They would become the "stars" of Prigoginian statistical mechanics. Their importance lies in the fact that, whenever it is possible to determine these variables by a canonical transformation of the initial phase space variables, one obtains a description with the following properties. The action variables $J_n(n = 1, 2, \ldots, N$, where N is the number of degrees of freedom of the system) are invariants of motion, whereas the angles α_n increase linearly in time, with frequencies $\omega_n(J)$, generally action-dependent. The integration of the equations of motion then becomes trivial. The systems for which such a canonical transformation exists are called *integrable systems*. But Poincaré demonstrated already at the end of the nineteenth century that for the majority of systems (for instance, even for three interacting particles) that it is mathematically impossible, in general, to determine (nontrivial) analytical invariants of motion. The impossibility of determining "trajectories" does not prevent, however, the study of ensembles of trajectories, thus using the methodology of statistical mechanics.

The approach of Brout and Prigogine[16] consisted of defining an "unperturbed state" in which the interactions are switched off, and for which, therefore, one could define actions and angles. The distribution function is then formally expanded in a power series, to all orders with respect to the interaction strength parameter λ. The use of these variables leads naturally to a Fourier representation with respect to the angle α_n. This Fourier representation opens the way to a **dynamics of correlations**, which will become the basis of many forthcoming works of the Brussels—and later, Austin—groups. In this representation, each Fourier component $\rho_{k_1, k_2, \ldots, k_n}(J)$ is associated with a certain type of correlation between $2, 3, \ldots$ particles (or degrees of freedom), corresponding to the number of nonzero wave vectors k_j. The component $\rho_{0,0,\ldots,0}(J)$, called the *vacuum of correlations* (i.e., the distribution of actions J), plays a privileged role. In this representation the Liouville equation is transformed into a coupled set of equations for the Fourier components. These equations are naturally interpreted as describing the coupled evolution of the correlations, consisting of *creations, transformations, and destructions of correlations* in time. This new language of dynamics of correlations replaces the traditional language of mechanics in terms of motion of a point along a trajectory.

2. *Projection Operators*

In C. George's method of projectors, the basic point is to distinguish between different "pieces" of the phase space distribution function ρ.[17]

---

[16]A detailed exposition of this first version of statistical mechanics is found in Prigogine's book *Nonequilibrium Statistical Mechanics* (LS.9).

[17]A detailed exposition is given in R. Balescu, *Equilibrium and Nonequilibrium Statistical Mechanics*, Wiley, New York, 1975.

A first distinction is made between the "vacuum of correlations" and the "true correlations." The former can be defined as the integral of the distribution function ρ over all angle variables. The set of all (normalizable) functions of J alone forms a subset of the whole functional space. Its complement is the subspace of correlations. The distribution function is thus written as

$$\rho = V\rho + C\rho \tag{4}$$

The separation can be formally realized by means of two *projection operators* V and C, with the following properties:

$$V + C = I, \qquad V^2 = V, \qquad C^2 = C, \qquad VC = CV = 0 \tag{5}$$

where I is the identity operator. On the other hand, the Liouville operator L is split into a "free motion" (or "unperturbed") part L_0, describing an integrable system, and an interaction term δL:

$$L = L_0 + \delta L \tag{6}$$

The decomposition (5) is related to the decomposition (6) by the commutation relation:

$$VL_0 - L_0V = 0 \tag{7}$$

The Fourier representation provides us with a particular realization of this decomposition: The vacuum $V\rho$ is realized by the Fourier component $\rho_{0,0,\ldots,0}(J)$ of the distribution function. The abstract decomposition (4) can, however, also be realized in more general ways. Equation (7) expresses the fact that in a system without interactions, one cannot create correlations out of the vacuum.

One may then try to generalize this operation, by defining another decomposition, different from the previous one. Two new projection operators Π and $\widehat{\Pi}$, were first "timidly" defined by C. George in 1967 and were later introduced definitively in the toolbox of statistical mechanics in 1969 by I. Prigogine, C. George, and F. Henin (MSN.60). The new objects possess the usual properties of projectors:

$$\Pi + \widehat{\Pi} = I, \qquad \Pi^2 = \Pi, \qquad \widehat{\Pi}^2 = \widehat{\Pi}, \qquad \Pi\widehat{\Pi} = \widehat{\Pi}\Pi = 0 \tag{8}$$

The projector Π is defined dynamically by the following relation, generalizing Eq. (7):

$$\Pi L - L\Pi = 0 \tag{9}$$

Π is an extension of the projector V; indeed, when the interactions vanish, $\delta L = 0$, Eq. (7) shows that $\Pi \to V$. The basic property (9) implies that the component $\overline{\rho} \equiv \Pi \rho$ evolves according to its own "**subdynamics**," independently of $\widehat{\rho} \equiv \widehat{\Pi} \rho$. One may therefore isolate the "coherent" part $\overline{\rho}$ of the distribution, which "contains" the irreversibility, from the "incoherent part" $\widehat{\rho}$, which tends to zero asymptotically for long times. The component $\overline{\rho}$ can in turn be decomposed: Its "vacuum" component, $V \overline{\rho} \equiv V \Pi \rho$, obeys an irreversible evolution equation and tends to the equilibrium distribution of actions, whereas $C \overline{\rho} \equiv C \Pi \rho$ appears to be a functional of the vacuum $V \overline{\rho}$. Thus, IF the operator Π exists and can be constructed,[18] irreversibility appears as an intrinsic property of an identifiable part of the distribution. The complementary part $\widehat{\rho}$ becomes irrelevant for sufficiently long times.

3. The Misra–Goldstein–Prigogine–Courbage (MPC, MGC) Formalism

We first consider a Hamiltonian, thus deterministic, system. Denoting by ω the set of all phase space coordinates of a point in phase space (which determines the instantaneous state of the system), the motion of this *point* is determined by a *canonical* transformation evolving in time, T_t, with $T_0 = I$. The function of time $T_t \omega$ thus represents the *trajectory* passing through ω at time zero. The evolution of the *distribution function* is obtained by the action on ρ of a *unitary* transformation U_t, related to T_t as follows:

$$(U_t \rho)(\omega) = \rho(T_{-t}\omega) \qquad (10)$$

The set of operators U_t forms a *group*:

$$U_t U_s = U_{t+s}, \qquad \forall t, \quad \forall s \qquad (11)$$

It is well known that the operator U_t is related to the Liouvillian by the relation $U_t = \exp(-iLt)$.

Consider, on the other hand, a *purely stochastic, Markovian process*. The evolution of a dynamical function $f(\omega)$ is now determined by a *transition probability* $P(\omega, t|\omega', 0)$ from the state ω' at time zero, to the state ω at time $t > 0$: This gives rise to a transformation W_t of f:

$$(W_t f)(\omega) = \int d\omega' P(\omega, t|\omega', 0) f(\omega'), \qquad t > 0 \qquad (12)$$

The evolution of a distribution function $\widetilde{\rho}(\omega)$ is given, as usual, by the adjoint operator: $\widetilde{\rho}_t(\omega) = (W_t^T \widetilde{\rho})(\omega)$. There is an essential difference between (10) and

---

[18]A nontrivial task in general, but that can be realized explicitly in the case of dilute gases, weakly coupled plasmas, and so on.

(12). In the deterministic case the value of the distribution function at time t depends on its value at a single point of the trajectory (initial condition); the latter point may be taken indifferently in the past or in the future of the time t. In the stochastic case, the value of $\widetilde{\rho}$ at a time $t > 0$ depends on the values taken by $\widetilde{\rho}$ at a time $t' = 0$ *in all points of phase space*; this is clearly expressed by the integral in Eq. (12). Indeed, every point ω' has a nonzero probability of arriving in ω at time t. The set of transformations W_t forms a *contracting semigroup*, rather than a group like the transformations U_t: this is the signature of *irreversibility*. These transformations conserve the positivity of the distribution functions (that must, at every time, represent a probability density) and admit the equilibrium distirbution as a stationary solution:

$$\text{(a) } \widetilde{\rho}(\omega) \geq 0 \rightarrow W_t^{\dagger} \widetilde{\rho}(\omega) \geq 0, \qquad \text{for} \quad t > 0$$

$$\text{(b) } W_t^{\dagger} \widetilde{\rho}_{eq} = \widetilde{\rho}_{eq}$$

MPC restrict themselves, moreover, to monotonous Markov processes:

$$\text{(b) } \| W_t^{\dagger} (\widetilde{\rho} - \widetilde{\rho}_{eq}) \| \rightarrow 0 \qquad \text{monotonously for} \quad t \rightarrow \infty$$

The authors then ask the following question: **Do there exist deterministic dynamical systems that are, in a precise sense, "equivalent" to a monotonous Markov process?** The question can be reformulated in a more operational way as follows: **Does there exist a similarity transformation Λ which, when applied to a distribution function ρ, solution of the Liouville equation, transforms the latter into a function $\widetilde{\rho}$ that can also be interpreted as a distribution function (probability density) and whose evolution is governed by a monotonous Markov process?** An affirmative answer to this question requires the following conditions on Λ (MPC):

$$\text{(i) } \rho \geq 0 \qquad \text{implies} \quad \Lambda\rho \geq 0$$

$$\text{(ii) } \int d\omega \, \rho = \int d\omega \, \Lambda\rho$$

$$\text{(iii) } \Lambda\rho_{eq} = \rho_{eq} \tag{13}$$

$$\text{(iv) } \Lambda U_t \rho = W_t^{\dagger} \Lambda\rho$$

$$\text{(v) } \Lambda \text{ possesses a densely defined inverse} \quad \Lambda^{-1}$$

The properties (i) and (ii) ensure that $\Lambda\rho$ has the same positivity and normalization properties as the initial ρ, and property (iii) ensures that the equilibrium distribution is not modified by Λ. Equation (iv) tells us what meaning should be attributed to the notion of "equivalence." Upon applying Λ to the deterministic evolution law $\rho \rightarrow U_t\rho$ of the distribution function, one obtains the stochastic evolution of the transformed distribution: $\Lambda\rho \rightarrow W_t^{\dagger}\Lambda\rho$. But we know that if Λ were a canonical transformation, the evolution of the transformed function would *necessarily* also be a unitary, deterministic transformation. Thus, in order to satisfy condition (iv), the transformation Λ must be **noncanonical**.

Condition (v) raises a subtle problem. If it were satisfied, we would be tempted to say that the transformation $\rho \rightarrow \Lambda\rho$ involves "no loss of memory" (as stated by MPC), because we could always reverse it. In order, however, to have a "complete equivalence" between the two representations, it is necessary that, upon reversion of the transformation we find again a true distribution function ρ. This implies, in particular, the preservation of positivity: $\Lambda^{-1}(\Lambda\rho) \geq 0$. But, in the second (MGC) paper it is shown that under these conditions there exists a *point transformation* such that $(\Lambda\rho)(\omega) = \rho(T\omega)$. The change of representation thus brings us back to a deterministic, canonical transformation [see Eq. (10)]. Thus, even if the inverse transformation Λ^{-1} exists mathematically, its result is physically unacceptable, because in an unstable dynamical system, trajectories have no meaning. Ten years later, Antoniou and Gustafson make the following statement about the inverse of Λ: "One may obtain abstract dilations, embedding probabilistic processes into larger deterministic dynamics. But their actual construction is computationally unattainable. . . . Due to the chaotic character of the exact endomorphism this is computationally impossible. . . . There is operational loss of information in going from deterministic to probabilistic dynamics." Thus, information is lost in going from ρ to $\Lambda\rho$; but this loss is much subtler than the one produced in any "coarse graining" operations, performed in the "traditional" treatments of statistical mechanics.

In conclusion, one should temper the vocabulary of the initial MPC paper: It is correct to state that the Λ-transformation leads to an evolution law "*similar to U_t*" (in the mathematical sense): $W_t^{\dagger} = \Lambda U_t \Lambda^{-1}$, but not to a physical "*equivalence*."

On the basis of the properties discussed above, the authors prove that $W_t^{\dagger}$, though being defined for all t, can only preserve positivity for $t > 0$, or for $t < 0$, but not for both together. Thus, the passage to the stochastic description implies the *breaking of the time symmetry*: The initial group of unitary transformations U_t splits up into *two distinct contracting semigroups* $W_t^{\dagger}$: for $t > 0$ and for $t < 0$. The second principle of thermodynamics appears now as a *selection rule:* Only one of the two semigroups describes the evolution toward the future of the dynamical system. It is at this level that the "loss of information" appears in the passage $U_t \rightarrow W_t^{\dagger}$.

A dynamical system is called "**intrinsically stochastic**" if it satisfies the conditions required for the existence of a Λ operator, and whose deterministic dynamics is thus "similar" to a stochastic evolution. The basic question raised at this stage is: "**Do there exist intrinsically stochastic dynamical systems?**" It appears that the answer is affirmative: The systems under consideration must be **dynamically unstable**. Physically, this implies that two trajectories, arbitrarily close together at a certain (initial) time, end up diverging exponentially from each other after a long time (this property is related to "*deterministic chaos*"). There exist various degrees of instability. MPC prove (with full mathematical rigor) that the "**mixing**" condition is necessary for the existence of Λ (we do not discuss here the details of the definitions). MGC showed later that a somewhat stronger condition (**K-flows** = Kolmogorov flows) is sufficient. In other words, the sentence "A K-flow is an intrinsically stochastic system" is a *mathematical theorem*. This relates the search for intrinsically stochastic systems to the ergodic theory, as developed in the second half of the twentieth century, especially by the Russian mathematicians Kolmogorov, Arnold, and Sinai.

An important consequence of the MPC theory is the existence of a "time operator" T. This operator does not commute with the Liouvillian: $LT - TL = i$. Its average value is interpreted by Prigogine as the "age" of the system, closely related to entropy. More generally, any positive, monotonously decreasing function, $M = M(T)$, is a Lyapounov function. On the other hand, the transformation Λ appears formally as a "square root": $\Lambda = M^{1/2}$. This property leads directly to an $\mathcal{H}$-theorem for intrinsically stochastic systems. Prigogine thus realized here the dream from his youth: the definition of a "thermodynamic time" (see Section I.C). But its properties are quite different from the one he imagined in 1945!

4. The Prigogine–Petrosky Formalism

The problem raised now is of a physical, rather than mathematical, nature: *Is it possible to realize the MPC program for a system of N particles* ($N \gg 1$) *enclosed in a volume V, large compared to the atomic scale?* The calculations are performed by taking N and V finite, anticipating the limit $N \to \infty$, $V \to \infty$, $N/V :$ *finite*, which is taken at the end of the calculations.

The objects of the theory are dynamical functions and distribution functions depending on the coordinates (q, p, or J, α) of phase space. In order to make the mathematical operations (in particular: scalar products) univocal, it is necessary to specify the nature of the functional space that is the framework of the dynamical objects that are being manipulated. The "simplest" choice is to adopt the Hilbert space, so common in quantum mechanics: this was the choice of PGHR (MSN.75). But this space is not adequate for the singular functions (Schwartz distributions) appearing in the theory. Antoniou and Prigogine (MSN.140) showed in 1993 that the objects of the theory had to be defined in a wider functional space, in particular, in a "**rigged Hilbert space**." An important property that is gained in this way is that the Hermitian operators (such as the

Liouvillian L) may possess eigenvectors ("Gamow vectors") corresponding to *complex eigenvalues*, a necessary condition for describing dissipative processes. This answers a traditional objection consisting of preventing a Liouvillian to describe dissipation, by calling on the fact that the latter may only possess real eigenvalues. The latter statement is only true in a Hilbert space.

Upon trying to solve the equations of motion (or the Liouville equation) by a series in powers of the interaction potential, it is found that all terms contain factors or products of factors of the form $\left[\sum_k n_k \omega_k(J)\right]^{-1}$, where $\omega_k(J)$ represents the generalized eigenfrequencies of the unperturbed Liouvillian, and n_k are integers. The important point is that whenever one of the sums in the denominators vanishes (**resonance**), the expression diverges. In an infinite system these divergences are always present, thus making the integration by the method of perturbation series impossible. This difficulty of the "small denominators" was raised already at the beginning of the twentieth century by H. Poincaré. In their review paper of 1996, Petrosky and Prigogine (PP) (MSN.155) attempted to *regularize* the divergent perturbation series.

In order to achieve this purpose, the authors add to the sum in each resonant denominator a term $\pm i\varepsilon$, where $\varepsilon > 0$ is a number forced to tend to zero at the end of the calculation. This procdure is well known in the case of a single resonant factor: It is an analytical continuation. The sign of ε is well-defined by the condition of causality, which requires that the term remain finite as $t \to \infty$. When several resonant factors are present (thus, in the higher-order terms of the perturbation series), the sign on ε is ambiguous. In order to make the sign univocal, PP turned back to the "rule of $i\varepsilon$" introduced by C. George in 1970 and also discussed in PGHR. I do not wish to include here a highly technical discussion regarding this rule. The terms of the perturbation series are, indeed, regularized by applying this rule. It is, nonetheless, true that that there exists no rigorous mathematical proof of the validity of this rule, which remains, at best, semi-intuitive.

I personally wonder whether the problem of resonances is, in fact, as deep as PP present it. H. Poincaré puts this problem in a precise mathematical framework: the search for **analytical** solutions (hence expandable in perturbation power series) of the equations of motion. It is in the coefficients of these expansions that the divergences due to small denominators appear. But why should the solutions of the equations of motion be analytical, after all? It is not impossible that, if the problem could be solved nonperturbatively, the problem would not be raised at all. [The success (in certain problems) of the renormalization procedures, consisting of grouping the divergent terms in partial series whose sum is convergent and nonanalytic, seems to be in favor of this view.] Note that in *La Fin des Certitudes* (1996, LG.7, p. 47) Prigogine cites a quite significant sentence of Max Born (1960): "It would be quite remarkable that Nature would have found the means to resist the progress of knowledge by hiding herself behind the rampart of the analytical difficulties of the N-body problem." Unfortunately, the previous remarks are purely formal and remain a mere conjecture. In order to calculate effectively the distribution function, we must necessarily, at present, resort to perturbation theory.

II.   SYSTEMATIC BIBLIOGRAPHY

I shall not give here the complete curriculum vitae of Ilya Prigogine, including all his functions and activities, the prizes and medals, and all the academic titles that were conferred to him around the world. This would lengthen this paper by at least 10 pages. Also, I shall not give here the the list of all his publications, whose number is around a thousand. All these data were already published in the book *Ilya Prigogine, Tempo, Determinismo, Divenire. Strumenti, Fatti e Documenti per una Ricognizione Storiografica*, a cura di Giuseppe Bozzolato, edizione Centro, Brugine, 1999 (in Italian). These data are conserved at the Université Libre de Bruxelles (Solvay Institutes for Physics and Chemistry); they were published on this basis by the *Istituto di documentazione e ricerca sull'opera di Ilya Prigogine*—Cisst, Via Roma 86a, 35020 Brugine, email: cisst@tin.it . They are also available (for the publications after 1964) at the site

www.crs4.it/CISST

Here I propose to offer a complete bibliography, *excluding interviews and articles published in daily newspapers*, of Prigogine's works, classified according to the subjects and the years of publication. Indeed, the existing bibliographies, listed in purely chronological order, mix all different subjects. I hope to facilitate in this way the access to individual publications. Note that the titles of every work is given in the original language of publication, accompanied, when necessary, by an English translation.

A.   Books

1.   Scientific Monographs

1939

LS.1. I. Prigogine, *Contributions à la Théorie des Electrolytes Forts* (Contributions to the Theory of Strong Electrolytes), Collection La Chimie Mathématique, Vol. V, Th. De Donder, éd., Gauthier-Villars, Paris (also published in Mémoires in-8°de la Classe des Sciences, *Acad. Roy. de Belgique*, tome **XVIII**).

1943

LS.2. I. Prigogine, Contribution à l'étude spectroscopique dans l'infra-rouge proche de la liaison d'hydrogène et la structure des solutions, (Contribution to the spectroscopic studies in the near infrared of the hydrogen bond and the structure of solutions), *Mémoires Acad. Roy. Belg., Cl. Sciences*, tome **XX**.

1944

LS.3. I. Prigogine et R. Defay, *Traité de Thermodynamique Conformément aux Méthodes de Gibbs et de Donder* (Treatise of Thermodynamics, According to the Methods of Gibbs and De Donder), Tome I. *Thermodynamique Chimique* (Chemical Thermodynamics), 1944; Tome II. *Thermodynamique Chimique, 1946*: these two books were republished as a single volume in 1950; Tome III. *Tension Superficielle et Adsorption* (Surface tension and adsorption), 1951, Desoer, Liège. Translated in English, German, Russian, and Japanese.

1947

LS.4. I. Prigogine, *Etude Thermodynamique des Phénomènes Irréversibles*, (Thermodynamic Study of Irreversible Phenomena) (Thèse d'agrégation de l'Enseignement supérieur, Université Libre de Bruxelles), Dunod, Paris et Desoer, Liège.

1953

LS.5. P. Mazur et I. Prigogine, Contribution à la thermodynamique de la matière dans un champ électromagnétique (Contribution to thermodynamics of matter in an electromagnetic field), *Mémoires Acad. Roy. Belg., Cl. Sciences*, **28**.

1955

LS.6. I. Prigogine, *Introduction to Thermodynamics of Irreversible Processes*, Charles C Thomas Publishers, Springfield, IL, 1955.

Second edition: John Wiley & Sons [Interscience Division], New York, 1962; Third edition, idem., 1967. Translations in Russian, Serbo-Croatian, French, Italian, and Spanish.

1957

LS.7. I. Prigogine (with the collaboration of V. Mathot and A. Bellemans), *The Molecular Theory of Solutions*, North Holland Publishers, Amsterdam, 1957.

1958

LS.8. I. Prigogine and R. Defay, *Chemical Thermodynamics* (translation by D. H. Everett of ref. LS-3), Jarrold & Sons, Norwich.

1962

LS.9. I. Prigogine, *Nonequilibrium Statistical Mechanics*, John Wiley & Sons [Interscience Division], New York.

1971

LS.10. P. Glansdorff et I. Prigogine, *Structure, Stabilité et Fluctuations*, Masson & Co., Paris.

LS.10a. P. Glansdorff and I. Prigogine, *Thermodynamic Theory of Structure, Stability and Fluctuations*, John Wiley & Sons, [Interscience], New York. Translations in Russian and Japanese

LS-11. I. Prigogine and R. Herman, *Kinetic Theory of Vehicular Traffic*, Elsevier, New York.

1977

LS.12. G. Nicolis and I. Prigogine, *Self-Organization in Nonequilibrium Systems: From Dissipative Structures to Order through Fluctuations*, John Wiley & Sons [Interscience], New York. Translations in Russian, Japanese, Italian, and Chinese.

1980

LS.13. I. Prigogine, *From Being to Becoming: Time and Complexity in the Physical Sciences*, Freeman & Co., San Francisco [French title: *Temps et devenir*, Masson, Paris]. Translations in eight languages.

1987

LS.14. G. Nicolis and I. Prigogine, *Exploring Complexity*, Freeman, New York. Translations in six languages.

1998

LS.15. D. Kondepudi and I. Prigogine, *Modern Thermodynamics, From Heat Engines to Dissipative Structures*, John Wiley & Sons, Chichester. Translations in five languages.

2. General Books

LG.1. I. Prigogine et I. Stengers, *La Nouvelle Alliance—Les Métamorphoses de la Science*, Gallimard, Paris. [English title: *Order Out of Chaos*, Bantam, New York, 1984]. Translations in 18 languages.

1988

LG.2. I. Prigogine, *La Nascita del Tempo* (The Birth of Time), Theoria, Roma.

1992

LG.3. I. Prigogine and I. Stengers, *Entre le Temps et l'Éternit é* (Betwen Time and Eternity), Librairie Arthème Fayard, Paris. Translations in five languages.

1993

LG.4. I. Prigogine, *Le Leggi del Caos* (The laws of Chaos), Laterza, Rome.

LG.5. I. Prigogine und I. Stengers, *Das Paradox der Zeit*: *Zeit, Chaos und Quanten* (The Paradox of Time: Time, Chaos and Quanta), Piper, München.

1994

LG.6. I. Prigogine, *Les Lois du Chaos* (The Laws of Chaos), Collection "Nouvelle Bibliothèque Scientfique," Flammarion, Paris.

1996

LG.7. I. Prigogine, *La Fin des Certitudes* (The End of Certitudes), Odile Jacob, Paris. Translations in 19 languages.

2001

LG.8. I. Prigogine, éd., *L'Homme Devant l'Incertain* (Man Facing the Uncertain), Odile Jacob, Paris.

B. Articles

1. Equilibrium Thermodynamics

1940

THE.1. I. Prigogine, Extension de l'équation de Saha au plasma non-isotherme (Extension of Saha's equation to a non-isothermal plasma), *Bull. Cl. Sci. Acad. Roy. Belg.* **26**, 1–11 (1940).

1941

THE.2. I. Prigogine, Thermodynamique et liaison d'hydrogène (Thermodynamics and the hydrogen bond), *Bull. Soc. Chim. Belg.* **50**, 153–171 (1941).

1942

THE.3. I. Prigogine et R. Hansen, Généralités sur l'introduction de grandeurs thermodynamiques dans l'expression des vitesses réactionnelles (Generalities on the introduction of thermodynamic quantities in the expression of reaction rates), *Bull. Cl. Sci. Acad. Roy. Belg.* **28**, 301–317 (1942).

1943

THE.4. I. Prigogine, Remarque sur la formule fondamentale de la pression osmotique (Remarks on the fundamental formula of osmotic pressure), *Bull. Soc. Chim. Belg.* **52**, 165–166 (1943).

THE.5. R. Defay et I. Prigogine, Systèmes monovariants et systèmes indifférents (Monovariant systems and indifferent systems), *Bull. Cl. Sci. Acad. Roy. Belg.* **29**, 525–535 (1943).

THE.6. I. Prigogine, Remarque sur la modération des titres molaires (Remarks on the moderation of molar titres), *Bull. Cl. Sci. Acad. Roy. Belg.* **29**, 695–699 (1943).

THE.7. I. Prigogine, Sur les phénomènes critiques de dissolution dans les systèmes ternaires (On the critical phenomena of solution in ternary systems), *Bull. Soc. Chim. Belg.* **52**, 115–123 (1943).

THE.8. I. Prigogine, Sur la variation de la température azéotropique sous l'influence de la pression (On the variation of the azeotropic temperature under the influence of pressure), *Bull. Soc. Chim. Belg.* **52**, 165–166 (1943).

1944

THE.9. I. Prigogine et R. Defay, Les phases superficielles idéales à adsorption mobile (The ideal superficial phases with moving adsorption), *Bull. Soc. Chim. Belg.* **53**, 115–140 (1944).

THE.10. I. Prigogine, Contribution à la théorie de l'état liquide (Contribution to the theory of liquid state), *J. Phys. Radium* **5**, 16–22 (1944).

THE.11. I. Prigogine, Contribution à la thermodynamique de l'azéotropie (Contribution to the thermodynamics of azeotropy), *J. Phys. Radium* **5**, 185 (1944).

1946

THE.12. R. Defay et I. Prigogine, Tension superficielle dynamique d'une surface parfaite (Dynamical surface tension of a perfect surface), *Bull. Cl. Sci. Acad. Roy. Belg.* **32**, 400–421 (1946).

THE.13. R. Defay et I. Prigogine, Théorie thermodynamique de la tension superficielle dynamique (Thermodynamic theory of dynamic surface tension), *J. Chim. Phys. Physico-chim. Biol.* **43**, 217–234 (1946).

1947

THE.14. I. Prigogine and R. Defay, On the number of independent constituents and the phase rule (On the number of independent constituents and the phase rule), *J. Chem. Phys.* **15**, 614 (1947).

THE.15. R. Defay et I. Prigogine, Méthode thermodynamique de Th. De Donder et méthode thermodynamique de Schottky, Ulich et Wagner (Thermodynamic method of Th. De Donder and thermodynamic method of Schottky, Ulich and Wagner), *Bull. Cl. Sci. Acad. Roy. Belg.* **33**, 222–232 (1947).

THE.16. I. Prigogine et R. Defay, La stabilité des transformations azéotropiques (Stability of azeotropic transformations), *Bull. Cl. Sci. Acad. Roy. Belg.* **33**, 694–703 (1947).

1949

THE.17. I. Prigogine, P. Van Rysselberghe, and J. L. Finck, Discussion of "On the second law from the standpoint of the equation of state," *J. Franklin Inst.* **247**, 497–503 (1949).

1954

THE.18. I. Prigogine et H. C. Mel, Sur la stabilité thermodynamique (On thermodynamic stability), *Bull. Cl. Sci. Acad. Roy. Belg.* **40**, 588–599 (1954).

1976

THE.19. R. Defay, Il. Prigogine, and A. Sanfeld, Surface thermodynamics, *J. Colloid Interface Sci.* **58**, 498–510 (1976).

2.   Theory of Solutions

1941

SOL.1. I. Prigogine, La structure de solutions d'électrolytes forts en solution concentrée (Structure of solutions of strong electrolytes in concentrated solution), *Bull. Soc. Chim. de Belg.* **50**, 89–98 (1941).

1945

SOL.2. I. Prigogine, La tension superficielle des mélanges binaires (Surface tension of binary mixtures), *Bull. Soc. Chim. Belges* **54**, 286–302 (1945).

1948

SOL.3. I. Prigogine, Surface tension of a regular solution, *Trans. Faraday Soc.* **44**, 626–628 (1948).

SOL.4. I. Prigogine, Etude spectroscopique de la loi d'action de masse (Spectroscopic study of the law of mass action), *J. Chim. Phys.* **45**, 17–21 (1948).

1949

SOL.5. I. Prigogine et R. Defay, Tension superficielle dynamique des solutions régulières (Dynamic surface tension of regular solutions), *J. Chim. Phys.* **46**, 367–372 (1949).

SOL.6. l. Prigogine, V. Mathot et A. Desmyter, Propriétés thermodynamiques des solutions associées (Thermodynamic properties of associated solutions), *Bull. Soc. Chim. Belges* **58**, 547–565 (1949).

1950

SOL.7. I. Prigogine, Sur la tension superficielle des solutions de molécules de dimensions différentes (On surface tension of solutions of molecules of different sizes), *J. Chim. Phys.* **47**, 35–40 (1950).

SOL.8. I. Prigogine et G. Garikian, Sur la thermodynamique statistique des solutions binaires (On the statistical thermodynamics of binary solutions), *Physica* **16**, 239–248 (1950).

SOL.9. I. Prigogine et V. Mathot, The influence of the shape of molecules on the thermodynamic properties of hydrocarbon mixtures, *J. Chem. Phys.* **18**, 765–766 (1950).

SOL.10. I. Prigogine et L. Saroléa, Sur la tension superficielle des solutions de molécules de dimensions différentes (II) (On surface tension of solutions of molecules of different sizes, II), *J. Chim. Phys.* **47**, 807–815 (1950).

1950

SOL.11. I. Prigogine, Le rôle de l'association dans la démixtion des solutions (The role of association in the unmixing of solutions), III-ème Congrès National des Sciences, Bruxelles, 1950.

SOL.12. I. Prigogine et R. Defay, Tension superficielle à la surface de séparation de deux solutions régulières (Surface tension at the separation surface of two regular solutions), *Bull. Soc. Chim. Belges* **59**, 255–262 (1950).

SOL.13. I. Prigogine and R. Defay, Surface tension of regular solutions, *Trans. Faraday Soc.* **46**, 199–204 (1950).

1951

SOL.14. P. Mazur et I. Prigogine, Sur l'hydrodynamique des mélanges liquides de ^{3}He et ^{4}He (On the hydrodynamics of liquid mixtures of ^{3}He and ^{4}He), *Physica* **17**, 680–693 (1951).

1952

SOL.15. I. Prigogine et V. Mathot, Application of the cell model to the statistical thermodynamics of solutions, *J. Chem. Phys.* **20**, 49–57 (1952).

SOL.16. I. Prigogine and J. Maréchal, The influence of difference in molecular size on the surface tension of solutions (IV), *J. Colloid Sci.* **7**, 122–127 (1952).

SOL.17. I. Prigogine, Thermodynamique statistique des solutions et phénomènes critiques de dissolution (Statistical thermodynamics of solutions and critical phenomena solution), C.R. 2-ème Réunion Chim. Phys., Paris, 1952, pp. 95–109.

SOL.18. I. Prigogine et L. Saraga, Stabilité et démixtion des solutions superficielles (Stability and unmixing of superficial solutions), C.R. 2-ème Réunion Chim. Phys., Paris, 1952, pp. 458–462.

SOL.19. I. Prigogine, Pression, diffusion et superfluidité dans les mélanges de gaz parfaits (Pressure, diffusion and superfluidity in mixtures of perfect gases), Colloquium Ultrasonore Trilingen, Cl. Sciences, Acad. Roy. Flamande, 1952, pp. 285–295.

1953

SOL.20. I. Prigogine and A. Bellemans, Statistical thermodynamics of solutions of molecules with spherical field of force, *Faraday Soc. Discussion* **15**, 80–92 (1953).

SOL.21. I. Prigogine, N. Trappeniers and V. Mathot, Statistical thermodynamics of r-mers and r-mer solutions, *Faraday Soc. Discussion* **15**, 93–107 (1953).

SOL.22. I. Prigogine and A. Bellemans, Application of the cell model to the statistical thermodynamics of solutions. Volume change of mixing in solutions of molecules slightly different in Size, *J. Chem. Phys.* **21**, 561–562 (1953).

SOL.23. I. Prigogine, *On the statistical theory of polymers and polymer solutions*, Proceedings, International Conference on Theoretical Physics, Kyoto and Tokyo, Sept. 1953, pp. 398–399.

SOL.24. I. Prigogine, Forces intermoléculaires et volume d'excès des solutions (Intermolecular forces and excess volume in solutions), *Bull. Soc. Chim. Belges* **62**, 125–137 (1953).

SOL.25. I. Prigogine, Forces intermoléculaires et propriétés thermodynamiques (Intermolecular forces and thermodynamic properties), *N. T. Natuurk.* **19**, 301–308 (1953).

SOL.26. I. Prigogine, N. Trappeniers and V. Mathot, On the application of the cell method to r-mer liquids, *J. Chem. Phys.* **21**, 559–560 (1953).

SOL.27. I. Prigogine, V. Mathot, and N. Trappeniers, On the statistical theory of r-mer solutions, *J. Chem. Phys.* **21**, 560–561 (1953).

SOL.28. I. Prigogine and J. Philippot, Sur la théorie moléculaire de l'hélium liquide. III. Les solutions de ^{3}He et de ^{4}He (On the molecular theory of liquid helium. III. The solutions of ^{3}He and ^{4}He), *Physica*, **19**, 235–240 (1953).

1954

SOL.29. I. Prigogine et S. Lafleur, Sur la mécanique statistique des mélanges, I (On statistical mechanics of mixtures, I), *Bull. Cl. Sci. Acad. Roy. Belg.* **40**, 481–496 (1954).

SOL.30. I. Prigogine et S. Lafleur, Sur la mécanique statistique des mélanges, II (On statistical mechanics of mixtures, II), *Bull. Cl. Sci. Acad. Roy. Belg.* **40**, 497–507 (1954).

SOL.31. I. Prigogine, On the theory of liquid helium, *Philos. Mag. Suppl.* **3**, 131–148 (1954).

1955

SOL.32. I. Prigogine and A. Bellemans, Statistical Theory of r-mer Solutions, *J. Polymer Sci.* **18**, 147–151 (1955).

1956

SOL.33. I. Prigogine, A. Bellemans, and A. Englert-Chwoles, Statistical thermodynamics of solutions, *J. Chem. Phys.* **24**, 518–527 (1956).

1957

SOL.34. A. Bellemans, C. Naar-Colin, and I. Prigogine, Statistical thermodynamics of r-mer mixtures, *J. Chem. Phys.* **26**, 712 (1957).

SOL.35. I. Prigogine, A. Bellemans, and C. Naar-Colin, Theorem of corresponding states for polymers, *J. Chem. Phys.* **26**, 710 (1957).

SOL.36. I. Prigogine, A. Bellemans, and C. Naar-Colin, *Theorem of corresponding states for polymers*, *J. Chem. Phys.* **26**, 751–755 (1957).

SOL.37. I. Prigogine and A. Bellemans, On isotope mixtures, Proceedings, Symposium on Isotope Separation, Amsterdam, 1957.

1958

SOL.38. I. Prigogine and A. Bellemans, Isotopic mixtures, *Il Nuovo Cimento* **9**, Suppl. **1**, 342–344 (1958).

SOL.39. I. Prigogine and A. Englert-Chwoles, On the statistical theory of the surface tension of binary mixtures, *Il Nuovo Cimento* **9**, Suppl. **1**, 347–355 (1958).

SOL.40. I. Prigogine and A. Englert-Chwoles, Sur la théorie statistique de la tension superficielle des mélanges binaires (On the statistical theory of the surface tension of binary mixtures), *J. Chim. Phys.*, p. 16 (1958).

3.　Nonequilibrium Thermodynamics: Linear Regime

1945

THL.1. I. Prigogine, Modération et transformation irréversibles des systèmes ouverts (Moderation and irreversible transformation in open systems), *Bull. Cl. Sci. Acad. Roy. Belg.* **31**, 600–606 (1945).

THL.2. I. Prigogine, Sources et courants d'entropie (Entropy sources and fluxes), *Comptes Rendus Acad. Sci.*, 1945.

1946

THL.3. I. Prigogine, Remarque sur le principe de réciprocité d'Onsager et le couplage des réactions chimiques (Remarks on Onsager's reciprocity principle and the coupling of chemical reactions), *Bull. Cl. Sci. Acad. Roy. Belg.* **32**, 30–35 (1946).

THL.4. R. Defay et I. Prigogine, Sur l'extension de la formule de Gibbs à la tension superficielle des surfaces non en équilibre (On the extension of Gibbs' formula to the surface tension of nonequilibrium surfaces), *Bull. Cl. Sci. Acad. Roy. Belg.* **32**, 36–51 (1946).

THL.5. R. Defay et I. Prigogine, Les potentiels chimiques latéraux d'une phase superficielle non en équilibre (The lateral chemical potentials of a nonequilibrium superficial phase), *Bull. Cl. Sci. Acad. Roy. Belg.* **32**, 176–184 (1946).

THL.6. R. Defay et I. Prigogine, Energie libre et Tension superficielle des surfaces non en équilibre (Free energy and surface tension of nonequilibrium surfaces), *Bull. Cl. Sci. Acad. Roy. Belg.* **32**, 335–350 (1946).

THL.7. I. Prigogine et J. M. Wiame, Biologie et thermodynamique des phénomènes irréversibles (Biology and thermodynamics of irreversible phenomena), *Experientia* **2**, 451–453 (1946).

1947

THL.8. I. Prigogine, Le rendement thermodynamique de la thermodiffusion (The thermodynamic efficiency of thermodiffusion), *Physica* **13**, 319–320 (1947).

1948

THL.9. I. Prigogine et A. Mertens, Sur l'hypothèse d'équilibre dans la théorie des vitesses absolues de réaction (On the equilibrium hypothesis in the theory of absolute reaction rates), in *Contribution à l'Etude de la Structure Moléculaire*, volume commémoratif Victor Henri, Desoer, Liège, 1948.

THL.10. I. Prigogine, P. Outer, and C. Herbo, Affinity and reaction rate close to equilibrium, *J. Phys. Colloid Chem.* **52**, 321–331 (1948).

THL.11. I. Prigogine, Remarque sur la dynamique des mélanges gazeux avec diffusion (Remarks on the dynamics of gaseous mixtures with diffusion), *Bull. Cl. Sci. Acad. Roy. Belg.* **34**, 789–794 (1948).

THL.12. I. Prigogine, Sur le rôle de la vitesse d'ensemble dans la diffusion (On the role of the bulk velocity in diffusion), *Bull. Cl. Sci. Acad. Roy. Belg.* **34**, 930–942 (1948).

THL.13. I. Prigogine, Quelques aspects de la thermodynamique des phénomènes irréversibles (Some aspects of thermodynamics of irreversible phenomena), Coll. Thermodynamique, Un. Internat. Phys., Bruxelles, 1948.

1949

THL.14. I. Prigogine et R. Buess, Sur la thermodiffusion dans le système eau-alcool éthylique (On the thermodiffusion in the system water-ethylic alcohol), *Physica* **15**, 465–466 (1949).

THL.15. I. Prigogine, Le domaine de validité de la thermodynamique des phénomènes irréversibles (On the domain of validity of the thermodynamics of irreversible phenomena), *Physica* **15**, 272–284 (1949).

1950

THL.16. I. Prigogine, L. de Brouckère, et R. Amand, Recherches sur la thermodiffusion en phase liquide (I) (Research on thermodiffusion in liquid phase (I)), *Physica* **16**, 577–598 (1950).

THL.17. I. Prigogine, L. de Brouckère, et R. Amand, Recherches sur la thermodiffusion en phase liquide (II) (Research on thermodiffusion in liquid phase (II)), *Physica* **16**, 851–860 (1950).

THL.18. I. Prigogine, Sur les fluctuations de l'équilibre chimique (On the fluctuations of chemical equilibrium), *Physica* **16**, 134–136 (1950).

1951

THL.19. I. Prigogine et P. Mazur, Thermodynamique des effets thermomagnétiques et galvano-magnétiques (Thermodynamics of the thermomagnetic and of the galvanomagnetic effects), *J. Phys.* **12**, 616–620 (1951).

THL.20. P. Mazur et I. Prigogine, Sur deux formulations de l'hydrodynamique et le problème de l'Hélium liquide II (On two formulations of hydrodynamics and the problem of liquid helium II), *Physica* **17**, 661–679 (1951).

THL.21. I. Prigogine, The equilibrium hypothesis in chemical kinetics, *J. Phys. Chem. Colloid Chem.* **55**, 765–772 (1951).

1952

THL.22. I. Prigogine, Thermodynamics of irreversible processes, *Appl. Mechanics Rev.* **5**, 193–195 (1952).

THL.23. I. Prigogine, Thermodynamique de la matière dans un champ électromagnétique (Thermodynamics of matter in an electromagnetic field), *J. Chim. Phys.*, **49**, 79–83 (1952).

THL.24. I. Prigogine et R. Buess, Distribution de matière et phénomènes de transport en présence de gradient de température et réaction chimique (I) (Distribution of matter and transport phenomena in presence of a temperature gradient and of chemical reactions (I)), *Bull. Cl. Sci. Acad. Roy. Belg.* **38**, 711–717 (1952).

THL.25. I. Prigogine et R. Buess, Distribution de matière et phénomènes de transport en présence de gradient de température et réaction chimique (II), (Distribution of matter and transport phenomena in presence of a temperature gradient and of chemical reactions (II)), *Bull. Cl. Sci. Acad. Roy. Belg.* **38**, 851–860 (1952).

THL.26. I. Prigogine, L. de Brouckère et R. Buess, Recherches sur la thermodiffusion en phase liquide, Thermodiffusion de l'eau (V) (Research on thermodiffusion in liquid phase. Thermodiffusion of water (V)), *Physica* **18**, 915–920 (1952).

1953

THL.27. I. Prigogine, On some aspects of Thermodynamics of Irreversible Processes, Proceedings International Conference on Theoretical Physics, Kyoto and Tokyo, Sept. 1953, pp. 475–487.

THL. 28. I. Prigogine, P. Mazur et R. Defay, Bilans thermodynamiques locaux dans les systèmes électrochimiques (Local thermodynamic balances in electrochemical systems), *J. Chim. Phys.* **50**, 146–155 (1953).

THL.29. I. Prigogine et P. Mazur, Sur l'extension de la thermodynamique aux phénomènes irréversibles liés aux degrés de libeerté internes (On the extension of thermodynamics to the irreversible phenomena related to internal degrees of freedom), *Physica* **19**, 241–254 (1953).

1954

THL.30. I. Prigogine, Sur la théorie variationnelle des phénomènes irréversibles (On the variational theory of irreversible phenomena), *Bull. Cl. Sci. Acad. Roy. Belg.* **40**, 471–483 (1954).

THL.31. I. Prigogine, Sur le problème de la vitesse des réactions chimiques (On the problem of chemical reaction rates), *Cahiers Phys.* **52**, 49–58 (1954).

1980

THL.32. P. Glansdorff et I. Prigogine, Théophile De Donder et la découverte de l'affinité (Th. De Donder and the discovery of affinity), *Bull. Soc. Math. Belg.* **31**, 41–46 (1980).

4. *Nonlinear Thermodynamics and Complex Systems*

1954

TNC.1. P. Glansdorff et I. Prigogine, Sur les propriétés différentielles de la production d'entropie (On the differntial properties of the entropy production), *Physica* **20**, 773–780 (1954).

1955

TNC.2. I. Prigogine et G. Mayer, Fluctuations dans les systèmes stationnaires de non-équilibre (Fluctuations in the stationary nonequilibrium systems), *Bull. Cl. Sci. Acad. Roy. Belg.* **41**, 22–29 (1955).

TNC.3. I. Prigogine et R. Balescu, Sur les propriétés différentielles de la production d'entropie, II (On the differntial properties of the entropy production, II), *Bull. Cl. Sci. Acad. Roy. Belg.* **41**, 917–928 (1955).

TNC.4. I. Prigogine, Thermodynamics of Irreversible Processes and Fluctuations, *Temperature* **2**, 215–232 (1955).

1956

TNC. 5. I. Prigogine et R. Balescu, Phénomènes cycliques dans la thermodynamique des phénomènes irréversibles (Cyclical phenomena in thermodynamics of irreversible phenomena), *Bull. Cl. Sci. Acad. Roy. Belg.* **42**, 256–265 (1956).

1958

TNC.6. I. Prigogine and R. Balescu, Cyclic Processes in Irreversible Thermodynamics, in Proceedings International Symposium on Transport Processes in Statistical Mechanics, Brussels, 1958, Interscience Publishers, New York, 1958, pp. 343–345.

1961

TNC.7. I. Prigogine, Evolutionsprobleme in der Thermodynamik irreversibler Prozesse (Evolution problems in thermodynamics of irreversible processes), *Z. Chemie* **1**, 203–208 (1961).

TNC.8. I. Prigogine and R. Balescu, Non-equilibrium thermodynamics, in Thermodynamics of Nuclear Materials, International Atomic Energy Agency, Vienna, 1962.

TNC.9. P. Glansdorf, I. Prigogine, and D. F. Hays, Variational properties of a viscous liquid at a nonuniform temperature, *Phys. Fluids* **5**, 144–149 (1962).

1963

TNC.10. P. Glansdorff and I. Prigogine, Generalized entropy production and hydrodynamic stability, *Phys.Lett.* **7**, 243–244 (1963).

1964

TNC.11. I. Prigogine, Steady states and variational principles, in *The Law of Mass Action*, Universitetsforlaget, Oslo, 1964, pp. 95–116.

TNC.12. P. Glansdorff and I. Prigogine, On a general evolution criterion in macroscopic physics, *Physica* **30**, 351–374 (1964).

1965

TNC.13. I. Prigogine, Steady states and entropy production, *Physica* **31**, 719–724 (1965).

TNC.14. I. Prigogine and P. Glansdorff, Variational properties and fluctuation theory, *Physica* **31**, 1242–1256 (1965).

1966

TNC.15. I. Prigogine, Evolution Criteria, Variational principles and fluctuations, in *Nonequilibrium Thermodynamics, Variational Techniques and Stability*, University of Chicago, 1966, pp. 3–16.

1967

TNC.16. I. Prigogine and G. Nicolis, On symmetry-breaking instabilities in dissipative systems, *J. Chem. Phys.* **46**, 3542–3550 (1967).

TNC.17. I. Prigogine, Dissipative structures in chemical systems, in *Fast Reactions and Primary Processes in Chemical Kinetics*, Nobel Symposium **5**, S. Claesson, ed., Interscience, New York, 1967, pp. 371–382.

TNC.18. R. Lefever, G. Nicolis and I. Prigogine, On the occurrence of oscillations around the steady state in systems of chemical reactions far from equilibrium, *J. Chem. Phys.* **47**, 1045–1047 (1967).

1968

TNC.19. I. Prigogine and R. Lefever, On symmetry-breaking instabilities in dissipative systems, II, *J. Chem. Phys.* **48**, 1695–1700 (1968).

1969

TNC.20. I. Prigogine, Structure, dissipation and life, in *Theoretical Physics and Biology*, M. Marois, ed., North Holland, Amsterdam, 1969.

TNC.21. I. Prigogine, R. Lefever, A. Goldbeter, and M. Herschkowitz-Kaufman, *Symmetry Breaking Instabilities in Biological Systems*, Nature **223**, 913–916 (1969).

TNC.22. P. Glansdorff and I. Prigogine, On the general theory of stability of thermodynamic equilibrium, in: *Problems of Hydrodynamics and Continuum Mechanics*, Society for Industrial and Applied Mathematics, Philadelphia, 1969.

1970

TNC.23. P. Glansdorff and I. Prigogine, Non-equilibrium stability theory, *Physica* **46**, 344–366 (1970).

1971

TNC. 24. I. Prigogine and R. Lefever, Termodinamica e Biologia (Thermodynamics and biology), in *Enciclopedia della Scienza e della Tecnica*, Mondadori, 1971.

TNC.25. I. Prigogine, Dissipative structures in biological systems, in *De la physique théorique à la biologie*, M. Marois, ed., Ed. CNRS, Paris, 1971.

TNC.26. I. Prigogine and A. Babloyantz, Coherent structures and thermodynamic stability, in *Chemical Evolution and the Origin of Life*, I, R. Buvet and C. Ponnamperuma, eds., North Holland, Amsterdam, 1971.

TNC.27. I. Prigogine and G. Nicolis, Biological order, structure and instabilities, *Q. Rev. Biophys.* **4**, 107–148 (1971).

TNC.28. I. Prigogine and R. Lefever, Thermodynamics, structure and dissipation, *Experientia, Suppl.* **18**, 101–126 (1971).

TNC.29. I. Prigogine, Entropy and dissipative structures, in *Lecture Notes in Physics*, Springer, **7**, 1–19 (1971).

TNC.30. G. Nicolis and I. Prigogine, *Fluctuations in Nonequilibrium Systems*, Proc. Nat Acad. Sciences, **68**, 2102–2107 (1971).

1972

TNC.31. R. S. Schechter, I. Prigogine, and J. R. Hamm, Thermal diffusion and convective stability, *Phys. Fluids* **15**, 379–386 (1972).

1973

TNC.32. I. Prigogine and G. Nicolis, Fluctuations and the mechanism of instabilities, Proceedings, 3rd International Conference: *From Theoretical Physics to Biology*, Versailles, 1971, Karger, Basel, 1973, pp. 89–109.

TNC.33. I. Prigogine et P. Glansdorff, L'écart à l'équilibre interprété comme une source d'ordre. *Structures dissipatives* (The departure from equilibrium interpreted as a source of order), *Bull. Cl. Sci. Acad. Roy. Belg.* **59**, 673–702 (1973).

TNC.34. I. Prigogine and R. Lefever, Theory of dissipative structures, Proceedings, *"Symposium on Synergetics,"* Schloss Elmau 1972, H. Haken, ed., Teubner, Stuttgart, 1973, pp.124–135.

1974

TNC.35. I. Prigogine et R. Lefever, Aspect thermodynamique et stochastique de l'évolution (Thermodynamic and stochastic aspects of evolution), Ecole de Roscoff, 1974, pp. 95–103.

TNC.36. I. Prigogine, G. Nicolis, and A. Babloyantz, Nonequilibrium problems in biological phenomena, *Ann. New York Acad. Sci.* **231**, 99–105 (1974).

TNC.37. P. Glansdorff, I. Prigogine, and G. Nicolis, The thermodynamic stability theory of nonequilibrium states, *Proc. Natl. Acad. Sci.* **71**, 197–199 (1974).

1975

TNC.38. I. Prigogine, Dissipative structures, dynamics and entropy, *Int. J. Quant. Chem. Symp.* **9**, 443–456 (1975).

TNC.39. I. Prigogine and R. Lefever, Stability and thermodynamic properties of dissipative structures in biological systems, in *Proceedings, 1st Aharon Katzir-Katchalsky Conference*, I. R. Miller, ed., Wiley, New York, 1975, pp. 26–57.

TNC.40. I. Prigogine, R. Lefever, J. S. Turner, and J. W. Turner, Stochastic theory of metastable states and nucleation, *Phys. Lett.* **51a**, 317–319 (1975).

TNC.41. I. Prigogine, G. Nicolis, R. Herman, and T. Lam, Stability, fluctuations and complexity, *Collect. Phenom.* **2**, 103–109 (1975).

TNC.42. I. Prigogine, G. Nicolis, and R. Lefever, *Models for Cellular Communication*, Institute De la Vie, 1975, pp. 50–81.

TNC.43. A. Babloyantz, J. Hiernaux, and I. Prigogine, Some remarks on thermodynamics and kinetics of self-organization, in *Molecular Biology of Nucleocytoplasmic Relationships*, Elsevier, Amsterdam, 1975, pp. 315–323.

TNC.44. G. Nicolis, I. Prigogine, and P. Glansdorff, On the mechanism of instabilities in nonlinear systems, *Adv. Chem. Phys.* **32**, 1–11 (1975).

TNC.45. I. Prigogine et M. Herschkowitz-Kaufman, Organisation spatiotemporelle et réactions chimiques (Spatio-temporal organization and chemical reactions), in *Reaction Kinetics in Heterogeneous Chemical Systems*, Proceedings, 25th International Meeting of Societé Chimie France, Dijon, 1974, Elsevier, Amsterdam 1975.

TNC. 46. I. Prigogine and R. Lefever, Stability and self-organization in open systems, *Adv. Chem. Phys.* **29**, 1–28 (1975).

1976

TNC.47. I. Prigogine, L'ordre par fluctuations et le système social (Order through fluctuations and the social system), *Rhein. Westf. Akad. Wiss., Vorträge*, no. 260, pp. 1–74, 1976.

TNC.48. G. Nicolis and I. Prigogine, Thermodynamic aspects and bifurcation analysis of spatio-temporal dissipative structures, in Proceedings, Faraday Symposium Chemical Society, no. 9, *Physical Chemistry of Oscillatory Phenomena*, 1975, pp. 7–20.

1977

TNC.49. I. Prigogine, L'ordre par fluctuations et le système social (Order through fluctuations and the social system), in *L'idée de Régulation dans les Sciences*, Maloine, Paris, 1976, pp. 153–191.

TNC.50. I. Prigogine, P. M. Allen, et J. L. Deneubourg, *Fluctuations et évolution de la complexité* (Fluctuations and evolution of complexity), in *"Modélisation et Maîtrise des Systèmes Techniques,"* Congrès AFCET, Versailles, éd. Hommes et Techniques, Paris 1977, pp. 26–41.

TNC.51. I. Prigogine, P. M. Allen, and R. Herman, *Long term trends and the evolution of complexity*, in *Goals in a Global Community*, Rapport Club de Rome, Vol. 1, E. Laszlo and J. Bierman, eds., Pergamon, New York, 1977, pp. 41–62.

1978

TNC.52. I. Prigogine and R. Lefever, *Coupling between difusion and chemical reactions*, in *16-e Conseil Solvay de Chimie*, 1976, Wiley, New York, 1978, pp. 1–53.

TNC.53. I. Prigogine and J. W. Turner, *Nonequilibrium phase transitions*, Proc. Karcher Symp. *Structural Aspects of Homogeneous, Heterogeneous and Biological Catalysis*, University Oklahoma, American Crystallographic Association, pp. 103—141.

TNC.54. I. Prigogine, New aspects of chemical kinetics and nonequilibrium phase transitions, Proceedings of Symposium *Structure and Dynamics in Chemistry*, Uppsala, 1977, P. Ahlberg and L.-O. Sundelöf, eds., pp. 172–186.

1980

TNC.55. I. Prigogine and R. Lefever, Stability problems in cancer growth and nucleation, *Comp. Biochem. Physiol.* **67B**, 389–393, 1980.

TNC.56. I. Prigogine, Nonequilibrium thermodynamics and chemical evolution, in *Aspects of Chemical Evolution*, 17th Solvay Conference Chemistry, G. Nicolis, ed., Wiley, New York 1980, pp. 43–62.

TNC.57. P. Glansdorff and I. Prigogine, Thermodynamics, nonequilibrium, in *Encyclopaedia of Physics*, Addison-Wesley, pp. 1029–1034.

1981

TNC.58. I. Prigogine and A. Goldbeter, Niet-evenwichtszelforganisatie in biochemische stelsels (Nonequilibrium self-organization in biochemical systems), *Farmaceut. Tijdschr. Belgie* **58**, 99–107, 1981.

TNC.59. D. K. Kondepudi and Il. Prigogine, Sensitivity of nonequilibrium systems, *Physica*, **107A**, 1–24, 1981.

TNC.60. I. Prigogine and A. Goldbeter, Nonequilibrium self-organization in biochemical systems, in *Biochemistry of Exercise*, Vol. **3**, J. Poortmans, ed., University Park Press, Baltimore, 1981.

TNC.61. G. Nicolis and I. Prigogine, Symmetry breaking and pattern selection in far from equilibrium systems, in *Proceedings, International Symposium Pierre Curie*, Paris, 1980, pp. 35–48.

TNC.62. G. Nicolis and I. Prigogine, Symmetry breaking and pattern selection in far from equilibrium systems, *Proc. Natl. Acad. Sci. USA* **78**, 659–663 (1981).

1982

TNC.63. I. Prigogine and P. M. Allen, The challenge of complexity, in *Self-Organization and Dissipative Structures*, W. C. Schieve and P. M. Allen, eds., University of Texas Press, Austin, 1982, pp. 3–39.

TNC.64. M. Malek-Mansour, G. Nicolis, and I. Prigogine, Nonequilibrium phase transitions in chemical systems, in *Thermodynamics and Kinetics of Biological Processes*, I. Lamprecht and A. I. Zotin, eds., W. de Gruyter, Berlin, 1982, pp. 75–103.

TNC.65. I. Prigogine and R. Lefever, On the spatio-temporal evolution of cellular tissues, in *Biological Structures and Coupled Flows*, A. Oplatka and M. Balaban, eds., Academic Press, New York, 1983, pp. 3–26.

1983

TNC.66. I. Prigogine, G. Nicolis, M. Mansour, and F. Baras, Fluctuations and explosive behavior in nonlinear chemical systems, in Proceedings, Symposium *Nonlinear Problems in Energy Engineering*, Argonne National Laboratory, 1983.

1984

TNC.67. I. Prigogine, Nonequilibrium Thermodynamics and Chemical Evolution: An Overview, in: *Aspects of Chemical Evolution*, G. Nicolis, ed., Wiley, New York, 1984.

1985

TNC.68. I. Prigogine and G. Nicolis, Self-organization in nonequilibrium systems: Towards a dynamics of complexity, in *Bifurcation Analysis*, M. Hatzewinkel, ed., Reidel, Dordrecht, 1985, pp. 3–12.

1986

TNC.69. R. Lefever and I. Prigogine, Sistema immunitario e stabilità. Un esempio di transizioni in fase di non-equilibrio in biologia (Immunitarian system and stability. An example of nonequilibrium phase transitions in biology), *Fondamenti* **5**, 179–192 (1986).

1989

TNC.70. I. Prigogine, From dynamical systems to socio-economic models, in *International Symposium on Evolutionary Dynamics and Nonlinar Economics*, University of Texas at Austin, 1989.

1990

TNC.71. P. Glansdorff et I. Prigogine, *Thermodynamique*, Dictionnaire encyclopédique Quillet, Suppl. J–Z, pp. 352–356.

TNC.72. I. Prigogine, Entropy revisited, in *Capillarity Today*, G. Pétré and A. Sanfeld, eds., Springer, New York, 1990, pp. 3–13.

1991

TNC.73. I. Prigogine, Theoretical physics and biology: An on-going dialogue, in *1st World Congress on Cellular and Molecular Biology*, Paris.

TNC.74. I. Prigogine, Schrödinger and the riddle of life, in *Molecular Theories of Cell Life and Death*, S. Ji, ed., Rutgers University Press, New Brunswick, 1991, pp. 238–242.

1997

TNC.75. G. Dewel, D. Kondepudi, and I. Prigogine, *Chemistry far from equilibrium—Thermodynamics, Order and Chaos*, New Chemistry, 1997.

TNC.76. J. P. Boon et Il. Prigogine, *Le temps dans la forme musicale* (Time in musical form), in *Le Temps et la Forme*, Darballay, E., ed., Genève, 1997.

TNC.77. I. Prigogine et J. P. Boon, Temps et complexité en physique et en musique, (Time and complexity in physics and in music), in *Le temps et la forme*, Darballay, E., ed., Genève, 1997.

TNC.78. D. Kondepudi and I. Prigogine, Thermodynamics, nonequilibrium, *Encyclopaedia Appl. Phys.* **21**, 311–337 (1997).

2000

TNC.79. I. Prigogine, La thermodynamique de la vie (Thermodynamics of life), La Recherche, Numéro spécial, *30 Ans de Science et de Recherche*, no. 331, mai 2000, pp. 38–41.

5. Vehicular Traffic

1960

TFL.1. I. Prigogine and F. C. Andrews, A Boltzmann-like approach for traffic flow, *Operations Res.* **8**, 789–797 (1960).

1961

TFL.2. I. Prigogine, A Boltzmann-like approach to the statistical theory of traffic flow, in *Theory of Traffic Flow*, Elsevier, Amsterdam, 1961.

1962

TFL.3. I. Prigogine, R. Herman, and R. L. Anderson, On individual and Collective Flow, *Bull. Cl. Sci. Acad. Roy. Belg.* **48**, 792–804 (1962).

TFL.4. I. Prigogine, P. Résibois, R. Herman, and R. L. Anderson, On a generalized Boltzmann-like approach for traffic flow, *Bull. Cl. Sci. Acad. Roy. Belg.* **48**, 805–814 (1962).

TFL.5. R. L. Anderson, R. Herman, and I. Prigogine, On the statistical distribution function theory of traffic flow, *Operations Res.* **10**, 180–196 (1962).

1972

TFL.6. I. Prigogine, R. Herman, and T. Lam, *Kinetic theory of vehicular traffic. Comparison with Data, Transportation Sci.* **6**(4) (1972).

1973

TFL.7. R. Herman, T. Lamm, and I. Prigogine, Multiline vehicular traffic and adaptive human behavior, *Science* **179**, 918–920 (1973).

1979

TFL.8. R. Herman and I. Prigogine, A two-fluid approach to town traffic, *Science*, **204**, 148–151 (1979).

1981

TFL.9. I. Prigogine, J. L. Deneubourg, A. De Palma, G. Engelen, et M. Sanglier, *Rapport final sur un modèle nonlinéaire de trafic urbain* (Final report on a nonlinear model of urban traffic), Secrétariat d'Etat à la Région Bruxelloise, et Institut International de Physique et de Chimie Solvay.

1996

TFL.10. I. Prigogine and R. Herman, Prologue, Proceedings of the Workshop *Traffic and Granular Flow*, World Scientific, Singapore.

6. Equilibrium Statistical Mechanics

1941

MSE.1. J. Géhéniau et I. Prigogine, Sur la mécanique statistique quantique, (On quantum statistical mechanics), *Bull. Cl. Sci. Acad. Roy. Belg.* **27**, 513–523 (1941).

1942

MSE.2. I. Prigogine et S. Raulier, Chaleur spécifique à volume constant des liquides mono-atomiques (Constant volume specific heat of monoatomic liquids), *Physica* 396–404 (1942).

MSE.3. I. Prigogine, Remarques sur la viscosité des liquides (Remarks on the viscosity of liquids), *Physica* **9**, 405–406 (1942).

1947

MSE.4. J. Géhéniau, M. Demeur, et I. Prigogine, Sur les abondances relatives des éléments (On the relative abundances of the elements), *Physica* **13**, 429–432 (1947).

1948

MSE.6. I. Prigogine et G. Garikian, Sur le modèle de l'état liquide de Lennard–Jones et Devonshire (On the Lennard-Jones and Devonshire model of the liquid state), *J. Chim. Phys.*, **45**, 273–289 (1948).

1950

MSE.7. I. Prigogine, Sur la perturbation de la distribution de Maxwell (On the perturbation of the Maxwell distirbution), *Physica* **16**, 51–00, (1950).

MSE.8. I. Prigogine, Remarque sur les ensembles statistiques dans les variables pression, température, potentiels chimiques (Remarks on the statistical ensembles in pressure, temperature and chemical potentials variables), *Physica* **16**, 133–136 (1950).

MSE.9. I. Prigogine et P. Janssens, Une généralisation de la méthode de Lennard–Jones–Devonshire pour le calcul de l'intégrale de configuration (A generalization of the Lennard-Jones and Devonshire method foir the calculation of the configuration integral), *Physica* **16**, 895–906 (1950).

1952

MSE.10. I. Prigogine et L. Saraga, Sur la tension superficielle et le modèle cellulaire de l'état liquide (On the surface tension and the cell model of liquid state), *J. Chim. Phys.* **49**, 399–407 (1952).

MSE.11. I. Prigogine, L. Saroléa, and L. Van Hove, On the combinatory factor in regular assemblies, *Trans. Faraday Soc.* **48**, 485–492 (1952).

MSE.12. I. Prigogine et J. Philippot, Théorie moléculaire du point lambda de l'hélium liquide, (Molecular theory of the lambda point of liquid helium), *Physica* **18**, 729–748 (1952).

1953

MSE.13. I. Prigogine et J. Philippot, Sur la théorie moléculaire de l'hélium liquide, II. Le coefficient de dilatation thermique de l'hélium liquide (Molecular theory of liquid helium, II. The coefficient of thermal dilatation of liquid helium), *Physica* **19**, 227–234 (1953).

MSE.14. I. Prigogine et J. Philippot, Sur la théorie moléculaire de l'Hélium liquide, IV. Le caractère coopératif de la transition du point lambda (Molecular theory of liquid helium, IV. The cooperative character of the lambda point transition), *Physica* **19**, 508–516 (1953).

1954

MSE.15. I. Prigogine, R. Bingen, et J. Jeener, Effets isotopiques et propriétés thermodynamiques en phase condensée, I (Isotope effects and thermodynamic properties in condensed phase, I), *Physica* **20**, 383–394 (1954).

MSE.16. I. Prigogine, R. Bingen, et J. Jeener, Effets isotopiques et propriétés thermodynamiques en phase condensée, II (Isotope effects and thermodynamic properties in condensed phase, II), *Physica* **20**, 516–520 (1954).

MSE.17. I. Prigogine, R. Bingen, et A. Bellemans, Effets isotopiques et propriétés thermody-namiques en phase condensée, III (Isotope effects and thermodynamic properties in condensed phase, III), *Physica* **20**, 633–654 (1954).

1964

MSE.18. I. Prigogine, The molecular theory of surface tension, in *Cavitation in Real Liquids*, R. Davies, ed., Elsevier, Amsterdam, 1964, pp. 147–163.

1980

MSE.19. I. Prigogine and A. Bellemans, Statistical mechanics of surface tension and adsorption, in *Adhesion and Adsorption of Polymers*, L. H. Lee, ed., Plenum, New York, 1980, pp. 5–14.

1984

MSE.20. I. Prigogine and A. Bellemans, Statistical mechanics of surface tension and adsorption, in *Adhesion and Adsorptions of Polymers*, L. H. Lee, ed., Plenum, New York, 1984, pp. 5–14.

2000

MSE.21. R. Balescu and I. Prigogine, Equilibrium statistical mechanics, in *The Legacy of Léon Van Hove*, A. Giovannini, ed., World Scientific Series in 20th Century Physics, Vol. 28, World Scientific, Singapore, 2000, pp. 119–122.

7. *Mécanique Statistique de Non-équilibre, Systèmes Dynamiques*

1949

MSN.1. I. Prigogine, Sur la perturbation de la distribution de Maxwell–Boltzmann par des réactions chimiques (On the perturbation of the Maxwell–Boltzmann distribution by chemical reactions), *Suppl. Nuovo Cimento* **6**, 289–295 (1949).

MSN.2. I. Prigogine and E. Xhrouet, On the perturbation of Maxwell distribution function by chemical reactons in gases, *Physica* **15**, 913–932 (1949).

MSN.3. I. Prigogine, Le domaine de validité de la thermodynamique des phénomènes irréver-sibles, (The domain of validity of thermodynamics of irreversible phenomena), *Physica* **15**, 272–284 (1949).

1953

MSN.4. I. Prigogine, On the statistical mechanics of irreversible processes, in *Proceedings International Conference on Theoretical Physics*, Kyoto and Tokyo, Sept. 1953, pp. 464–470.

MSN.5. G. Klein et I. Prigogine, Sur la mécanique statistique des phénomènes irréversibles, I (On statistical mechanics of irreversible phenomena, I), *Physica* **19**, 74–00 (1953).

MSN.6. G. Klein et I. Prigogine, Sur la mécanique statistique des phénomènes irréversibles, II (On statistical mechanics of irreversible phenomena, II), *Physica* **19**, 89–00 (1953).

MSN.7. G. Klein et I. Prigogine, Sur la mécanique statistique des phénomènes irréversi-bles, III (On statistical mechanics of irreversible phenomena, III), *Physica* **19**, 1053–1071 (1953).

1955

MSN.8. I. Prigogine et R. Bingen, Sur la mécanique statistique des phénomènes irréversibles, IV (On statistical mechanics of irreversible phenomena, IV) *Physica* **21**, 299–311 (1955).

MSN.9. F. Waelbroeck, S. Lafleur, et I. Prigogine, Conductibilité thermique des gaz réels (Thermal conductivity of real gases), *Physica* **21**, 667–675 (1955).

1956

MSN.10. R. Brout and I. Prigogine, Statistical mechanics of irreversible processes, Part V: Anharmonic Forces, *Physica*, **22**, 35–47 (1956).

MSN.11. R. Brout and I. Prigogine, Statistical mechanics of irreversible processes, Part VI: Thermal conductivity of crystals, *Physica* **22**, 263–272 (1956).

MSN.12. R. Brout and I. Prigogine, Statistical mechanics of irreversible processes, Part VII: General theory of weakly coupled systems, *Physica* **22**, 621–636 (1956).

MSN.13. I. Prigogine, On the statistical mechanics of irreversible processes, *Can. J. Phys.* **34**, 1236–1245 (1956).

1957

MSN.14. I. Prigogine et R. Balescu, Sur la théorie moléculaire du mouvement brownien (On the molecular theory of Brownian motion), *Physica* **23**, 556–568 (1957).

MSN.15. I. Prigogine and J. Philippot, On irreversible processes in non-uniform systems, *Physica* **23**, 569–584 (1957).

MSN.16. I. Prigogine and F. Henin, On the general perturbational treatment of irreversible processes, *Physica* **23**, 585–596 (1957).

MSN.17. I. Prigogine et F. Henin, Equation de Liouville et section efficace (Liouville equation and scattering cross section), *Bull. Acad. Roy. Belg. (Cl. Sci.)* **43**, 814–827 (1957).

1958

MSN.18. I. Prigogine, Aspects nouveaux de la mécanique classique (New aspects of classical mechanics), *Bull. Cl. Sci. Acad. Roy. Belg.* **44**, 1066–1081 (1958).

MSN.19. I. Prigogine and R. Brout, Irreversible processes in weakly coupled systems, Proceedings International Symposium on *Transport Processes in Statistical Mechanics*, part II, Brussels, 1958, Interscience Publishers, New York, 1958, p. 25–32.

MSN.20. I. Prigogine and F. Henin, On the transport equation for dilute gases, *Physica* **24**, 214–230 (1958).

MSN.21. I. Prigogine and P. Résibois, On the approach to equilibrium of a quantum gas, *Physica* **24**, 795–816 (1956).

MSN.22. I. Prigogine and M. Toda, On irreversible processes in quantum mechanics, *Mol. phys.* **1**, 48–62 (1958).

1959

MSN.23. I. Prigogine and T. A. Bak, Diffusion and chemical reaction in a one-dimensional condensed system, *J. Chem. Phys.* **31**, 1368–1370 (1959).

MSN.24. I. Prigogine and S. Ono, On the transport equation in quantum gases, *Physica* **25**, 171–178 (1959).

MSN.25. I. Prigogine and R. Balescu, Irreversible processes in gases, I: The diagram technique, *Physica* **25**, 281–301 (1959).

MSN.26. I. Prigogine and R. Balescu, Irreversible processes in gases, II: The equations of evolution, *Physica* **25**, 302–323 (1959).

MSN.27. I. Prigogine and B. Leaf, On the field-matter interaction in classical electrodynamics, I., *Physica* **25**, 1067–1079 (1959).

1960

MSN.28. I. Prigogine and R. Balescu, Irreversible processes in gases, III: inhomogeneous systems, *Physica* **26**, 145–159 (1960).

MSN.29. I. Prigogine, R. Balescu, P. Résibois, and F. Henin, On the dynamical derivation of equilibrium statistical mechanics, *Physica* **26**, S 36–48 (1960).

MSN.30. I. Prigogine, R. Balescu, and I. Krieger, On the decay of long-range Correlations, *Physica* **26**, 145–159 (1960).

MSN.31. I. Prigogine and F. Henin, On the general theory of the approach to equilibrium, I. Interacting normal modes, *J. Math. Phys.* **1**, 349–371 (1960).

MSN.32. I. Prigogine and P. Résibois, Analytical invariants in N-body systems, *Bull. Cl. Sci. Acad. Roy. Belg.* **46**, 53–60 (1960).

MSN.33. I. Prigogine, *Superfluidité et Equation de Transport Quantique* (Superfluidity and Quantum Transport Equation), P. Résibois, ed., Monographie no. 6, Institut Interuniversitaire des Sciences Nucléaires, 1960.

1961

MSN.34. I. Prigogine and P. Résibois, On the kinetics of the approach to equilibrium, *Physica* **27**, 629–646 (1961).

1962

MSN.35. I. Prigogine, Irréversibilité et Corrélations (Irreversibility and Correlations), Revue de Métaphysique et de Morale, Libr. A. Colin, Paris, 1962, pp. 228–236.

MSN.36. I. Prigogine and P. Résibois, On the statistical theory of superfluidity, *Physica* **28**, 1–14 (1962).

1963

MSN.37. I. Prigogine and Ph. De Gottal, Correlations in a non-isothermal plasma, *Physica* **29**, 706–711 (1963).

MSN.38. I. Prigogine and G. Severne, On the validity of Onsager's reciprocity relations for strongly coupled systems, *Phys. Lett.* **6**, 173–176 (1963).

MSN.39. I. Prigogine and G. Severne, The general theory of transport processes and the correlation function method, *Phys. Lett.* **6**, 177–179 (1963).

MSN.40. I. Prigogine, Remarks on scattering theory, in *12-ème Conseil de Chimie Solvay*, Interscience Publishers, New York, 1963, pp. 481–499.

1964

MSN.41. I. Prigogine et P. Résibois, *Temps et Irréversibilité en Physique Statistique* (Time and Irreversibility in Statistical Physics), Atti Congr. Lagrangiano, Ac. Scienze, Torino (1964).

MSN.42. I. Prigogine, P. Résibois and G. Severne, The general theory of transport processes and the correlation function method, II, *Phys. Lett.* **7**, 317–318 (1964).

1965

MSN.44. I., Prigogine, Some aspects of relativistic statistical mechanics, in Proceedings, Symposium on *Statistical Mechanics and Thermodynamics*, J. Meixner, ed., Aachen 1964, North Holland Publishers, Amsterdam, 1965, pp. 20–35.

MSN.45. Ph. De Gottal and I. Prigogine, Relativistic effects in statistical hydrodynamics, *Physica* **31**, 677–687 (1965).

MSN.46. F. Henin, I. Prigogine, P. Résibois, and M. Watabe, Kinetic equations, quasiparticles and renormalisation, *Phys. Lett.* **16**, 253–254 (1965).

MSN.47. I. Prigogine, G. Nicolis, and J. Misguich, Local equilibrium approach to transport processes in dense media, *J. Chem. Phys.* **43**, 4516–4521 (1965).

1966

MSN.48. I. Prigogine and F. Henin, On the entropy of strongly interacting systems, *Phys. Lett.* **20**, 255–257 (1966).

MSN.49. F. Henin, I. Prigogine, C. George and F. Mayné, Kinetic equations and quasiparticle description, *Physica* **32**, 1828–1872 (1966).

MSN.50. I. Prigogine, F. Henin, and C. George, Entropy and quasiparticle description of anharmonic lattices, *Physica* **32**, 1873–1900 (1966).

MSN.51. I. Prigogine and F. Mayné, On the application of non-equilibrium statistical mechanics to quantum field theory, *Phys. Lett.* **21**, 42–44 (1966).

1967

MSN.52. I. Prigogine, Quantum theory of Dissipative Systems, Nobel Symposium **5**, S. Claesson, ed., Interscience, New York, 1967, pp. 99–129.

MSN.53. I. Prigogine and F. Henin, *Kinetic Equation, Quasiparticles and Entropy*, in Proceedings, IUPAP Meeting on *Statistical Mechanics, Foundations and Applications*, Copenhagen 1966, Benjamin, New York, 1967.

1968

MSN.54. I. Prigogine, F. Henin, and C. George, Dissipative processes, quantum states and entropy, *Proc. Natl. Acad. Sci.* **59**, 7–14 (1968).

MSN.55. I. Prigogine, Dissipative processes, quantum states and field theory, in XIV-ème Conseil de Physique Solvay, *Fundamental Problems in Elementary Particle Physics*, Interscience Publishers, New York, 1968, pp. 155–195.

1969

MSN.56. I. Prigogine, Quantum theory of dissipative processes and nonequilibrium thermo-dynamics, *J. Phys. Soc. Japan* **26**, Supplement, 208–211 (1969).

MSN.57. I. Prigogine, *Quantum Statistical Mechanics of Systems with an Infinite Number of Degrees of Freedom, Contemporary Phys.* **1**, 315–331 (1969).

MSN.58. I. Prigogine and F. Henin, Kinetic Theory and Subdynamics, in *Problemy Teoreticheskoi Fiziki*, Nauka, Moscou, 1969.

MSN.59. I. Prigogine, Quantum states and dissipative processes, *Adv. Chem. Phys.* **15**, 11–35 (1969).

MSN.60. I. Prigogine, C. George, and F. Henin, Dynamical and statistical descriptions of N-body systems, *Physica* **45**, 418–434 (1969).

MSN.61. I. Prigogine, C. George, and F. Henin, Dynamics of systems with large number of degrees of freedom and generalized transformation theory, *Proc. Natl. Acad. Sci.* **65**, 789–796 (1970).

MSN.62. I. Prigogine, Dynamic foundations of thermodynamics and statistical mechanics, in *A Critical Review of Thermodynamics*, E. B. Stuart, B. Gal-Or, and A. Brainard, eds., Mono Book Corp., Baltimore, 1970, pp. 1–18.

MSN.63. I. Prigogine, C. George, F. Henin, P. Mandel, and J. W. Turner, Physical particle representation and generalized transformation theory, *Proc. Natl. Acad. Sci.* **66**, 709–715 (1970).

MSN.64. I. Prigogine, G. Nicolis, and P. Allen, Eyring's theory of viscosity of dense media and nonequilibrium statistical mechanics, in *Chemical Dynamics, Papers in Honor of H. Eyring*, Hirshfelder, ed., Wiley, New York, 1971.

MSN.65. I. Prigogine, C. George, and J. Rae, Classical dynamics as an eigenvalue problem, *Physica* **56**, 25–42 (1971).

MSN.66. I. Prigogine and C. George, Quantization as a problem in eigenprobabilities, *Physica* **56**, 329–344 (1971).

1972

MSN.67. C. George, I. Prigogine, and L. Rosenfeld, The macroscopic level of quantum mechanics, *Nature* **240**, 25–27 (1972).

MSN.68. C. George, I. Prigogine, and L. Rosenfeld, The macroscopic level of quantum mechanics, *Det Kong. Danske Vid. Selskab* **38**, 1–44 (1972).

MSN.69. A. P. Grecos and I. Prigogine, Kinetic and ergodic properties of quantum systems—The Friedrichs Model, *Physica* **59**, 77–96 (1972).

MSN.70. A. P. Grecos and I. Prigogine, Dissipative Properties of Quantum Systems, *Proc. Natl. Acad. Sci.* **69**, 1629–1633 (1972).

1973

MSN.71. I. Prigogine, Irreversibility as a symmetry-breaking process, *Nature* **246**, 67–71 (1973).

MSN.72. I. Prigogine, *Irreversibility in the Many-Body Problem*, J. Biel and J. Rae, eds., Book review, *Nat. Phys. Sci.* **241**, 104 (1973).

MSN.73. I. Prigogine and A. P. Grecos, Irreversibility and dissipativity of quantum systems, in *Cooperative Phenomena*, H. Haken and M. Wagner, eds., Springer, Berlin, 1973, pp. 373–380.

MSN.74. I. Prigogine, The statistical interpretation of nonequilibrium entropy, *Acta Phys. Austriaca*, Suppl. **X**, 401–450 (1973).

MSN.75. I. Prigogine, C. George, F. Henin, and L. Rosenfeld, [**PGHR**], A unified formulation of dynamics and thermodynamics, *Chem. Scripta* **4**, 5–32 (1973).

MSN.76. F. Mayné and I. Prigogine, Scattering theory and subdynamics, *Physica* **63**, 1–32 (1973).

1974

MSN.77. I. Prigogine, Microscopic aspects of entropy and the statistical foundations of nonequilibrium thermodynamics, Proceedings, International Symposium on *Foundations of Continuum Thermodynamics*, Bussaco, Delgado Domingo, M. N. R. Nina and J. H. Whitelaw, eds., Lisboa, 1974, pp. 81–112.

MSN.78. I. Prigogine and F. Mayné, *Entropy, Dynamics and Scattering Theory*, Ecole d'été de Sitgès, Garrido, ed., 1974, pp. 35–73.

MSN.79. C. George and I. Prigogine, Quantum mechanics of dissipative systems and noncanonical formalism, *Int. J. Quantum Chem. Symp.* **8**, 335–346 (1974).

MSN.80. F. Henin and I. Prigogine, Entropy, Dynamics and Molecular Chaos, *Proc. Natl. Acad. Sci.* **71**, 2618–2622 (1974).

MSN.81. W. C. Schieve and I. Prigogine, The role of subdynamics in kinetic theory, in *Rarefied Gas Dynamics*, Academic Press, San Francisco, 1974, pp. 19–35.

1975

MSN.82. A. P. Grecos and I. Prigogine, On the theory of subdynamics and its relation to scattering processes, in *Théories cinétiques classiques et relativistes*, Coll. Internat. CNRS, no. 236, 1975, pp. 98–111.

MSN.83. I. Prigogine and A. P. Grecos, Quantum theory and dissipativity, in Proceedings, International Research Symposium on *Statistical Physics with special sessions on Topics Related to Bose Statistics*, University of Calcutta, **12**, Suppl. 1, 177–184 (1975).

1976

MSN.84. I. Prigogine, A. P. Grecos, and Cl. George, Kinetic theory and ergodic properties, *Proc. Natl. Acad. Sci. USA* **73**, 1802–1805 (1976).

1977

MSN.85. I. Prigogine and A. P. Grecos, Kinetic theory and ergodic properties in quantum mechanics, in *75 Jahre Quantenmechanik*, Akademie-Verlag, Berlin, 1977, pp. 57–68.

MSN.86. I. Prigogine, F. Mayné, C. George, and M. de Haan, Microscopic theory of irreversible processes, *Proc. Natl. Acad. Sci. USA* **74**, 4152–4156 (1977).

MSN.87. I. Prigogine and A. P. Grecos, On the relation of dynamics to statistical mechanics, *Celestial Mech.* **16**, 489–507 (1977).

MSN.88. I. Prigogine and C. George, New quantization rules for dissipative systems, *Int. J. Quantum Chem.* **12**, Suppl. 1, 177–184 (1977).

1978

MSN.89. I. Prigogine and A. Grecos, The dynamical theory of irreversible processes, Proceedings, International Conference *Frontiers of Theoretical Physics*, Calcutta, 1977, F. C. Auluck, L. S. Kothari, V. S. Nanda, eds., pp. 51–62.

MSN.90. C. George, F. Henin, F. Mayné, and I. Prigogine, New quantum rules for dissipative systems, *Hadronic J.* **1**, 520–573 (1978).

MSN.91. I. Prigogine and A. Grecos, On the dynamical theory of irreversible processes and the microscopic interpretation of nonequilibrium entropy, Proceedings, *13th IUPAP Conference on Statistical Physics, Ann. Israel Phys. Soc.* **2**, 84–97, 1978.

MSN.92. I. Prigogine, M. Theodosopoulou, and A. Grecos, On the derivation of linear irreversible thermodynamics for classical fluids, *Proc. Natl. Acad. Sci. USA*, **75**, 1632–1636 (1978).

1979

MSN.93. I. Prigogine and A. P. Grecos, Topics in nonequilibrium statistical mechanics, in *Problems in the Foundations of Physics*, LXXII Corso, Soc. Ital. Fisica.

MSN.94. I. Prigogine, The microscopic theory of irreversible processes, in Proceedings, *11th Symposium Rarefied Gas Dynamics*, Vol. 1, CEA, Paris, 1979, pp. 1–27.

MSN.95. C. George and I. Prigogine, Coherence and randomness in quantum theory, *Physica* **99A**, 369–382 (1979).

MSN.96. A. P. Grecos and I. Prigogine, Irreversible processes in quantum theory, in *Group Theoretical Methods in Physics*, Lecture Notes in Physics, Vol. 94, Springer, 1979, pp. 229–237.

MSN.97. B. Misra, I. Prigogine, and M. Courbage, From deterministic dynamics to probabilistic descriptions, *Proc. Natl. Acad. Sci. USA* **76**, 3607–3611 (1979).

MSN.98. B. Misra, I. Prigogine, and M. Courbage [**MPC**], From deterministic dynamics to probabilistic descriptions, *Physica* **98A**, 1–26 (1979).

MSN.99. B. Misra, I. Prigogine, and M. Courbage, Lyapounov variable: Entropy and measurement in quantum mechanics, *Proc. Natl. Acad. Sci. USA* **76**, 4678–4772 (1979).

MSN.100. G. Nicolis and I. Prigogine, Irreversible processes at nonequilibrium steady states and Lyapounov functions, *Proc. Natl. Acad. Sci. USA* **76**, 6060–6061 (1979).

1980

MSN.101. B. Misra and I. Prigogine, On the foundations of kinetic theory, *Suppl. Progr. Theor. Phys.* **69**, 101–110 (1980).

MSN.102. I. Prigogine, Entropy, time and kinetic description, in *Order and Fluctuations in Equilibrium and Nonequilibrium Statistical Mechanics*, 17th International Solvay Conference on Physics, G. Nicolis, G. Dewel, and J. W. Turner, eds., Wiley, 1980, pp. 35–75.

1983

MSN.103. I. Prigogine and C. George, The second law as a selection principle: The microscopic theory of dissipative processes in quantum systems, *Proc. Natl. Acad. Sci. USA* **80**, 4590–4594 (1983).

MSN.104. I. Prigogine and M. Courbage, Intrinsic randomness and intrinsic irreversibility in classical dynamical systems, *Proc. Natl. Acad. Sci. USA* **80**, 2412–2416 (1983).

MSN.105. B. Misra and Il. Prigogine, Irreversibility and nonlocality, *Lett. Math. Phys.* **7**, 421–429 (1983).

MSN.106. I. Prigogine and B. Misra, Time, probability and dynamics, in *Long Time Prediction in Dynamics*, W. Horton, L. E. Reichl, and A. G. Szebehely, eds., Wiley, New York, 1983, pp. 21–43.

1984

MSN.107. I. Prigogine, Irreversibility and space-time structure, in Proceedings, International Conference on *Fluctuations and Sensitivity in Nonequilibrium Systems*, Austin, 1984, W. Horsthemke and K. D. K. Kondepudi, eds., Springer, Berlin, 1984, pp. 2–9.

1985

MSN.108. C. George, F. Mayné, and I. Prigogine, *Scattering theory in superspace, Adv. Chem. Phys.* **61**, 223–299 (1985).

1986

MSN.109. I. Prigogine, Irreversibility, stochasticity and non-locality in classical dynamics, in *Quantum Implications*, B. Hiley and D. Peat, eds., Routledge & Keagan, London, 1986.

MSN.110. Y. Elskens and I. Prigogine, From instability to irreversibility, *Proc. Natl. Acad. Sci. USA* **83**, 5756–5760 (1986).

1987

MSN.111. I. Prigogine and T. Petrosky, Intrinsic irreversibility in quantum theory, *Physica* **147A**, 33–47 (1987).

MSN.112. T. Petrosky and Il. Prigogine, Poincaré's theorem and unitary transformations for classical and quantum systems, *Physica*, **147A**, 439–460 (1987).

MSN.113. I. Prigogine and T. Petrosky, An alternative to quantum theory, *Physica* **147A**, 461–486 (1987).

MSN.114. I. Prigogine and T. Petrosky, *Limits to Quantum Theory* (in honor of Prof. Tomita), Kyoto University, 1987.

MSN.115. I. Prigogine, A new microscopic level of irreversibility? Meeting: *Spatial Inhomogeneities and Transient Behavior in Chemical Kinetics*, Bruxelles, 1987.

MSN.116. I. Prigogine, Irréversibilité et mécanique quantique (Irreversibility and quantum mechanics), *J. Chim. Phys.* **84**, 1275–1278 (1987).

MSN.117. I. Prigogine, Un itinéraire dans un siècle troublant (An itinerary in a troubled century), *Rev. Sci. Morales et Polit.* **1987**, 603–618.

MSN.118. I. Prigogine, *Irreversibility and space–time structure, Res Mechanica* **21** (1987).

1988

MSN.119. Il. Prigogine, E. Kestemont, and M. Mareschal, Velocity correlations and irreversibility: A molecular dynamics approach, in *From Chemical to Biological Organization*, M. Markus, S. C. Müller, and G. Nicolis, eds., Springer, Berlin, 1988, pp. 22–26.

1989

MSN.120. I. Prigogine and T. Petrosky, Nonequilibrium statistical mechanics: Beyond the Van Hove limit, in *Festschrift in Honor of Léon Van Hove*, World Scientific, Singapore, 1989.

MSN.121. I. E. Antoniou and I. Prigogine, Intrinsic irreversibility in classical and quantum mechanics, in *The Concept of Probability*, E. I. Bitsakis and C. A. Nicolaides, eds., Kluwer, 1989, pp. 223–234.

MSN.122. I. Prigogine, The microscopic meaning of irreversibility, *Z. Phys. Chem.* **270**, 477–490 (1989).

1990

MSN.123. I. Prigogine, E. Kestemont, and M. Mareschal, *The Approach to Equilibrium and Molecular Dynamics, in Microscopic Simulations of Complex Flows*, M. Mareschal, ed., Plenum, New York, 1990.

MSN.124. T. Petrosky and I. Prigogine, Laws and events: The dynamical basis of self-organization, *Can. J. Phys.* **68**, 670–682 (1990).

MSN.125. I. Prigogine, The microscopic meaning of irreversibility, *Z. Phys. Chem.* **270**, 477–490 (1989).

MSN.126. T. Petrosky, S. Tasaki, and I. Prigogine, Quantum zeno effect, *Physica,* **A 170**, 306–325 (1990).

MSN.127. T. Petrosky, S. Tasaki, and I. Prigogine, Quantum zeno effect, *Phys. Lett.* **A 151**, 109–113 (1990).

MSN.128. T. Petrosky, I. Prigogine, and S. Tasaki, Quantum Theory of Nonintegrable systems, *Physica* **A 173**, 175–242 (1990).

MSN.129. I. Prigogine, Time, dynamics and chaos: Integrating Poincaré's nonintegrable systems, in *"XXVI Nobel Conference"*, J. Holte, ed., Gustavus Adolphus College, St. Peter, Minnesota, pp. 55–88.

1991

MSN.130. T. Petrosky and I. Prigogine, Alternative formulation of classical and quantum dynamics for non-integrable systems, *Physica* **A 175**, 146–209 (1991).

MSN.131. I. Prigogine, T. Y. Petrosky, H. H. Hasegawa, and S. Tasaki, Integrability and chaos in classical and quantum mechanics, *Chaos, Solit. Fractals* **1**, 3–24 (1991).

MSN.132. H. Hasegawa, T. Petrosky, I. Prigogine, and S. Tasaki, Quantum mechanics and the direction of time, *Found. Phys.* **21**, 263–281 (1991).

1992

MSN.133. I. Prigogine, T. Petrosky, H. Hasegawa, and S. Tasaki, Integration of non-integrable systems, in *Solitons and Chaos*, I. Antoniou and F. Lambert, eds., Springer, Berlin, 1992.

MSN.134. I. Prigogine and T. Petrosky, Quantum chaos—Towards the formulation of an alternate quantum theory, Proceedings, *II International Wigner Symposium*, H. D. Doebner, W. Scherer, and F. Schroeck Jr., eds., World Scientific, Singapore, 1992.

MSN.135. T. Petrosky and I. Prigogine, Integration of Poincaré's nonintegrable systems, new formulation of classical and quantum dynamics, Proceedings, *9th Symposium on Energy Engineering Sciences*, Argonne National Laboratory, Argonne, Ill. 1992.

MSN.136. I. Prigogine and I. Antoniou, From microscopic irreversibility to macroscopic ireversibility. Discussion remarks and comments to R. L. Stratonovich, *Z. Phys. Chem.* **170**, 219–221 (1992).

MSN.137. I. Prigogine, Dissipative processes in quantum theory, *Phys. Rep.* **219**, 93–108 (1992).

MSN.138. T. Petrosky and I. Prigogine, Complex spectral representations and singular eigenvalue problem of Liouvillian in quantum scatering, in Symposium *Quantum Physics and the Universe*, Waseda University, Tokyo, Japan, 1992.

MSN.139. T. Petrosky and I. Prigogine, Complex spectral representations and quantum chaos, in *Research Trends in Physics: Chaotic Dynamics and Transport in Fluids and Plasmas*, Institute of Advanced Studies, La Jolla, 1992.

1993

MSN.140. I. Antoniou and I. Prigogine, Intrinsic irreversibility and integrability of dynamics, *Physica* **A 192**, 443–464 (1993).

MSN.141. I. Prigogine, From classical chaos to quantum chaos, *Vistas in Astron.* **37**, 7–25 (1993).

MSN.142. T. Petrosky and I. Prigogine, Poincaré resonances and the limits of quantum mechanics, *Phys. Lett.* **A 182**, 5–15 (1993).

MSN.143. T. Petrosky and I. Prigogine, Poincaré resonances and the limits of trajectory dynamics, *Proc. Natl. Acad. Sci. USA* **90**, 9393–9397 (1993).

MSN.144. R. Passante, T. Petrosky, and I. Prigogine, Virtual transitions, self-dressing and indirect spectroscopy, *Optics Commun.* **99**, 55–60 (1993).

1994

MSN.145. T. Petrosky and I. Prigogine, Quantum chaos, complex spectral representations and time symmetry breaking, *Chaos, Solitons and Fractals* **4**, 311–359 (1994).

MSN.146. I. Prigogine and T. Petrosky, Poincaré's resonances and extension of classical and quantum mechanics, in Proceedings, 12th Symposium *Energy Engineering Sciences*, Argonne National Laboratory, 1994, pp. 8–16.

MSN.147. H. Hasegawa, E. Luschei, and I. Prigogine, Generalized spectral representations of chaotic maps displaying intermittency and anomalous diffusion, Proceedings, 12th Symposium, *Energy Engineering Sciences*, Argonne National Laboratory, 1994, pp. 249–255.

MSN.148. I. Prigogine, Why irreversibility? The formulation of classical and quantum mechanics for nonintegrable systems, Lecture Notes, Department of Physics, Keio University, Yokohama, 1994.

1995

MSN.149. I. Prigogine, Classical and quantum mechanics of unstable dynamical systems, in Proceedings, International Conference on *Dynamical Systems and Chaos*, Y. Aizawa, S. Saito, and K. Shiraiwa, eds., World Scientific, Singapore, Vol. 2, 1995.

MSN.150. I. Prigogine, Why irreversibility? The formulation of classical and quantum mechanics for non-integrable systems, *Int. J. Quantum Chem.* **53**, 105–118 (1995).

MSN.151. I. Prigogine, Why irreversibility? The formulation of classical and quantum mechanics for non-integrable systems, *Int. J. Bifurcation and Chaos* **5**, 3–16 (1995).

MSN.152. R. Passante, T. Petrosky, and I. Prigogine, Long-time behaviour of self-dressing and indirect spectroscopy, *Physica* **A 218**, 437–456 (1995).

MSN.153. I. Prigogine and T. Petrosky, *Poincaré Resonances and the Extension of Classical and Quantum Mechanics, in Nonlinear, deformed and irreversible quantum systems*, H. D. Doebner, V. K. Dobrev, and P. Nattermann, eds., World Scientific, Singapore, pp. 3–21, 1995.

MSN.154. I. Prigogine, Science of chaos or chaos in science: A rearguard battle, *Physicalia Mag.* **17**, 213–218 (1992).

1996

MSN.155. T. Petrosky and I. Prigogine, Poincaré resonances and the extension of classical dynamics, in Special issue: *Time symmetry breaking in classical and quantum mechanics, Chaos, Solitons and Fractals* **7**, 441–497 (1996).

MSN.156. I. Prigogine, Time, chaos and the laws of Nature, in *Law and Prediction in the Light of Chaos Research*, P. Weingartner and G. Schurz, eds., Springer, Berlin, 1996.

MSN.157. I. Antoniou, I. Prigogine, and S. Tasaki, New spectral representations of mixing dynamical systems, in Proceedings, *World Congress of Nonlinear Analysts*, 1992, V. Lakshmikantham, ed., W. de Gruyter, Berlin, 1996.

MSN.158. I. Prigogine and T. Petrosky, Extension of classical dynamics—the case of anharmonic lattices, in *Ivanenko Memorial Volume in Gravity, Particles and Space–Time*, P. Pronin and G. Sardanshvili, eds., World Scientific, Singapore, 1996.

MSN.159. I. Prigogine and T. Petrosky, Chaos, time symmetry breaking and the extension of classical and quantum mechanics, in Proceedings, *"El Escorial" course on Foundations of Quantum Physics*, Ed. Complutense, Madrid, pp. 183–215.

1997

MSN.160. T. Petrosky and I. Prigogine, The Liouville space extension of quantum mechanics, *Adv. Chem. Phys.* **XCIX**, 1–120 (1997).

MSN.161. I. Prigogine, From Poincaré's divergences to quantum mechanics with broken time symmetry, *Z. Naturforschung* **52a**, 37–45 (1997).

MSN.162. I. Prigogine, Non-linear science and the laws of Nature, *J. Franklin Inst.* **334 B**, 745–758 (1997).

MSN.163. I. Prigogine, Non-linear science and the laws of Nature, *Int. J. Bifurcation and Chaos*, **7**, 1917–1926 (1997).

MSN.164. T. Petrosky and I. Prigogine, The extension of classical dynamics for unstable Hamiltonian systems, *Computers Math. Apic.* **34**, 1–44 (1997).

MSN.165. I. Prigogine and D. Driebe, Time, chaos and the laws of nature, in *Nonlinear Dynamics, Chaotic and Complex Systems*, E. Infeld, R. Zelazny, and A. Galkowski, eds., Cambridge University Press, Cambridge, 1997, pp. 206–223.

1998

MSN.166. I. Antoniou, M. Gadella, I. Prigogine, and G. Pronko, Relativistic Gamow vectors, *J. Math. Phys.* **39**, 2295–3018 (1998).

MSN.167. I. Prigogine and T. Petrosky, Semigroup representation of the Vlasov equation, *J. Plasma Phys.* **59**, 611–618.

MSN.168. I. Prigogine, Irreversibility and nonlocality, in Proc. *XXIst Solvay Confeence in Physics*, Nov. 1998, Japan

MSN.169. T. Petrosky and I. Prigogine, Irreversibility, Probability and Laws of Nature, in *Fronteras da Ciencia*, Proceedings, Coimbra, Oct. 1998.

1999

MSN.170. I. Prigogine, Laws of Nature, Probability and time symmetry breaking, *Physica* **A 263**, 528–539 (1999).

MSN.171. I. Prigogine and T. Petrosky, Laws of nature, probability and time symmetry breaking, in *Generalized Functions, Operator Theory and Dynamical Systems*, I. Antoniou and G. Lumer, eds., Chapman & Hall, London, pp. 99–110, 1999.

MSN.172. I. Prigogine and I. Antoniou, Laws of nature and time symmetry breaking, in *Tempos in Science and Nature: Structures, Relations and Complexity*, Ann. *New York Acad. Sci.* **879**, 8–28 (1999).

MSN.173. I. Prigogine, Non-locality and superluminosity, in Proceedings *Bogolyubov Conference*, Moscow, Sept. 1999.

MSN.174. T. Petrosky and I. Prigogine, Extension of classical dynamics: emergence of irreversibility and stochasticity, in *Fundamental and Applications of Complex Systems*, G. Zgrablich, ed., Nueva Ed. Univ., San Luis, pp. 429–445, 1999.

MSN.175. I. Prigogine, T. Petrosky, and G. Ordoñez, *Report: The Extension of Classical and Quantum Mechanics, as Well as of Field Theory to Non-integrable Systems.* EC report, Contract PSS*0992, 1999.

2000

MSN.176. T. Petrosky and I. Prigogine, Thermodynamic limit, Hilbert space and breaking of time symmetry, *Chaos, Solitons and Fractals* **11**, 373–382 (2000).

MSN.177. I. Antoniou, I. Prigogine, V. Sadovnichii, and S. A. Shkarin, Time operator for diffusion, *Chaos, Solitons & Fractals* **11**, 465–467 (2000).

MSN.178. E. Karpov, G. Ordoñez, T. Petrosky, I. Prigogine, and G. Pronko, Causality, delocalization, and positivity of energy, *Phys. Rev.* **A 62**, 02103 (2000).

MSN.179. T. Petrosky, G. Ordoñez, and I. Prigogine, Quantum transitions and nonlocality, *Phys. Rev.* **A 62**, 042106 (2000).

MSN.180. T. Petrosky and I. Prigogine, Limits to causality and delocalization in classical field theory, in *Evolution Equations and Their Applications in Physical and Life Sciences*, G. Lumer and L. Weis, eds., Marcel Decker, New York, 2000.

MSN.181. E. Karpov, I. Prigogine, T. Petrosky, and G. Pronko, Friedrichs Model with virtual transitions: Exact solution and indirect spectroscopy, *J. Math. Phys.*, **41**, 118–131 (2000).

MSN.182. T. Petrosky and I. Prigogine, Non-equilibrium statistical mechanics, in *The Legacy of Léon Van Hove*, Vol. 28, A. Giovannini, ed., World Scientific Series in 20th Century Physics, World Scientific, Singapore, 2000, pp. 135–138.

2001

MSN.183. G. Ordoñez, T. Petrosky, and I. Prigogine, Quantum transitions and dressed unstable states, *Phys. Rev.* **A 63**, 052106 1–23 (2001).

MSN.184. G. Ordoñez, T. Petrosky, E. Karpov, and I. Prigogine, Explicit construction of a time superoperator for quantum unstable systems, *Chaos, Solitons and Fractals* **12**, 2591–2601 (2001).

MSN.185. T. Petrosky, G. Ordoñez, and I. Prigogine, Space–time formulation of quantum transitions, *Phys. Rev.* **A 64**, 062101 1–21 (2001).

MSN.186. I. Prigogine, Dynamics of correlations—A formulation for both integrable and non-integrable dynamical systems, Appendix to the new edition of *From Being to Becoming*, 2001.

MSN.187. I. Antoniou, M. Gadella, E. Karpov, I. Prigogine, and G. Pronko, Gamov algebras, *Chaos, Solitons and Fractals* **12**, 2757–2775 (2001).

MSN.188. I. Prigogine, T. Petrosky, and G. Ordoñez, Time symmetry breaking and stochasticity in Hamiltonian physics, in Proceedings, *XXIV Workshop on High Energy Physics and Field Theory*, Protvino, State Research Centre of Russia, Institute of High Energy Physics, 2001, pp. 204–220.

MSN.189. I. Prigogine, E. Karpov, T. Petrosky, and G. Ordoñez, Time operator in the Friedrichs model, in Proceedings, *XXIV Workshop on High Energy Physics and Field Theory*, Protvino, State Research Centre of Russia, Institute of High Energy Physics, 2001, pp. 274–281.

2002

MSN.190. I. Prigogine, Introductory remarks, in *Dynamical Systems and Irreversibility*, Proceedings, XXI Solvay Conference of Physics, *Adv. Chem. Phys.* **122**, pp. xxi–xxii, 2002.

MSN.191. I. Prigogine, Dynamics of correlations, A formalism for both integrable and nonintegrable dynamical systems, in *Dynamical Systems and Ireversibility*, Proceedings, XXI Solvay Conference on Physics, *Adv. Chem. Phys.* **122**, 261–275, 2002.

MSN.192. I. Prigogine, E. Karpov, G. Ordoñez, and T. Petrosky, Quantum transitions in interacting fields, *Phys. Rev.* **A 66**, 012109 (2002).

2003

MSN.193. I. Prigogine, Chemical kinetics and dynamics, in *Chemical Explanation—Characteristics, Development, Autonomy, Ann. New York Acad. Sci.* **988**, 128–132 (2003).

MSN.194. E. Karpov, G. Ordoñez, T. Petrosky, and I. Prigogine, Microscopic entropy and nonlocality, in Proceedings International Workshop *Quantum Physics and Communication*, Dubna 2002, *Particles and Nuclei Lett.* **1** (116), 8–15 (2003).

MSN.195. T. Petrosky, G. Ordoñez, and I. Prigogine, Radiation damping in classical systems: The role of nonintegrability, *Phys. Rev.* **A 68**, 1–22 (2003).

MSN.196. I. Prigogine, S. Kim, G. Ordoñez, and T. Petrosky, Stochasticity and time symmetry breaking in Hamiltonian dynamics, in Proceedings, XXII Solvay Conference Physics, *The Physics of Communication*, Delphi, 2001, World Scientific, Singapore, 2003, pp. 1–22.

MSN.197. I. Prigogine, G. Ordoñez and T. Petrosky, Microscopic entropy flow and entropy production in resonance scattering, in Proceedings, XXII Solvay Conf. on Physics, *The Physics of Communication*, Delphi, 2001, World Scientific, Singapore, 2003, pp. 370–388.

MSN.198. I. Prigogine and G. Ordoñez, Acceleration and entropy: A macroscopic analogue of the twin paradox, *Chaos and Complexity Lett.* (to appear, 2003).

2004

MSN.199. I. Prigogine, Microscopic entropy, in: Proceedings, Workshop Complexity: Microscopic and macroscopic aspects, Fond. Les Treilles, 2002, *Int. J. Quant. Chem.* **98** (2004).

8. Gravitation and Cosmology

1966

GRA.1. I. Prigogine, Entropy and gravitation, *Nature* **209**, 602–603 (1966).

GRA.2. I. Prigogine and G. Severne, On the statistical mechanics of gravitational plasmas, *Physica* **32**, 1376–1396 (1966).

1968

GRA.3. I. Prigogine and G. Severne, Nonequilibrium Statistical Mechanics and gravitational interactions, *Bull. Astronomique* **3**, 273–287 (1968).

1982

GRA.4. C. M. Lockhart, B. Misra, and I. Prigogine, Geodesic instability and internal time in relativistic cosmology, *Phys. Rev.* **D 25**, 921–929, 1982.

1986

GRA.5. J. Géhéniau and I. Prigogine, The birth of time, *Found. Phys.* **16**, 437–443, 1986.

GRA.6. I. Prigogine and J. Géhéniau, Entropy, matter and cosmology, *Proc. Natl. Acad. Sci. USA* **83**, 6245–6249, 1986.

GRA.7. I. Prigogine, *Entropy and Cosmology*, Conf. mondiale Inst. Internat. Froid, Paris, 1986.

GRA.8. E. Gunzig, J. Géhéniau, and I. Prigogine, Entropy and cosmology, *Nature* **330**, 621–624, 1987.

1988

GRA.9. I. Prigogine, J. Géhéniau, E. Gunzig, and P. Nardone, Thermodynamics and cosmology, *Gen. Relativ. Gravit.* **21**, 767–776, 1988 (5th award, Gravity Research Foundation, 1988).

GRA.10. I. Prigogine, J. Géhéniau, E. Gunzig, and P. Nardone, Thermodynamics of cosmological matter creation, *Proc. Natl. Acad. Sci. USA* **85**, 7428–7432, 1988.

1992

GRA.11. I. Prigogine, *Cosmologie*, Conférence Petrofina, Bruxelles.

1993

GRA.12. S. Tasaki, P. Nardone, and I. Prigogine, Resonance and instability in a cosmological model, *Vistas in Astronomy* **37**, 645–648.

1996

GRA.13. M. Castagnino, E. Gunzig, P. Nardone, I. Prigogine, and S. Tasaki, Quantum cosmology and large poincaré systems, in *Quantum physics, Chaos theory and Cosmology*, American Institute of Physics, Woodbury, NY, pp. 3–20.

1998

GRA.14. I. Prigogine et T. Petrosky, Le vide quantique et le dilemme hamiltonien (The quantum vacuum and the Hamiltonian dilemma), in *Le vide—Univers de tout et de rien*, E. Gunzig et S. Diner, eds., Ed. Complexe, pp. 290–297.

9. Varia

1942

DIV.1. G. Pry et I. Prigogine, Sur le calcul des niveaux énergétiques à l'aide de la méthode de Wentzel–Kramers–Brillouin, *Acad. Roy. Belg.* **28**, 652–659 (1942).

DIV.2. G. Pry et I. Prigogine, Rayons et nombres de coordination de quelques réseaux simples, *Bull. Cl. Sci. Acad. Roy. Belg.* **28**, 866–873 (1942).

1947

DIV.2a. I. Prigogine et G. Garikian, Sur le calcul des niveaux énergétiques par la méthode de Wentzel–Kramers–Brillouin et son application à l'hydrogène liquide (On the calculation of the energy levels by the Wentzel–Kramers–Brillouin method and its application to liquid hydrogen), *J. Phys.* 330–332 (1947).

1960

DIV.3. I. Prigogine and B. Leaf, On the field-matter interaction in classical electrodynamics, II, *Bull. Cl. Sci. Acad. Roy. Belg.* **46**, 915–928 (1960).

1961

DIV.4. I. Prigogine and F. Henin, Radiation damping and the equation of motion in classical electrodynamics, *Physica* **27**, 982–984 (1961).

DIV.5. I. Prigogine, On the motion of a charged particle, *J. Nucl. Energy, Part C: Plasma Phys.* **2**, 184–187 (1961).

1962

DIV.6. I. Prigogine, Universal constants, time scales and self-interaction, *Bull. Cl. Sciences, Acad. Roy. Belg.* **48**, 1322–1332 (1962).

DIV.7. I. Prigogine and F. Henin, Motion of a relativistic charged particle, *Physica* **28**, 667–688 (1962).

1963

DIV.8. I. Prigogine and F. Henin, Motion of a relativistic charged particle. II, *Physica* **29**, 286–292 (1963).

1964

DIV.9. I. Prigogine and F. Henin, On the structure of elementary particles in classical electrodynamics, in *Nucleon Structure*, R. Hofstadter and Schiff eds., Stanford University Press, Palo Alto, CA, 1964, pp. 334–336.

1965

DIV.10. I. Prigogine and F. Henin, On a reformulation of the classical electron theory, *Mém. Cl. Sci. Acad. Roy. Belg.* **35**, (1965).

10. Science Policy

1980

POL.1. I. Prigogine, *Science et technique: Vers une nouvelle dimension européenne*? (Science and technique: Toward a new European dimension?), Le Soir, Carte blanche, 08/12/1980.

POL.2. I. Prigogine, Science et société dans l'Europe en mutation (Science and society in changing Europe), in *La recherche—Développement dans la CEE: Vers une Nouvelle Phase de la Politique Commune*, Strasbourg, 1980, Commission des Comm. Eur.

POL.3. I. Prigogine, On science policy, *Commun. Cognition* **13**, 133–140 (1980).

1981

POL.3a. I. Prigogine, *Science et technique: Vers une nouvelle dimension européenne*? (Science and technique: Toward a new European dimension?), Infordata, Sept. 1981.

1982

POL.4. I. Prigogine, Wissenschaft und Gesellschaft im Wandel Europas (Science and society in changing Europe), *Merkur* **5**, 480 (1982).

POL.5. A. Danzin et I. Prigogine, *Quelle science pour demain? Recherche et besoins humains, la science come facteur de mutation* (What kind of science for tomorrow? Research and human needs, science as a factor of change), Courrier UNESCO, Février 1982, 4–9.

1985

POL.6. I. Prigogine, *Science, civilisation et démocratie* (Science, civilisation and democracy), 6-e Conference Parlem. Scientifique, Conseil de l'Europe, Strasbourg, 1985.

1987

POL.7. I. Prigogine, *Projet d'Assemblée Européenne des Sciences* (Project of a European Assembly of Science), Bureau du CODEST, 1987.

POL.8. I. Prigogine, *Quelle politique scientifique en Belgique?* (What kind of sicence policy in Belgium?), Colloque du Parti Libéral, 1987.

1988

POL.9. I. Prigogine, *Rapport pour une Assemblée Européenne des Sciences* (Report for a European Assembly of Science), C.E.E., 1988.

1989

POL.10. I. Prigogine, *L'universitá nel mondo contemporaneo* (The University in the contemporary world), Alma Mater Studiorum, **II**, 21–35, Bologna, 1989.

POL.11. I. Prigogine, *Environmental Ethics in a time of Bounded Rationality*, 6th CEC Summit Conference on Bioethics, Brussels, 1989 (CEC Luxembourg 1990).

POL.12. I. Prigogine, Universities in a changing context, *H en H*, **36**, 33–43 (1989).

1990

POL.13. I. Prigogine, *European science and technology for sustainable development, in respect to the values of life*, Presidenza Italiana CE, 1990.

1993

POL.14. I. Prigogine and G. Nicolis, *The Fourth European Framework Programme and Research on Complex Systems.*

1994

POL.15. I. Prigogine, *Les universités dans un monde en mutation*, The University: Facing up to its European responsibilities, Pisa (publié en 1996).

POL.16. I. Prigogine, *Complex systems, an innovative approach*, 4th Community RTD Framework Programme, European Commission, 1994.

1996

POL.17. I. Prigogine, *Projet pour l'appel d'Heidelberg* (Project for the call of Heidelberg), juillet 1996.

11. General Papers, Philosophy

1937

GEN.1. I. Prigogine (Prigoshin), Essai de philosophie physique (Essay of physical philosophy), *Cahiers du Libre Examen* **1** (1), 19 (1937).

GEN.2. I. Prigogine (Prigoshin), *Le problème du déterminisme* (The problem of determinism), *Cahiers du Libre Examen* **1** (3), 15–17 (1937).

GEN.3. H. Bolle et I. Prigogine, J. Prigoshin, ed., L'évolution, *Cahiers du Libre Examen* **2** (6), 17–18 (1937).

1950

GEN.4. I. Prigogine, *Irréversibilité, évolution et finalité* (Irreversibility, evolution and finality), Revue de Université de Bruxelles, juillet–août 1950, 1–13.

1951

GEN.5. I. Prigogine, Probabilités et irréversibilité (Probabilities and irreversibility), *Soc. Belge de Logique et de Philos. des Sci.* 109–116 (1951).

1956

GEN.6. I. Prigogine, *Les Particules Elémentaires* (The Elementary Particles), Revue de l'Université de Bruxelles, juin–septembre 1956, pp. 1–13.

1963

GEN.7. I. Prigogine, *Symboles en Physique* (Symbols in Physics), Cahiers International De Symbolisme, no. 3, 1963.

1964

GEN.8. I. Prigogine, Hommage à Edmond Bauer, *J. Chim. Phys.*, 976–977 (1964).

1965

GEN.9. I. Prigogine, *Quelques remarques sur la structure de la Physique* (A few remarks on the structure of physics), *Rev. Int. Philos.* **73–74**, 335–341 (1965).

1967

GEN.10. I. Prigogine, temps, structure et entropie (Time, structure and entropy), *Bull. Cl. Sci. Acad. Roy. Belg.* **53**, 273–287 (1967).

1969

GEN.11. I. Prigogine, Unité et pluralité du monde physique (Unity and plurality of the physical world), *Synthèses* **276**, 12–20 (1969).

GEN.12. I. Prigogine, *La fin de l'atomisme* (The end of atomism), *Bull. Cl. Sci. Acad. Roy. Belg.* **55**, 1110–1117 (1969).

1970

GEN.13. I. Prigogine, Processus irréversibles et structures disipatives (Irreversible processes and dissipative structures) (Editorial), *Entropie* **34–35** (1970).

1971

GEN.14. I. Prigogine, Time, structure and entropy, in *Time in Science and Philosophy*, J. Zeman, ed., Elsevier, Amsterdam, 1971, pp. 89–99.

GEN.15. I. Prigogine, Unity of physical laws and levels of description, in *Interpretations of Life and Mind*, M. Grene, ed., Routledge & Kegan Paul, London, 1971, pp. 1–13.

GEN.16. J. Améry und I. Prigogine, Die tragische Philosophie Jacques Monods (The tragic philosophy of Jacques Monod), *Merkur* **25**, 1108–115.

1972

GEN.17. I. Prigogine, La thermodynamique de la vie (Thermodynamics of life), *La Recherche* **3**, 547–562 (1972).

GEN.18. I. Prigogine, Structure, entropy and quantum theory, *Nova Acta Leopoldina* **37**, 139–150 (1972).

GEN.19. I. Prigogine, G. Nicolis, and A. Babloyantz, Thermodynamics of Evolution, I, *Physics Today* **25** (11), 23–28 (1972).

GEN.20. I. Prigogine, G. Nicolis, and A. Babloyantz, Thermodynamics of Evolution, II, *Physics Today* **25** (12) (1972).

GEN.21. I. Prigogine, In memoriam A. Katzir-Katchalsky, *Bull. Cl. Sci. Acad. Roy. Belg.* **58**, 1018–1020 (1972).

GEN.22. I. Prigogine, *La naissance du temps* (The birth of time), *Bull. Cl. Sci. Acad. Roy. Belg.* **58**, 1189–1215 (1972).

1973

GEN.23. I. Prigogine et al., Can thermodynamics explain biological order?, *Impact Sci. Soc.* **23**, 159–179 (1973).

GEN.24. I. Prigogine et al., *La thermodynamique peut-elle expliquer l'ordre biologique?*, (Is thermodynamics able to explain biological order?), *Impact: Science et Société* **23**, 175–195 (1973).

GEN.25. I. Prigogine, *Time, Irreversibility and Structure*, in *The Physicist's Conception of Nature* J. Mehra, ed., Reidel, Dordrecht, 1973, pp. 561–593.

GEN.26. I. Prigogine, Measurement process and the macroscopic level of quantum mechanics, in *The Physicist's Conception of Nature*, J. Mehra, ed., Reidel, Dordrecht, 1973, pp. 697–701.

GEN.27. I. Prigogine, *The Physicist's Conception of Nature*, J. Mehra, ed., Reidel, Dordrecht, 1973 (présentation d'ouvrage), *Bull. Cl. Sci. Acad. Roy. Belg.* **59**, 1217–1218 (1973).

1974

GEN.28. P. Glansdorff et I. Prigogine, Entropie, structure et dynamique (Entropy, structure and dynamics), in *Sadi Carnot et l'Essor de la Thermodynamique*, CNRS, Paris, 1974.

GEN.29. I. Prigogine, *Entropie et Dynamique* (Entropy and dynamics), *Entropie* **57**, 5–11 (1974).

GEN.30. I. Prigogine, Léon Rosenfeld et les fondements de la physique moderne (Léon Rosenfeld and the foundations of modern physics), *Bull. Cl. Sci. Acad. Roy. Belg.* **60**, 841–854 (1974).

1975

GEN.31. I. Prigogine, Physique et métaphysique (Physics and metaphysics), in Colloque *Connaissance Scientifique et Philosophie*, Académie. Royal de Belgique, 1973.

GEN.32. I. Prigogine, Comments on the problem of the unity of Science, in *The Centrality of Science and Absolute Values*, Proceedings, 4th International Conference on the Unity of Science, The International Cultural Foundation, 1975.

GEN.33. I. Prigogine, G. Nicolis, and A. Babloyantz, *Termodynamika ewolucji* (in Polish) (Thermodynamics of evolution), *Postepy Fizyki* **26**, 253–281 (1975).

GEN.34. I. Prigogine et I. Stengers, Nature et créativité (Nature and creativity), *La Revue de l'AUPELF* **13**, 47–72 (1975).

1976

GEN.35. I. Prigogine, *Science et Société* (Science and society), *Bull. ULB et UAE*, no. 5, 24–27 (1976).

GEN.36. I. Prigogine, Order through fluctuation: Self-organization and social system, in *Evolution and Consciousness, Human Systems in Transition*, E. Jantsch and C. Waddington, eds., Addison-Wesley, Reading, MA, 1976, pp. 93–133.

1977

GEN.37. I. Prigogine, Wandlungen der Wissenschaft-Kultur und Wissenschaft heute (Transformations of science-culture and science today), *Wirtschaft und Wissenschaft* **3**, 22–32.

GEN.38. I. Prigogine et I. Stengers, La nouvelle Alliance, 1-ère partie, De la dynamique à la thermodynamique: La progressive ouverture de la physique au monde des processus naturels (see GEN.39), *Scientia* **71**, 287–304 (1977).

GEN.39. I. Prigogine and I. Stengers, The New Alliance, part 1, From dynamics to thermodynamics: Physics, the gradual opening towards the world of natural processes, *Scientia* **71**, 319–332 (1977).

GEN.40. I. Prigogine et I. Stengers, La nouvelle Alliance, 2-ème partie, L'élargissement de la dynamique, vers une science humaine de la nature (see GEN.41), *Scientia* **71**, 617–630 (1977).

GEN.41. I. Prigogine and I. Stengers, The New Alliance, part 2, An extended dynamics: Towards a human science of nature, *Scientia* **71**, 319–332 (1977).

GEN.42. I. Prigogine, Remarques introductives (Introductory remarks), *Revue de l'Univ. De Bruxelles*, no. 2, 180–186 (1977).

GEN.43. I. Prigogine, Métamorphoses de la science; culture et science aujourd'hui, *Revue de Univ. De Bruxelles*, no. 2, 233–262 (1977).

GEN.44. I. Prigogine, Physics and metaphysics, *Adv. Biol. Med. Phys.* **16**, 241–265 (1977).

1978

GEN.45. I. Prigogine et I. Stengers, *Neptuniens et Vulcaniens. Essais sur la transdisciplinarité* (Neptunians and Vulcanians. Essay on transdisciplinarity), in *Hommage à François Perroux*, Presses Universitaire Grenoble, 1978, pp. 44–55.

GEN.46. I. Prigogine, Entropie, fluctuations et dynamique (Entropy, fluctuations and dynamics), *Sciences et Techniques*, janvier 1978, pp. 34–37.

GEN.47. I. Prigogine, Zeit, Struktur und Fluktuationen (Structure and fluctuations) (Nobel-Vortrag), *Angew. Chemie* **90**, 704–715 (1978).

GEN.48. I. Prigogine, Time, structure and fluctuations, *Science* **201**, 777–785 (1978).

GEN.49. I. Prigogine, *Métamorphoses de la Science, Culture et Science aujourd'hui* (Metamorphosis of Science, Culture and Science Today), Séminaire Ecologie Quantitative, 3-e session, E4, tome II, pp. 345–368 1978.

GEN.50. I. Prigogine, I. Stengers, e G. Nicolis, *Controllo/Retroazione* (Control/feedback), Enciclopedia, III, *Divino-Fame*, Einaudi, Torino, 1978, pp. 978–1016.

GEN.51. I. Prigogine, e I. Stengers, *Energia*, Enciclopedia, V, *Divino-Fame*, Einaudi, Torino, 1978, pp. 411–438.

GEN.52. I. Prigogine e I. Stengers, *Equilibrio/squilibrio* (Equilibrium/nonequilibrium), Enciclopedia, V, *Divino-Fame*, Einaudi, Torino, 1978, pp. 523–546.

1979

GEN.53. I. Prigogine e I. Stengers, *Interazione* (Interaction), Enciclopedia, V, *Imitazione-Istituzioni*, Einaudi, Torino, 1978, pp. 863–873.

GEN.54. I. Prigogine, Irreversibility and randomness, *Astrophys. Space Sci.* **65**, 371–381 (1979).

GEN.55. I. Prigogine, I. Stengers, et S. Pahaut, *La dynamique, de Leibniz à Lucrèce* (Dynamics, from Leibniz to Lucretius), *Critique* **35**, 35–55 (1979).

GEN.56. I. Prigogine, The microscopic significance of irreversibility and the emergence of a new time, *Scientia*, 1979, pp. 173–184 (version française: 185–196).

GEN.57. I. Prigogine et I. Stengers, *Les deux cultures aujourd'hui* (The two cultures today), Nouvelle Revue Française, I: mai 1979, 42–54, II: juin 1979, 41–48.

1980

GEN.58. I. Prigogine et I. Stengers, Le problème de l'invention et la philosophie des sciences (The problem of invention and the philosophy of science), *Rev. Int. Philosphie* **34**, 5–25 (1980).

GEN.59. I. Prigogine, *Einstein: Triumphs and Conflicts*, in *Albert Einstein, Four commemorative Lectures*, The Humanities Research Center, University Texas Austin, 1979.

GEN.60. I. Prigogine, Vers la réconciliation science et humanisme (Toward the reconciliation of science and humanism), *Lumière* **316**, 25–27 (1980).

GEN.61. I. Prigogine, Time, dynamics and probability, *Phys. Mag.* **2**, 21–29 (1980).

GEN.62. I. Prigogine e I. Stengers, *Interazione, Ordine/disordine* (Interaction, Order/disorder), Enciclopedia X, *Opinione-Probabilitá,* Einaudi, Torino, 1978, pp. 87–106.

GEN.63. I. Prigogine e I. Stengers, *Or ganizzazione* (Organization), Enciclopedia X., *Opinione-Probabilitá,* Einaudi, Torino, 1978, pp. 177–198.

GEN.64. I. Prigogine, Vers la réconciliation science et humanisme (II): La vie naît de la mort (Toward the reconciliation of science and humanism (II): life is born from death), *Lumière,* 1980, pp. 22–24.

GEN.65. I. Prigogine, Probing into time, *Discovery,* Sept. 1980, University of Texas, Austin, pp. 4–7.

GEN.68. I. Prigogine, L'ordre à partir du chaos? (Order out of chaos?), *Prospective et Santé* **13**, 29–39 (1980).

GEN.69. I. Prigogine, *Zeit, Entropie und der Evolutionsbegriff in der Physik* (Entropy and the concept of evolution in physics), Mannheimer Forum 80/81, Böhringer, Mannheim, 1980, 9–44.

GEN.70. I. Prigogine, Temps, évolution et destin (Time, evolution and destiny), Actes 1-er Colloque International, *Relations actuelles entre les Sciences, les Arts et la Philosophie,* Maison de l'UNESCO, Paris, 1980.

GEN.71. I. Prigogine, La transparence et l'obstacle (Transparency and obstacle), in *Science et littérature: une même question, Bull. Acad. Roy. Langue et Litt. Franç.* **58**, 238–249 (1980).

GEN.72. I. Prigogine, *Loi, histoire et désertion* (Science, History and Desertion), *Le Débat,* **6**, Nov. 1980, 122–130.

1981

GEN.73. I. Prigogine, Einstein: Triomphe et conflits (Einstein: triumph and conflicts), in *Albert Einstein, 1879–1955,* Acad. Roy. Belg., 1981.

GEN.74. I. Prigogine and G. Dewel, *Pierre Résibois, 1936–1979, J. Stat. Phys.* **24**, 7–19 (1981).

GEN.75. I. Prigogine e I. Stengers, Semplice/complesso (Simple/complex), Enciclopedia XII. *Ricerca-Socializzazione,* Einaudi, Torino, 1981, pp. 715–730.

GEN.76. I. Prigogine e I. Stengers, Sistema (System), Enciclopedia XII. *Ricerca-Socializzazione,* Einaudi, Torino, 1981, pp. 993–1023.

GEN.77. I. Prigogine e I. Stengers, Soglia (Threshold), Enciclopedia XIII, *Societá-Tecnica,* Einaudi, Torino, 1981, pp. 78–93.

GEN.78. I. Prigogine en F. Boenders, Pluralisme en eenmaking (Pluralism and unification), *Nieuw Vlaams Tijdschr.* **34/6**, 914–930 (1981).

GEN.79. I. Prigogine, In memory of Erich Jantsch, in *The Evolutionary Vision,* AAAS Selected Symposium 61, Westview Press, Boulder, CO. 1981, pp. xiii–xiv.

GEN.80. I. Prigogine, Time, irreversibility and randomness, in *The evolutionary vision,* AAAS Selected Symposium 61, Westview Press, Boulder, CO, 1981, pp. 73–81.

GEN.82. I. Prigogine e I. Stengers, *Vincolo* (Bond), Enciclopedia XIV, *Tema/motivo-Zero,* Einaudi, Torino, 1981, pp. 1064–1080.

1982

GEN.83. I. Prigogine, Tempo, Entropia, Dinamica (Time, entropy, dynamics), *Critica Marxista* **6**, 35–49 (1982).

GEN.84. I. Prigogine and I. Stengers, Dynamics from Leibniz to Lucretius (Dynamics, from Leibniz to Lucretius), postface to the book *Hermes* of M. Serres, Johns Hopkins University Press, Baltimore, 1982, pp. 137–155.

GEN.85. A. P. Grecos and Il. Prigogine, On the problem of irreversibility in theoretical physics, in Proceed. 6th International Congress on Logic, Methodology and Philosophy of Science, Hannover 1979, North Holland, Amsterdam, 1982, pp. 429–439.

1983

GEN.86. I. Prigogine, *La Chimie et l'Aléatoire* (Chemistry and randomness), 7-ème Conference on Education en Chimie, Montpellier, 1983, pp. 87–98.

GEN.87. I. Prigogine, Probing into time, in *Biological Foundations and Human Nature*, M. Balaban, ed., Academic Press, New York, 1983, pp. 47–80.

GEN.88. I. Prigogine, *Einstein, Triomphe et Conflit* (in Russian) (Einstein: Triumph and Conflicts), Nauka, Moscow, 1983, pp. 109–123.

GEN.89. I. Prigogine, *Entropy and Evolution*, VWNL Cahier Het Leven, 1983, 23–48.

GEN.90. I. Prigogine, Man's dialogue with nature, *Perkins J.* **Summer**, 4–14 (1983).

GEN.91. I. Prigogine, *La lecture du complexe* (The reading of complexity), in Actes Coll. *Innovation et Société*, UNESCO—Acad. Europ. Sciences, Arts et Cult. 1983, pp. 27–37.

GEN.92. I. Prigogine et G. Nicolis, La chimie de l'aléatoire (The chemistry of randomness), *Chimie Nouvelle* **1**, 3–8 (1983).

GEN.93. I. Prigogine, *Man's New Dialogue with Nature*, Lecture at the conferring ceremony of Honda Prize, 1983.

GEN.94. I. Prigogine, *Between Time and Eternity, Nehru and Einstein*, Jawaharlal Nehru Lecture 1983, New Delhi, 1983.

GEN.95. I. Prigogine, *Zeit und Werden: Das Problem der Irreversibilität* (Time and becoming: The problem of irreversibility), Cerisy-la Salle, 1983.

1984

GEN.96. I. Prigogine, Review of *The Historical Development of Quantum Theory by J. Mehra and H. Rechenberg, Found. Phys.* **14**, 275–277 (1984).

GEN.97. I. Prigogine, *En guise d'introduction: Un monde à découvrir* (By way of introduction: A world to be discovered), *Nouvelles Sci. Technol.* **2**, 5–10 (1984).

GEN.98. I. Prigogine, L'avenir de la recherche (The future of research), *Nouvelles Sci. Technol.* **2**, 65–67 (1984).

GEN.99. I. Prigogine, *Only an Illusion*, in Tanner Lectures on Human Values, Jawaharlal Nehru University, Dec. 1982.

GEN.100. I. Prigogine, Order out of chaos, in Proceedings Stanford International Symposium, *Disorder and Order*, 1981, P. Livingston, ed., ANMA Libri 1984, pp. 41–60.

GEN.101. I. Prigogine, The microscopic theory of irreversible processes, in Proceedings, International Symposium, *Self-Organization: Autowaves and Structures far from Equilibrium*, 1983, V. I. Krinsky, ed., Springer, Berlin 1984, pp. 22–28.

GEN.102. I. Prigogine, The rediscovery of time, *Zygon* **19**, 433–447 (1984).

GEN.103. I. Prigogine et S. Pahaut, Redécouvrir le temps (Rediscovering time), in *L'art et le temps*, Societé Des Exposition Du Palais des Beaux Arts, Bruxelles, 1984, pp. 23–33.

1985

GEN.104. I. Prigogine, *A Major Influence, Indira Gandhi*, G. Parthasarathi and Sharada Prasad, eds., Vikas Publishers, 1985.

GEN.105. I. Prigogine et I. Stengers, *Hasard et Nécessité* (Randomness and necessity), Encyclopaedia Universalis, Symposium 85, pp. 324–330.

GEN.106. P. Glansdorff et I. Prigogine, *Thermodynamique: Lois Fondamentales* (Thermodynamics: Fundamental laws), Encyclopaedia Universalis, pp. 1163–1166.

GEN.107. P. Glansdorff et I. Prigogine, *Thermodynamique: Processus irréversibles non linéaires* (Thermodynamics: Nonlinear irreversible processes), Encyclopaedia Universalis, 1177–1178.

GEN.108. I. Prigogine, The rediscovery of time; in Proceedings, IBM Conference *Science and Complexity*, London 1985, Science Reviews Ltd., pp. 11–25.

GEN.109. I. Prigogine, *Exploring Complexity*, Seminar: *Perspective of the technological society toward the 21-st century, Japan Economic Journal*, Tokyo 1985.

GEN.110. I. Prigogine, *Redécouvrir le Temps* (Rediscovering time), Institute Supérieur de' Architecture V. Horta, Dec. 1985, pp. 2–6.

GEN.111. I. Prigogine, *Affronter la complexité* (Facing complexity), Centre d'Action Laïque, 1985.

GEN.112. I. Prigogine, *Coping with the irrational*, Lecture at meeting: *Science and Culture*, University of Barcelona, 1985.

GEN.113. I. Prigogine, *A new rationality?*, in Symposium, *Laws of Nature and Human Conduct: Specificities and Unifying Themes*, Brussels, 1985.

GEN.114. I. Prigogine, *New perspectives on complexity*, in *The Science and Practice of Complexity*, United Nations University, 1985.

1986

GEN.115. I. Prigogine and Y. Elskens, Irreversibility and the resurrection of the dead, *Nature* **320**, 661 (1986).

GEN.116. I. Prigogine, Natur, Wissenschaft und neue Rationalität (Nature, science and new rationality), *Dialektik* **12**, 11–37 (1986).

GEN.117. I. Prigogine and I. Stengers, *Eine offene Wissenschaft* (An open science), *Denkanstösse* **1**, 140–144 (1986).

GEN.118. I. Prigogine and I. Stengers, Von Euklid zu Aristoteles, (From Euclid to Aristoteles), in *Lust an der Natur*, R. Böhm and K. Meschkowski, eds., Piper, 1986.

GEN.119. I. Prigogine and G. Nicolis, Science in an age of transition (in Japanese), *Exploration of Nature* **5**, 1–9 (1986).

GEN.120. I. Prigogine and I. Stengers, *Devenir et Irréversibilité* (Becoming and irreversibility), Encyclopédie Philosophique, Paris 1986.

GEN.121. I. Prigogine, The open universe, in *The Infinite in the Sciences*, Rome, 1986.

GEN.122. I. Prigogine, Origine della complessità (Origin of complexity), lecture given at Darwin College, Cambridge, UK, 1986.

GEN.123. I. Prigogine, Science, civilization and democracy: Values, systems, structures and affinities, in *VIth Parliamentary and Scientific Conference*, Council of Europe, Tokyo, 1986.

GEN.124. I. Prigogine, *De l'univers clos de la mécanique à l'univers ouvert de l'entropie* (From the closed universe of mechanics to the open universe of entropy), Institut Royal Supérieur de Défense, Contact, No. 75, 1986.

GEN.125. I. Prigogine, The challenge of chemistry, *Chimica Oggi*, **Oct.**, 11–15 (1986).

GEN.126. I. Prigogine, *The Rediscovery of Time*, The World View of Contemporary physics, Colorado University, 1986.

GEN.127. I. Prigogine, *Nouvelles perspectives sur la complexité* (New perspectives of complexity), in Colloque *Science et pratique de la complexité*, Montpellier 1984, Université Nations Unies, 1986.

GEN.128. I. Prigogine, *The message of entropy*, in Workshop, *Patterns, Defects and Microstructures in Nonequilibrium Systems*, Austin, Texas 1986.

GEN.129. I. Prigogine, Life and physics: New perspectives, in *150-ème Anniversaire de la Naissance d'Alfred Nobel*, 1986.

GEN.130. I. Prigogine, A new rationality?, in Symposium Solvay Institute—Honda Foundation, Brussels 1985, *Mondes en développement*, no. 54–55, 19–39 (1986).

GEN.131. I. Prigogine, *Science in an Age of Transition*, Conférence VUB, 1986.

GEN.132. I. Prigogine, *Science et Culture* (Science and Culture), XXV-ème anniversaire de AUPELF, Université Paris–Sorbonne, 1986.

GEN.133. I. Prigogine, Die Botschaft der Entropie (The message of entropy), *Neue Zürcher Zeit.*, 3 Déc. (1986).

GEN.134. I. Prigogine y S. Pahaut, Redescubrir el tiempo (Rediscovering the time), *El Pascante*, no. 4 (1986).

1987

GEN.135. I. Prigogine, *Capter l'éphémère: A propos d'Arcimboldo* (Capturing the ephemera: About Arcimboldo), in Manifesto *Ecce Arcimboldo*, Palazzo Grassi, Venezia, 1987.

GEN.136. I. Prigogine og I. Stengers, *At forsta det nyue, hen imod en ny kohaerens* (in Swedish), Paradigma, juin 1987.

GEN.137. I. Prigogine, *La Naissance du Temps* (The Birth of Time), Conférence, Progetto Culturale Montedison, Roma, 1987.

GEN.138. I. Prigogine, *The Ethical Value of Science in a World of Limited Predictability*, Conférence ICTB, Venezia, 1987.

GEN.139. I. Prigogine, *Dix ans après* (Ten years later), in Conference, *Spatial Inhomogeneities and Transient Behavior in Chemical Kinetics*, Bruxelles, 1987.

GEN.140. I. Prigogine, The meaning of entropy, in *Evolutionary Epistemology*, W. Callebaut and R. Pinxten, eds., Reidel, Dordrecht, 1987; also: *Krisis*, No. 5–6, 112–128 (1987).

1988

GEN.141. I. Prigogine, Die physikalisch-chemischen Wurzeln des Lebens (The physico-chemical roots of life), in *Die Herausforderung der Evolutionsbiologie*, H. Meier, ed., Piper, 1988.

GEN.142. I. Prigogine, Quel regard sur le monde? (What glance on the world?), in *Science et Culture au XXe Siècle*, Conference Des lauréats du Prix Nobel, Paris, 1988.

GEN.143. I. Prigogine, Une nouvelle alliance de la science et de la culture, (A new alliance of science and culture), *Le Courrier de l'Unesco*, mai 1988.

GEN.144. I. Prigogine, Encapsulating time?, in *Molecules in Physics, Chemistry and Biology*, J. Maruani, ed., Kluwer, 1988.

GEN.145. I. Prigogine, *Science and Technology: New Perspectives*, Conference at Deutsche Aktionsgemeinschaft Bildung-Erfindung-Innovation., Berlin 1988.

GEN.146. I. Prigogine, *The Origins of Complexity*, Opening MERIT, Maastricht, 1988.

GEN.147. I. Prigogine, *Irréversibilité et Structure de l'Espace-Temps* (Irreversibility and the Structure of Space–Time), Congress Philosophie, Dijon, 1988.

GEN.148. I. Prigogine, *L'homme et la nature* (Man and nature), Conférence, Lisboa, 1988.

GEN.149. I. Prigogine, La redécouverte du temps (The rediscovery of time), *L'Homme* **108**, 5–26 (1988).

GEN.150. I. Prigogine, *The challenge of chemistry*, Message to the Chemical Society of Japan, 1988.

GEN.151. I. Prigogine, K. Chemla, et S. Pahaut, *Réflexions sur l'histoire des sciences en Chine* (*Reflexions on the history of sciences in China*), in Catalogue, *Chine, Ciel et Terre*, Musées Royale, Art d' et d', Histoire, Bruxelles, 1988.

1989

GEN.152. (I.P.) La physique et la redécouverte du temps (Physics and the rediscovery of time), (after Ilya Prigogine's lecture), in Colloque *Le Temps*, Ph. Grotard, D. Thieffry, eds., Cercle de philosophie, U.L.B., 1989.

GEN.153. I. Prigogine, What is entropy? *Naturwissenschaften* **76**, 1–8 (1989).

GEN.154. I. Prigogine, The rediscovery of time, in *Logic, Methodology and Philosophy of Science*, VIII, 1989, pp. 29–45.

GEN.155. I. Prigogine, *Etica della Conoscenza scientifica* (Ethics of scientific knowledge), Symposium ICTB, Venezia, Enciclopedia Italiana, 1989.

GEN.156. I. Prigogine, The philosophy of instability, *Futures*, Aug. 1989, pp. 396–400.

GEN.157. I. Prigogine, e I. Stengers, Disordine sovrano (Supreme disorder), *Epoca*, no. 2023, juillet 1989.

GEN.158. I. Prigogine, On ne peut pas parler d'une immobilité de l'homme (One cannot speak of man's immobility), *Dynasteurs*, juin 1989.

GEN.159. I. Prigogine, Le chaos et l'enthousiasme (Chaos and enthusiasm), *Sci. Technol.*, 19 Oct. 1989.

GEN.160. I. Prigogine, The rediscovery of time, in *Logic, Methodology and Philosophy of Science*, VIII, J. E. Fenstad, ed., Elsevier, Amsterdam, 1989, pp. 29–46.

GEN.161. I. Prigogine, What is entropy? *Naturwissenschaften* **76**, 1–8 (1989).

1990

GEN.162. I. Prigogine, What is time?, in *Metaphysics as Foundation*, I. Leclercq Festschrift, Suny Press, 1990.

GEN.163. I. Prigogine, *The New Environmental Science*, Fond. San Paolo, Torino, 1990.

GEN.164. I. Prigogine, Les objets nomades et la bibliothèque (The nomadic objects and the library), in *1-ère Conference Européen Automatisation et les Réseaux de Bibliothèques*, Bruxelles, 1990.

GEN.165. I. Prigogine, *Crise de la Société, Crise de la Science: Perspectives Nouvelles* (Crisis of Society, Crisis of Science: New Perspectives), Conférence donnée à Etudes et Expansion, Liège.

GEN.166. I. Prigogine, *The Arrow of Time*, in *Ecological Physical Chemistry*, C. Rossi and E. Tiezzi, eds., Elsevier, Amsterdam, 1990, pp. 1–24.

GEN.167. I. Prigogine, Loi, histoire et... désertion (Law, history and... desertion), in *La Querelle du Déterminisme*, Le Débat, Gallimard, Paris, 1990, pp. 102–112.

GEN.168. I. Prigogine et I. Stengers, La querelle du déterminisme, six ans après (The quarrel of determinism: six years later), in *La Querelle du Déterminisme*, Le Débat, Gallimard, Paris, pp. 247–265.

GEN.169. I. Prigogine et I. Stengers, *Hasard et nécessité* (Randomness and necessity), La connaissance en devenir, Encyclopaedia Universalis, pp. 712–719.

1991

GEN.170. I. Prigogine, *Le paradoxe du temps* (The paradox of time), Les Cahiers du MURS, no. 26, 4-e trimestre, 1991.

GEN.171. I. Prigogine, Temps, Dynamique, Chaos (Time, dynamics and chaos), in Colloque International, *L'Homme Face à la Science*, Criterion, Paris, 1991.

GEN.172. I. Prigogine, *Irréversibilité et structure de l'espace–temps* (Irreversibility and the structure of space–time), in *L'espace et le temps*, XXII-e Congrès De l'Association des Société de Philosophie de Langue Française, Dijon, 1991.

GEN.173. I. Prigogine, Science et créativité (Science and creativity), in *Irréalisme et Art Moderne*, Mélanges Ph. Roberts-Jones, 1991.

GEN.174. I. Prigogine, I. Stengers, und S. Pahaut, Die Dynamik, von Leibniz zu Lukrez (Dynamics, from Leibniz to Lucretius), in *Anfänge*, Merve Verlag, Berlin, 1991, pp. 19–62.

1992

GEN.175. I. Prigogine, *Le Temps et le Problème des Deux Cultures* (Time and the Problem of Two Cultures), Conférence Ansaldo, Gênes, 1992.

GEN.176. I. Prigogine, *Tijd en het probleem van de twee beschavingen* (Time and the problem of two cultures), Humanistisch Verbond, Gent, 1992.

GEN.177. I. Prigogine, *Science, Raison et Passion* (Science, Reason and Passion), Proceedings, Spoleto Scienza, Fond. Sigma Tau, 1992.

GEN.178. I. Prigogine, *End of Science?*, University of Arkansas for Medical Sciences, Little Rock, AR, 1992.

GEN.179. I. Prigogine, *The End of Science?*, Murata Symposium, Tokyo, 1992.

GEN.180. I. Prigogine, *Vers un humanisme scientifique* (Toward a scientific humanism), *Ist Italiano per gli studi Filosofici*, Seminario di scienze, Napoli, 1992.

GEN.181. I. Prigogine, *El redescubrimiento del tiempo* (The rediscovery of time), *Editorial Archipiélago* **10–11** (1992).

1993

GEN.182. I. Prigogine, Temps et devenir (Time and becoming), *Bull. Cl. Lett., Acad. Roy. Belg.*, 6-ème série, **IV**, 7–12 (1993).

GEN.183. I. Prigogine, Temps et devenir, A propos de l'histoire du temps, (Time and becoming, about the history of time), *4-ème Fesival du Film Scientifique du Québec*, Musée de la Civilisation, Québec.

GEN.184. I. Prigogine, Laws of nature—The search for certainty, in *Zum Naturbegriff der Gegenwart*, Projekt "Natur im Kopf" Stuttgart, 1993.

GEN.185. I. Prigogine, Creativity in the Sciences and the Humanities. A study in the relation between the two cultures, in *The Creative Process*, L. Gustafsson, S. Howard, and L. Niklasson, eds., Swedish Ministry of Education, Stockholm, 1993.

GEN.186. I. Prigogine, Uncertainty: The key to the science of the Future?, *New York Times*, Article pour le Club de Rome.

GEN.187. I. Prigogine, El redescubrimiento del tiempo / 2 (The rediscovery of time), *Archipiélago* **12**, 87–96 (1993).

GEN.188. I. Prigogine, *Exploring Complexity*, Conférence: Dieci Nobel per il futuro, Milano, 1993.

GEN.189. I. Prigogine, *The laws of Chaos*, Annales des Mines, Dec. 1993, pp. 61–62.

1994

GEN.190. I. Prigogine, Réinventer le temps (Reinventing time), in Erasme International, *Les Enigmes du temps*, hors série no. 96, 1994.

GEN.191. I. Prigogine, *Peace keeping and peace building: A commentary*, UNESCO, Venise.

GEN.192. I. Prigogine, Mind and matter, beyond the cartesian dualism, in *Origins: Brain and Self-Organization*, K. Pribram, ed., Lawrence Erlbaum Publishers, Hillsdale, NJ, 1994.

GEN.193. I. Prigogine, Science, reason and passion, *World Futures* **40**, 35–43 (1994).

GEN.194. I. Prigogine, Time, chaos and the laws of nature, Lecture, Université Complutense, Madrid, 1994.

GEN.195. I. Prigogine, Time, chaos and the two cultures, in International Meeting: *With Darwin Beyond Descartes, The Historical Concept of Nature and the Overcoming of the Two Cultures*, University of Pavia, 1994.

GEN.196. I. Prigogine, *Prologue* to *La Nueva Pagina* de Federico Mayor, Zaragoza, Ediciones UNESCO, 1994.

GEN.197. I. Prigogine, Résonances et domaines du savoir (Resonances and the domain of knowledge), dans *La thérapie familiale en changement* (M Elkaïm, dir.), Collect. "Les empêcheurs de penser en rond," Synthelabo, pp. 215–226, 1994.

GEN.198. I. Prigogine, Le due culture di oggi (Today's two cultures), *Acque e terre*, **V** (1), 37–39 (1994).

GEN.199. I. Prigogine, *A (very) brief history of certainty*, in Special issue: *Prigogine Solves Time Paradox, Brain/Mind*, **19** 1–8 (1994).

GEN.200. I. Prigogine, Ereigniss und Gesetz. Das Zusammenspiel von Ordnung und Unordnung im Universum (Event and law: The interaction of order and disorder in the universe), in *Strukturen des Chaos*, T. Schabert and E. Hornung, eds., W. Fink Verlag, München, 1994.

GEN.201. I. Prigogine, Zeitpfeil und Naturgesetze, Eine neue Deutung der Irreversibilität von Zeit (The arrow of time and the laws of nature, a new interpretation of time irreversibility), *Aus Forschung und Medizin* **9**, 65–74 (1994).

GEN.202. I. Prigogine, El fin de la ciencia? (The end of science?), Nuevos paradigmas Cultura y Subjectividad, Paidos, Buenos Aires, pp. 36–60, 1994.

GEN.203. I. Prigogine, De los relojes a las nubes (From clocks to clouds), Nuevos paradigmas Cultura y Subjectividad, Paidos, Buenos Aires, pp. 395–419, 1994.

GEN.204. I. Prigogine, El Universo y el Tiempo (Universe and time), Conference, Museo de La Plata, 1994.

GEN.205. I. Prigogine, L'échec des certitudes scientifiques et de la symétrie entre passé et futur (The failure of scientific certainties and of the symmetry between past and future), *Scienza e Storia, Bull. CISST* **10**, 31–41 (1994).

GEN.206. I. Prigogine, *Time, chaos and the two cultures*, Inauguration of the J. C. Polanyi Chair in Chemistry, University of Toronto, 1994.

1995

GEN.207. I. Prigogine, The rediscovery of time, in *Federico Mayor Amicorum Liber*, Bruylant, Bruxelles, 1995, pp. 1091–1095.

GEN.208. I. Prigogine, Le désordre créateur (The creative disorder), BIC, Dossier *Les sciences de l'homme*, no. 27, Institut du Management, Paris, 1995, pp. 71–77.

GEN.209. I. Prigogine, Que sont les Instituts Solvay? (What are the Solvay Institutes), text written on the occasion of the striking of the medal, May 1995.

GEN.210. I. Prigogine, The end of certitudes, in Conférence, *Babu Jagjivan Ram Memorial*, Déc. 1995.

GEN.211. I. Prigogine et I. Stengers, Was heisst es, die Welt zu verstehen?, (What means: understanding the world), *Denkanstösse '96*, Sept. 1995, Piper, pp. 72–75.

GEN.212. I; Prigogine, Préface to *Paroles d'homme* de A. Haulot, 1995.

GEN.213. I. Prigogine, The converging of Western and Eastern viewpoints on science and nature, Lecture, MOA Museum of Art, Atami, Japan, 1995.

GEN.214. I. Prigogine, Rencontre entre les différents points de vue occidentaux et orientaux concernant la science et la nature (see GEN.213), Version française du précédent.

GEN.215. I.Prigogine, The rediscovery of value and the opening of economics, Lecture, University of Texas at Austin, Oct. 1995.

GEN.216. I. Prigogine, Qu'est-ce que je ne sais pas? (What do I not know?), *Rencontres philosophiques*, UNESCO, 1995.

GEN.217. I. Prigogine, *Allocution au Conseil Solvay 1995*, Palais Royal de Laeken.

GEN.218. I. Prigogine, La fin des certitudes (The end of certitudes), *Bull. Coll. Royal des Doyens d'honneur du Travail*, **5** (1995).

GEN.219. I. Prigogine, Lecture at the occasion of the honorary degree, University of Valladolid, 1995.

GEN.220. I. Prigogine, *La fin des certitudes* (The end of certitudes), Conférence, Académie Royal Belgique, 1995.

1996

GEN.221. I. Prigogine, *The Rediscovery of Time*, 75 years of physics at Hoza street, Warsaw University, 1996.

GEN.222. I. Prigogine, *Time, Chaos and the Laws of Nature*, 50th anniv. Conf. of the Korean Chem. Society, The Korean Chem. Society, 1996.

GEN.223. I. Prigogine, *Chaos and the End of Certitudes—A New Paradigm*, Conférences à Ehwa Women's University, Seoul, et Warsaw University, 1996.

GEN.224. I. Prigogine, La ultima frontera (The last frontier), *El Pais*, 05/05/96.

GEN.225. I. Prigogine, *Time's Arrow Suspended in Flight*, Scientific and Medical Networks, 1996.

GEN.226. I. Prigogine, La quête de la certitude (The quest of certitude), in *Temps cosmique et histoire humaine*, Annales Institut Philosophie ULB, Vrin, Paris, 1996.

GEN.227. I. Prigogine, La fin des certitudes (The end of certitudes), *37e Symp. Int. Teilhard de Chardin*, Revue du XXIe siècle **2**, 9–24 (1996).

GEN.228. I. Prigogine, L'avenir en construction (The future in construction), *Sciences et Avenir*, hors-série, mai 1996.

GEN.229. I. Prigogine, *Zeitpfeil und instabile dynamische Systeme* (The arrow of time and dynamically unstable systems), Zürcher Zeit. 1996.

GEN.230. I. Prigogine, *Science, Reason and Passion, Leonardo* **29**, 39–42 (1996).

GEN.231. I. Prigogine, *The rediscovery of time, J. of Intellectual World*, 10/27/96.

GEN.232. I. Prigogine, La science au service de la planète (Science at the service of the planet), *Socialisme*, Oct. 1996.

GEN.233. I. Prigogine, Chasing the arrow of time, On Campus, The University of Texas at Austin, 20/06/96.

GEN.234. I. Prigogine, Le réenchantement du Monde (The reenchantment of the world), in *La Société en Quête de Valeurs—Pour Sortir de l'Alternative Entre Scepticisme et Dogmatisme*, Colloque Institut du management d'EDF et de GDF, Maxima, 1996.

GEN.235. I. Prigogine, Science en transition, *Bull. Mém. Acad. Roy. Méd. Belg.* **15**, 481–490 (1996).

GEN.236. I. Prigogine, Dos relogios as nuvens, in *Novos paradigmas, Cultura e subjetividade*, D. Schnitman, ed., Artes Medicas, Porto Alegre, 1996, pp. 257–273.

GEN.237. I. Prigogine, *The Venice Deliberations: Transformations in the Meaning of "Security,"* UNESCO 1996.

1997

GEN.238. I. Prigogine, *Rencontre de deux cultures: Humaniste et scientifique* (The meeting of two cultures: humanistic and scientific), Les Raisons de l'IRE, no. 10, fev–avril 1997.

GEN.239. I. Prigogine, Ping Chen, and Kehong Wen, Instability, complexity and bounded rationality in economic change, in IMPACT, *How IC^2 Institute Research Affects Public Policy and Business Practices* W. W. Coper, S. Thore, D. Gibson, and F. Phillips, eds., Quorum Boks, Westport, CT, 1997, pp. 209–218.

GEN.240. I. Prigogine, Zeit, Chaos und Naturgesetze (Time, chaos and laws of nature), in *Die Wiederentdeckung der Zeit*, A. Gimmler, M. Sandbothe and W. Zimmerli, eds., Primus Verlag, Darmstadt, 1997, pp. 79–94.

GEN.241. I. Prigogine, The rediscovery of value and the opening of economics, in *Value, Architecture and Design in America*, M. Benedikt, ed., University of Texas at Austin, 1997.

GEN.242. I. Prigogine, *Boulding Lecture*, Séoul National University, July 1997.

GEN.243. I. Prigogine, El desorden creador (The creative disorder), *Iniciativa Socialista* **9**, 48–50 (1997).

GEN.244. I. Prigogine, O fim da ciência? (The end of science?), in *Novos Paradigmas, Cultura e Subjetividade*, D. Schnitman, ed., Artes Medicas, Porto Alegre, 1997, pp. 26–40.

GEN.245. I. Prigogine, Nature in construction—perspectives, Presented at *From Energy to Information: Representation in Science, Art and Literature*, University of Texas, Austin, 1997.

1998

GEN.246. I. Prigogine, Zeit, Chaos und die zwei Kulturen (Time, chaos and the two cultures), in *Selbstorganisation*, Band 8: *Evolution und Irreversibilität*, J. Krug and L. Pohlmann, eds., Duncker & Humblot, Berlin, 1998, pp. 13–23.

GEN.247. I. Prigogine, Pluralité des futurs et fin des certitudes (Plurality of futures and the end of certitudes), in *Le XXI-ème siècle aura-t-il lieu?*, UNESCO.

GEN.248. I. Prigogine, Pluralité des futurs et fin des certitudes (Plurality of futures and the end of certitudes), *Le Monde*, 15/09/98, *L'Humanité*, 23/09/98.

GEN.249. I. Prigogine, Nell'universo delle probabilitá un solo punto fermo: l'incertezza (In the uiverse of probabilities, a single fixed point: incertitude), *Telema* **IV**, 17–18 (1998).

GEN.250. I. Prigogine, De l'être au devenir (From being to becoming), *III-ème Congrès International d'Ontologie*, San Sebastian, Oct. 1998.

GEN.251. J. P. Boon et I. Prigogine, Le temps dans la forme musicale (Time in the musical form), in *Le Temps et la Forme: Pour une Épistémologie de la Connaissance Musicale*, E. Darbellay, ed., Droz, Genève, 1998.

GEN.252. I. Prigogine, *The Networked Society*, IST 98, Vienna, Dec. 1998.

GEN.253. I. Prigogine, *Science et valeurs* (Science and values), Association Descartes, Vérone, Mai 1998.

GEN.254. I. Prigogine, Time and the laws of nature (unpublished).

GEN.255. I. Prigogine, Is the future given?, in *The Logic of Growth*, The Crafoord Memorial Symposium, Lund University, Sept. 1998.

GEN.256. I. Prigogine, *Science for the XXI-st Century*, lecture, Cercle Gaulois, 1998.

1999

GEN.257. I. Prigogine, Einstein and Magritte. A study of creativity, Conférence, *Einstein Meets Magritte*, CLEA, VUB, mai 1995 ; publié en 1999 in *Einstein Meets Magritte*, Kluwer, Dordrecht, pp. 99–105.

GEN.258. I. Prigogine, *Time and Laws of Nature*, in *Visione del mondo nella storia della scienza*, I.P.E., Napoli, 1999.

GEN.259. I. Prigogine, *Science for the XXI-st Century*, Acad. Roy. Sci. Outre-mer, 1999, pp. 12–18.

GEN.260. I. Prigogine, Le sommet de l'iceberg (The top of the iceberg), *Horizons* **33**, 8–10 (1999).

GEN.261. I. Prigogine, Le temps a-t-il un sens? (Does time have a meaning?), in *La Science: Dieu ou Diable?*, G. Pessis-Pasternak, ed., Odile Jacob, Paris, 1999.

GEN.262. I. Prigogine and J. Attali, Diriger dans un monde toujours plus complexe (How to manage in a more and more complex world), BMMA, 9/02/99, ULB.

GEN.263. I. Prigogine, Les jeux ne sont pas faits, in *Lettres aux Générations Futures*, UNESCO, 1999, pp. 133–140.

GEN.264. I. Prigogine, Préface, 11-ème rapport du Commissaire Géneral aux Réfugiés et aux Apatrides, 1999.

GEN.265. I. Prigogine, La riscoperta del valore e lo sviluppo della scienza economica (The rediscovery of value and the development of economic science), *Intersezioni*, numéro spécial, 1999.

GEN.266. I. Prigogine, La flèche du temps (The arrow of time), *Le Figaro*, 22/12/99.

GEN.267. I. Prigogine, Vers un monde moins arbitraire (Toward a less arbitrary world), *Le Nouvel Observateur*, 23–29/12/99.

GEN.268. I. Prigogine, Introduction au Séminaire de Padoue, *Scienza, Storia, Societá: Il pensiero di Ilya Prigogine e la sua influenza nella cultura del Novecento*, Scienza e Storia, Brugine, 1999.

GEN.269. I. Prigogine, *Is future given? Revisiting classical and quantum mechanics*, Lecture, Universita degli Studii, Salerno, May 1999.

GEN.270. I. Prigogine, Aspects de la temporalité: l'événement (Aspects of temporality: The event), 2-ème séminaire Ilya Prigogine: *Penser la Science—Qu'est-ce qu'un événement?*, ULB, 1999.

2000

GEN.271. I. Prigogine, Créativité de la Nature—Créativité humaine (Creativity of nature—creativity of man), in *L'Homme qui Marche*, Editions du Regard, Paris, 2000.

GEN.272. I. Prigogine, Le futur est-il donné? (Is future given?), in *Mondialisation et sociétés multiculturelles*, Ricciardielli, S. Urban, and K. Nanopoulos, eds., PUF, Paris, 2000.

GEN.273. I. Prigogine, Carta para as futuras geraçoes (in Portuguese), Folha de S. Paolo—*Mais!*, 30/01/2000.

GEN.274. I. Prigogine, The networked society (Festschrift honoring I. Wallerstein), *J. World-Systems Res.* **VI**, Fall–Winter 2000.

GEN.275. I. Prigogine, De l'être au devenir (From being to becoming), III. Congreso Internacional de Ontologia, Physis, No. 1, 2000, pp. 31–39.

GEN.276. I. Prigogine, La riscoperta del valore e lo sviluppo della scienza economica (The rediscovery of value and the development of economic science), *Intersezioni* **XX**, no. 2, Agosto 2000, pp. 297–304.

GEN.277. I. Prigogine, The arrow of time, in *The chaotic Universe*, V. Gurzadyan, ed., *Proc. 2-d ICRA Network Workshop*, Rome, Pescara, World Scientific, Singapore, 2000.

GEN.278. I. Prigogine, L'apport de l'Ecole de Thermodynamique et de Mécanique statistique de Bruxelles (The contribution of the Brussels school of thermodynamics and statistical mechanics), article rédigé dans le cadre d'un ouvrage sur les activités de l'ULB à l'approche de l'an 2000 (non publié).

2001

GEN.279. I. Prigogine, Vingt ans après (Twenty years later), in *L'Homme Devant l'Incertain*, I. Prigogine, ed., Odile Jacob, Paris, 2001, pp. 7–10.

GEN.280. I. Prigogine, *Le futur n'est pas donné* (Future is not given), in *L'homme devant l'incertain*, Il. Prigogine, ed., Odile Jacob, Paris, 2001, pp. 13–30.

GEN.281. I. Prigogine, L'humanisation de l'homme, créativité de la nature, créativité de l'homme (The humanisation of man, the creativity of nature, the creativity of man), in *L'Humanité de l'Homme*, J. Sojcher, ed., Cercle d'Art, Paris, 2001, pp. 11–21 + Entretien avec J. Sojcher, pp. 187–189.

GEN.282. I. Prigogine, The arrow of time and the end of certainty, in *Keys to the 21-st Century*, J. Bindé, ed., Unesco Publ., 2001.

GEN.283. I. Prigogine, Science et culture, in *Scienza e Storia*, Rivista del Centro Int. Di Storia dello Spazio e del Tempo, 2001, pp. 149–156.

GEN.284. I. Prigogine, Preface to *The Soviet Union and the Inner Side of the Nobel Prizes* by A. M. Blokh, Izd. Gumanistika, St. Petersburg.

GEN.285. I. Prigogine, Some reflection on time and universe, Proceedings, XXIV Workshop on High Energy Physics and Field Theory, Protvino, State Research Centre of Russia, Institute High Energy Physics, 2001, pp. 4–6.

2002

GEN.286. I. Prigogine, The end of science?, in *New Paradigms, Culture and Subjectivity*, D. Schnittman and J. Schnitman, eds., Ampton Press, 2002, pp. 21–36.

GEN.287. I. Prigogine, Dynamical roots of time symmetry breaking, *Philos. Trans. Roy. Soc. London*, **A 360**, 299–301 (2002).

GEN.288. I. Prigogine, A new page, Proceedings, International Forum 2000—World Expo Hanover, International Forum 2000 Executive Council, pp. 6–9.

GEN.289. I. Prigogine, *Is future given? Changes in our description of nature*, Lecture, National Technical University, Athens, 2002.

GEN.290. I. Prigogine, Lecture at the homage session at the occasion of the 25th anniversary of the Nobel Prize of chemistry award, Assoc. Diplômés Fac. Sc. (AScBr) ULB, 22 Nov. 2002.

GEN.291. I. Prigogine, *Marcel Grossmann Award talk*, in Proc. 9th Marcel Grossmann meeting, World Scientific, Singapore, 2002, pp. xx–xxii.

GEN.292. I. Prigogine, *Le Libre Examen dans un Monde en Mutation* (The free thought in a changing world), Conférence, ULB, 16/11/02.

GEN.293. I. Prigogine, Naïveté relative de chacun (Everyone's relative naivety), in *Paul Valéry à tous les points de vue*, P. Gifford, R. Pickering, and J. Schmidt-Radefeld, eds., L'Harmattan, Paris, 2002.

2003

GEN.294. I. Prigogine, *Kosci nie zostaly rzucone* (in Polish), *U podloza globalnych zagrozen, dylematy rozwoju*, Warsaw University.

GEN.295. I. Prigogine, Dynamical roots of the time symmetry breaking, in *Advanced Topics in Theoretical Chemical Physics*, 3–6, J. Maruani et al., eds., 2003, Kluwer, Dordrecht, pp. 3–6.

GEN.296. I. Prigogine, Is future given? in *Globalization and Multicultural Societies*, M. Ricciardelli, S. Urban, and K. Nanopoulos, eds., Notre Dame University Press, 2003.

GEN.297. I. Prigogine, Surprises in half of a century, in *Uncertainty and Surprise*, Red McCombs School of Business, University of Texas at Austin, April 2003.

GEN.298. I. Prigogine, Foreword, in *Essays Commemorating the Founding of the International Institute of Integration Research*, Daiseion-Ji (to appear). **[Last (handwritten) manuscript of Ilya Prigogine.]**

TIME ASYMMETRY IN NONEQUILIBRIUM STATISTICAL MECHANICS

PIERRE GASPARD

*Center for Nonlinear Phenomena and Complex Systems,
Université Libre de Bruxelles, B-1050 Brussels, Belgium*

CONTENTS

Special Volume in Memory of Ilya Prigogine: Advances in Chemical Physics, Volume 135,
edited by Stuart A. Rice

I. INTRODUCTION

Natural phenomena are striking us every day by the time asymmetry of their evolution. Various examples of this time asymmetry exist in physics, chemistry, biology, and the other natural sciences. This asymmetry manifests itself in the dissipation of energy due to friction, viscosity, heat conductivity, or electric resistivity, as well as in diffusion and chemical reactions. The second law of thermodynamics has provided a formulation of their time asymmetry in terms of the increase of the entropy. The aforementioned irreversible processes are fundamental for biological systems which are maintained out of equilibrium by their metabolic activity.

These phenomena are described by macroscopic equations that are not time-reversal symmetric in spite of the fact that the motion of the particles composing matter is ruled by the time-reversal symmetric equations of Newton or Schrödinger. This apparent dichotomy has always been very puzzling. Recently, a new insight into this problem has come from progress in dynamical systems theory. The purpose of dynamical systems theory is to analyze the trajectories of dynamical systems and to describe their statistical properties in terms of invariant probability measures. This theory has been developed to understand, in particular, how deterministic systems ruled by ordinary differential equations such as Newton's equation can generate random time evolutions, the so-called chaotic behaviors. New concepts such as the Lyapunov exponents, the Kolmogorov–Sinai entropy per unit time, the Pollicott–Ruelle resonances, and others have been discovered which are important additions to ergodic theory [1]. These new concepts have been used to revisit the statistical mechanics of irreversible processes, leading to advances going much beyond the early developments of ergodic theory. Thanks to these advances, it is nowadays possible to understand in detail how a system ruled by time-reversal symmetric Newton equations can generate a relaxation toward the state of thermodynamic equilibrium and thus to explain how the time asymmetry can appear in the statistical description. Furthermore, new relationships have been derived about the properties of the fluctuations in nonequilibrium systems—in particular, for the dissipated work [2]. These new relationships are based on the description of nonequilibrium systems in terms of paths or histories, description to which dynamical systems theory has greatly contributed.

Very recently, a new concept of time-reversed entropy per unit time was introduced as the complement of the Kolmogorov–Sinai entropy per unit time in order to make the connection with nonequilibrium thermodynamics and its entropy production [3]. This connection shows that the origin of entropy production can be attributed to a time asymmetry in the dynamical randomness of nonequilibrium steady states. Dynamical randomness is the property that the paths or histories of a fluctuating system do not repeat themselves and present random time evolutions. Examples are stochastic processes such as coin tossing or dice games of chance. The paths or histories of these processes can be depicted in space–time plots, representing the time evolution of their state. The disorder of the state of the system at a given time is characterized by the thermodynamic entropy or the entropy per unit volume. In contrast, the dynamical randomness can be characterized by the concept of entropy per unit time, which is a measure of disorder along the time axis instead of the space axis. Under nonequilibrium conditions, a time asymmetry appears in the dynamical randomness measured either forward or backward in time with the Kolmogorov–Sinai entropy per unit time or the newly introduced time-reversed entropy per unit time. The difference between both quantities is precisely the entropy production of nonequilibrium thermodynamics [3]. This new result provides an interpretation of the second law of thermodynamics in terms of temporal ordering out of equilibrium, which has far reaching consequences for biology as will be explained below.

These new methods of nonequilibrium statistical mechanics can be applied to understand the fluctuating properties of out-of-equilibrium nanosystems. Today, nanosystems are studied not only for their structure but also for their functional properties. These properties are concerned by the time evolution of the nanosystems and are studied in nonequilibrium statistical mechanics. These properties range from the electronic and mechanical properties of single molecules to the kinetics of molecular motors. Because of their small size, nanosystems and their properties such as the currents are affected by the fluctuations which can be described by the new methods.

The plan of this chapter is the following. Section II gives a summary of the phenomenology of irreversible processes and set up the stage for the results of nonequilibrium statistical mechanics to follow. In Section III, it is explained that time asymmetry is compatible with microreversibility. In Section IV, the concept of Pollicott–Ruelle resonance is presented and shown to break the time-reversal symmetry in the statistical description of the time evolution of nonequilibrium relaxation toward the state of thermodynamic equilibrium. This concept is applied in Section V to the construction of the hydrodynamic modes of diffusion at the microscopic level of description in the phase space of Newton's equations. This framework allows us to derive *ab initio* entropy production as shown in Section VI. In Section VII, the concept of Pollicott–Ruelle resonance is also used to obtain the different transport coefficients, as well as the rates of various kinetic processes in the framework of the escape-rate theory. The time asymmetry in the dynamical randomness of nonequilibrium systems and the fluctuation theorem for the currents are presented in Section VIII. Conclusions and perspectives in biology are discussed in Section IX.

II. HYDRODYNAMICS AND NONEQUILIBRIUM THERMODYNAMICS

A. Hydrodynamics

At the macroscopic level, matter is described in terms of fields such as the velocity, the mass density, the temperature, and the chemical concentrations of the different molecular species composing the system. These fields evolve in time according to partial differential equations of hydrodynamics and chemical kinetics.

A fluid composed of a single species is described by five fields: the three components of the velocity, the mass density, and the temperature. This is a drastic reduction of the full description in terms of all the degrees of freedom of the particles. This reduction is possible by assuming the local thermodynamic equilibrium according to which the particles of each fluid element have a Maxwell-Boltzmann velocity distribution with local temperature, velocity, and density. This local equilibrium is reached on time scales longer than the intercollisional time. On shorter time scales, the degrees of freedom other than the five fields manifest themselves and the reduction is no longer possible.

On the long time scales of hydrodynamics, the time evolution of the fluid is governed by the five laws of conservation of mass, momenta, and energy:

$$\partial_t \rho = -\boldsymbol{\nabla} \cdot (\rho v) \tag{1}$$

$$\partial_t \rho v = -\boldsymbol{\nabla} \cdot (\rho v v + \boldsymbol{P}) \tag{2}$$

$$\partial_t \rho e = -\boldsymbol{\nabla} \cdot (\rho e v + \boldsymbol{J}_q) - \boldsymbol{P} : \boldsymbol{\nabla} v \tag{3}$$

where ρ is the mass density, v the fluid velocity, and e is the specific internal energy or internal energy per unit mass [4]. The pressure tensor $\boldsymbol{P}$ decomposes into the hydrostatic pressure P and a further contribution due to the viscosities:

$$P_{ij} = P \, \delta_{ij} + \Pi_{ij} \tag{4}$$

with the components $i, j = x, y, z$. Both the viscous pressure tensor Π and the heat current density $\boldsymbol{J}_q$ are expressed in terms of the gradients of the velocity and the temperature by the phenomenological relations

$$\Pi_{ij} = -\eta \left(\nabla_i v_j + \nabla_j v_i - \frac{2}{3} \boldsymbol{\nabla} \cdot v \, \delta_{ij} \right) - \zeta \boldsymbol{\nabla} \cdot v \, \delta_{ij} \tag{5}$$

$$J_{q,i} = -\kappa \nabla_i T \tag{6}$$

where η is the shear viscosity, ζ the bulk velocity, and κ the heat conductivity.

The hydrodynamic equations rule in particular the relaxation of the fluid toward its state of global thermodynamic equilibrium, in which the velocity vanishes while the hydrostatic pressure and the temperature become uniform. The approach to the global thermodynamic equilibrium can be described by linearizing the hydrodynamic equation around the equilibrium state. These linearized equations can be solved by using the principle of linear superposition. Accordingly, the general solution is the linear combination of special solutions which are periodic in space with a wavenumber k giving the direction of the spatial modulations and their wavelength $\lambda = 2\pi/\|k\|$:

$$\rho(r, t) = \tilde{\rho}_k \, e^{ik \cdot r + st} + \rho_0 \tag{7}$$

$$v(r, t) = \tilde{v}_k \, e^{ik \cdot r + st} \tag{8}$$

$$T(r, t) = \tilde{T}_k \, e^{ik \cdot r + st} + T_0 \tag{9}$$

These spatially periodic solutions are known as the hydrodynamic modes. The decay rates $-s$ of these solutions are obtained by solving an eigenvalue problem for the five linearized hydrodynamic equations. These decay rates define the so-called dispersion relations of the five hydrodynamic modes. The dispersion relations can be expanded in powers of the wavenumber k as shown in Table I. The sound modes are propagative because their dispersion relations include the imaginary term $\pm i U_s k$ with the sound velocity

$$U_s = \sqrt{\left(\frac{\partial P}{\partial \rho}\right)_s} \tag{10}$$

The shear and thermal modes are not propagative. All the hydrodynamic modes are damped with a relaxation rate proportional to the square of the wavenumber so that their damping vanishes as the wavelength of the mode tends to infinity, which has its origin in the fact that these five modes are associated with the five

TABLE I
The Five Hydrodynamic Modes of a One-Component Fluid

Mode	Dispersion Relation	Multiplicity
Longitudinal sound modes	$s \simeq \pm i U_s k - \Gamma k^2$	2
Shear modes	$s \simeq -\frac{\eta}{\rho_0} k^2$	2
Thermal mode	$s \simeq -\frac{\kappa}{\rho_0 c_P} k^2$	1

Notation: k is the wavenumber, U_s the sound velocity, Γ the damping coefficient (11), η the shear viscosity, ρ_0 the uniform mass density, κ the heat conductivity, and c_P the specific heat capacity at constant pressure [4].

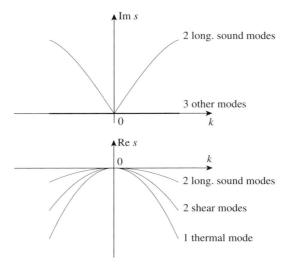

Figure 1. Schematic dispersion relations of the five hydrodynamic modes of a fluid with one component.

conserved quantities: mass, linear momenta, and energy. The damping coefficient of the sound modes is given by

$$\Gamma = \frac{1}{2\rho_0}\left[\left(\frac{1}{c_V} - \frac{1}{c_P}\right)\kappa + \frac{4}{3}\eta + \zeta\right] \tag{11}$$

where c_V and c_P are the specific heat capacities at constant volume and pressure, respectively [4].

The dispersion relations of the five hydrodynamic modes are depicted in Fig. 1. Beyond the hydrodynamic modes, there may exist kinetic modes that are not associated with conservation laws so that their decay rate does not vanish with the wavenumber. These kinetic modes are not described by the hydrodynamic equations but by the Boltzmann equation in dilute fluids. The decay rates of the kinetic modes are of the order of magnitude of the inverse of the intercollisional time.

A major preoccupation of nonequilibrium statistical mechanics is to justify the existence of the hydrodynamic modes from the microscopic Hamiltonian dynamics. Boltzmann equation is based on approximations valid for dilute fluids such as the Stosszahlansatz. In the context of Boltzmann's theory, the concept of hydrodynamic modes has a limited validity because of this approximation. We may wonder if they can be justified directly from the microscopic dynamics without any approximation. If this were the case, this would be great progress since it would prove that the irreversible relaxation toward the equilibrium state is ruled by decaying modes that are truly intrinsic to the microscopic Hamiltonian dynamics. Since the hydrodynamic modes decay exponentially, their microscopic construction would introduce time asymmetry into the statistical description and prove that irreversible processes can be deduced from the underlying Hamiltonian dynamics. We shall show in the following that dynamical systems theory has brought this extraordinary conceptual progress.

B.   Reaction–Diffusion Processes

Similar considerations concern the irreversible processes of diffusion and reaction in mixtures [5]. A system of M different molecular species is described by the three components of velocity, the mass density, the temperature, and $(M-1)$ chemical concentrations and is ruled by $M+4$ partial differential equations. The $M-1$ extra equations govern the mutual diffusions and the possible chemical reactions

$$\sum_{a=1}^{M} v_a^{<\rho} \, X_a \underset{w_{-\rho}}{\overset{w_{+\rho}}{\rightleftharpoons}} \sum_{a=1}^{M} v_a^{\rho>} \, X_a \tag{12}$$

$(\rho = 1, 2, \ldots, R)$ between the M molecular species $\{X_a\}_{a=1}^{M}$. The stoichiometric coefficients

$$v_a^{\rho} \equiv v_a^{\rho>} - v_a^{<\rho} = -v_a^{-\rho} \tag{13}$$

give the number of molecules of the species a produced during the reaction ρ. The speed w_ρ of the reaction ρ is defined as the number of reactive events per unit time and is determined by the chemical concentrations of the molecules participating in the reaction. Each reaction ρ is balanced by a reversed reaction $-\rho$, unless the reaction ρ is fully irreversible in which case the reversed reaction has vanishing speed $w_{-\rho} = 0$. The chemical concentrations can be represented by the numbers of molecules per unit volume—that is, the particle densities $\{n_a\}_{a=1}^{M}$.

For an isothermal process, the chemical concentrations obey reaction–diffusion equations

$$\partial_t n_a = -\boldsymbol{\nabla} \cdot (n_a \boldsymbol{v} + \boldsymbol{J}_a) + \sum_{\rho=1}^{R} v_a^{\rho} w_\rho \tag{14}$$

with $a = 1, 2, \ldots, M$. The diffusive currents $\boldsymbol{J}_a$ are defined with respect to the center of mass of each fluid element so that they satisfy the constraint

$$\sum_{a=1}^{M} m_a \boldsymbol{J}_a = 0 \tag{15}$$

where m_a is the mass of the molecule of species a. This constraint shows that the number of independent diffusive currents is limited to $M - 1$. Accordingly, only $M - 1$ mutual diffusions can exist in a fluid with M components. Mutual diffusion is absent in a one-component fluid and becomes possible in a binary mixture.

A simple example of reaction is the isomerization

$$A \underset{w_-}{\overset{w_+}{\rightleftharpoons}} B \tag{16}$$

ruled by the coupled equations

$$\partial_t n_A + \boldsymbol{\nabla} \cdot \boldsymbol{J}_A = -w_+ + w_- \tag{17}$$

$$\partial_t n_B + \boldsymbol{\nabla} \cdot \boldsymbol{J}_B = +w_+ - w_- \tag{18}$$

We notice that the total mass is conserved because the mass density $\rho = m_A n_A + m_B n_B$ obeys the continuity equation

$$\partial_t \rho + \boldsymbol{\nabla} \cdot \boldsymbol{J} = 0 \tag{19}$$

with the mass current density $\boldsymbol{J} = m_A \boldsymbol{J}_A + m_B \boldsymbol{J}_B$.

In a dilute mixture, the reaction speeds are proportional to the chemical concentrations

$$w_+ \simeq \kappa_+ n_A \tag{20}$$

$$w_- \simeq \kappa_- n_B \tag{21}$$

If cross-diffusion due to the chemical reaction is neglected, the diffusive currents are proportional to the gradients of concentrations

$$\boldsymbol{J}_A \simeq -\mathcal{D}_A \boldsymbol{\nabla} n_A \tag{22}$$

$$\boldsymbol{J}_B \simeq -\mathcal{D}_B \boldsymbol{\nabla} n_B \tag{23}$$

and we get the coupled reaction–diffusion equations

$$\partial_t n_A = \mathcal{D}_A \nabla^2 n_A - \kappa_+ n_A + \kappa_- n_B \tag{24}$$

$$\partial_t n_B = \mathcal{D}_B \nabla^2 n_B + \kappa_+ n_A - \kappa_- n_B \tag{25}$$

These equations are linear and their general solution is given by the linear combination of spatially periodic modes

$$n_A(\boldsymbol{r}, t) = \tilde{n}_{A,k}\, e^{i\boldsymbol{k}\cdot\boldsymbol{r}+st} + n_{A,0} \tag{26}$$

$$n_B(\boldsymbol{r}, t) = \tilde{n}_{B,k}\, e^{i\boldsymbol{k}\cdot\boldsymbol{r}+st} + n_{B,0} \tag{27}$$

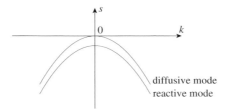

Figure 2. Schematic dispersion relations of the diffusive and reactive modes of the reaction of isomerization $A \rightleftharpoons B$ taking place in a solvent at rest.

The system admits two kinds of modes. The diffusive mode has the dispersion relation

$$s = -\mathcal{D}k^2 + O(k^4) \tag{28}$$

with the diffusion coefficient

$$\mathcal{D} = \frac{\kappa_+ \mathcal{D}_B + \kappa_- \mathcal{D}_A}{\kappa_+ + \kappa_-} \tag{29}$$

The diffusive mode is associated with the conservation (19) of the total mass since its dispersion relation vanishes with the wavenumber. Besides, we find a reactive mode with the dispersion relation

$$s = -\kappa_+ - \kappa_- - \mathcal{D}^{(r)} k^2 + O(k^4) \tag{30}$$

which does not vanish with the wavenumber. The associated reactive diffusion coefficient is given by

$$\mathcal{D}^{(r)} = \frac{\kappa_+ \mathcal{D}_A + \kappa_- \mathcal{D}_B}{\kappa_+ + \kappa_-} \tag{31}$$

The dispersion relations of the two modes are depicted in Fig. 2. The reactive mode is one of the kinetic modes existing beside the hydrodynamic modes such as the diffusive mode. Here also, we may wonder if these modes can be justified from the microscopic dynamics.

C. Nonequilibrium Thermodynamics

The second law of thermodynamics asserts that the total entropy S of a system may change in time because of exchanges with its environment and internal entropy production which is vanishing at equilibrium and positive out of equilibrium [5]

$$\frac{dS}{dt} = \frac{d_e S}{dt} + \frac{d_i S}{dt} \quad \text{with} \quad \frac{d_i S}{dt} \geq 0 \tag{32}$$

The second law can be expressed in terms of the local balance equation for the entropy:

$$\partial_t \rho s = -\boldsymbol{\nabla} \cdot \boldsymbol{J}_s + \sigma_s \tag{33}$$

The entropy is given in terms of the specific entropy s by $S = \int \rho s\, d^3r$ so that the entropy flux out of the system is related to the entropy current density by

$$\frac{d_e S}{dt} = \int \boldsymbol{J}_s \cdot d\Sigma \tag{34}$$

and the entropy production to the entropy source by

$$\frac{d_i S}{dt} = \int \sigma_s\, d^3r \geq 0 \tag{35}$$

The entropy production is due to the contributions of the different irreversible processes taking place in the system. The entropy source is given by the sum [5]

$$\sigma_s = \sum_\alpha A_\alpha J_\alpha \geq 0 \tag{36}$$

of the affinities or thermodynamic forces A_α [6] multiplied by the currents J_α associated with each one of these irreversible processes listed in Table II. In the linear regime, Onsager has assumed that the currents are proportional to the affinities [7]:

$$J_\alpha \simeq \sum_\beta L_{\alpha\beta} A_\beta \tag{37}$$

The microscopic justification of the second law of thermodynamics has always been a major problem in nonequilibrium statistical mechanics. Recent

TABLE II
Different Irreversible Processes with Their Affinity or Thermodynamic Force A_α and Their Current J_α.

Irreversible Process (α)	Affinity A_α	Current J_α
Shear viscosity (η)	$-\frac{1}{2T}\left(\nabla_i v_j + \nabla_j v_i - \frac{2}{3}\boldsymbol{\nabla} \cdot \boldsymbol{v}\,\delta_{ij}\right)$	$\Pi_{ij} - \frac{1}{3}\,\mathrm{tr}\Pi\,\delta_{ij}$
Bulk viscosity (ζ)	$-\frac{1}{3T}\boldsymbol{\nabla} \cdot \boldsymbol{v}\,\delta_{ij}$	$\frac{1}{3}\,\mathrm{tr}\Pi\,\delta_{ij}$
Heat conduction (κ)	$\boldsymbol{\nabla}\frac{1}{T}$	$\boldsymbol{J}_q$
Diffusion ($\mathcal{D}$)	$-\boldsymbol{\nabla}\frac{\mu_a}{T}$	$\boldsymbol{J}_a$
Reaction (L)	$-\frac{1}{T}\sum_{a=1}^M \nu_a^\rho \mu_a$	w_ρ

Notation: T is the temperature, v_i the fluid velocity, Π_{ij} the viscous pressure tensor, $\boldsymbol{J}_q$ the heat current density, μ_a its chemical potential, $\boldsymbol{J}_a$ the current density of molecular species a, ν_a^ρ the stoichiometric coefficient (13), and w_ρ the speed of reaction ρ.

work has shown that it is possible to derive *ab initio* the entropy production thanks to the construction of the hydrodynamic modes [8, 9]. As shown by Boltzmann, the entropy is an extramechanical quantity that characterizes the disorder of the system at a given time. It is therefore related to the probability distribution describing the statistical state of the system. The characterization of the disorder requires the counting of the number of possible microstates corresponding to a given probability distribution. This counting is possible if discrete coarse-grained states are defined for this purpose as proposed by Gibbs in 1902 [10]. This coarse graining has been further justified by quantum mechanics which showed that the discrete states must be identified with phase-space cells of volume $\Delta^f r \Delta^f p = (2\pi\hbar)^f$, as done in the Sackur–Tetrode formula. In this way, a natural definition can be given to the entropy even in nonequilibrium states. Gibbs coarse-grained entropy has been in agreement with the experimental measures of entropy for a century. As explained below, it is possible to derive the entropy production (36) of nonequilibrium thermodynamics thanks to the recent advances in dynamical systems theory and, in particular, the construction of the hydrodynamic modes at the microscopic level of description.

III. MICROREVERSIBILITY AND TIME ASYMMETRY

A major preoccupation in nonequilibrium statistical mechanics is to derive hydrodynamics and nonequilibrium thermodynamics from the microscopic Hamiltonian dynamics of the particles composing matter. The positions $\{r_a\}_{a=1}^N$ and momenta $\{p_a\}_{a=1}^N$ of these particles obey Newton's equations or, equivalently, Hamilton's equations:

$$\frac{dr_a}{dt} = +\frac{\partial H}{\partial p_a} \tag{38}$$

$$\frac{dp_a}{dt} = -\frac{\partial H}{\partial r_a} \tag{39}$$

Since the Hamiltonian function H is an even function of the momenta, Hamilton's equation are symmetric under time reversal:

$$\Theta(r_1, r_2, \ldots, r_N, p_1, p_2, \ldots, p_N, t) = (r_1, r_2, \ldots, r_N, -p_1, -p_2, \ldots, -p_N, -t) \tag{40}$$

Furthermore, the phase-space volumes are preserved during the Hamiltonian time evolution, according to Liouville's theorem. We denote by

$$\Gamma = (r_1, r_2, \ldots, r_N, p_1, p_2, \ldots, p_N) \in \mathcal{M} \tag{41}$$

a point in the phase space $\mathcal{M}$ of positions and momenta and by

$$\Gamma_t = \Phi^t(\Gamma_0) \tag{42}$$

the unique solution of Hamilton's equations starting from the initial condition Γ_0.

The time-reversal symmetry of the Hamiltonian dynamics, also called the microreversibility, is the property that if the phase-space trajectory

$$\mathcal{C} = \{\Gamma_t = \Phi^t(\Gamma_0) : t \in \mathbb{R}\} \tag{43}$$

is a solution of Hamilton's equations, then its time reversal

$$\Theta(\mathcal{C}) = \{\tilde{\Gamma}_{t'} = \Phi^{t'} \circ \Theta(\Gamma_0) : t' \in \mathbb{R}\} \tag{44}$$

is also a solution of Hamilton's equations. It is not often emphasized that, typically, the trajectory and its time reversal are physically distinct trajectories

$$\mathcal{C} \neq \Theta(\mathcal{C}) \tag{45}$$

and that it is rather exceptional that they coincide (as in the case of a harmonic oscillator). For a free particle of Newton's equation $d^2x/dt^2 = 0$, the trajectory $x = v_0 t + x_0$ is physically distinct from its time reversal $x = -v_0 t + x_0$ unless $v_0 = 0$. This remark shows that the solutions of Hamilton's equations do not necessarily have the time-reversal symmetry of the set of equations itself. This phenomenon is well known under the name of spontaneous symmetry breaking and is common in condensed matter physics. In this regard, we notice that it is a major historical development that Newton and his followers have conceptually separated the actual trajectory of a system of interest from the fundamental equation of motion which rules all the possible trajectories given by its solutions. The microreversibility is the time-reversal symmetry of the set of all the possible solutions of Newton's equations but this does not imply the time-reversal symmetry of the unique trajectory followed by the Universe. Most of the solutions of typical Newton's equations break the time-reversal symmetry of the equations. In this perspective, the Newtonian scheme appears to be compatible with the possibility of irreversible time evolutions if irreversibility is understood as the property of the trajectory (as in pre-Newtonian science) and not with the fundamental equations of motion. Today, we are familiar with such phenomena of spontaneous symmetry breaking according to which the solution of an equation may have a lower symmetry than the equation itself. It is our purpose to show that irreversibility can be understood in a similar way as a spontaneous breaking of the time-reversal symmetry at the statistical level of description.

IV. POLLICOTT–RUELLE RESONANCES AND TIME-REVERSAL SYMMETRY BREAKING

Modern dynamical systems theory has shown that the solutions of Newton's equations may be as random as a coin tossing probability game. A mechanism allowing such a dynamical randomness is the sensitivity to initial conditions of chaotic dynamical systems. This mechanism can conciliate causality and determinism with the existence of random events. Causality says that each effect has a cause and determinism that a unique cause can be associated to each effect once a system is described in its globability. In constrast, we observe in nature many random events as effects without apparent causes. Maxwell and Poincaré have judiciously pointed out that there exist systems with sensitivity to initial conditions in which small causes can lead to big effects. If the cause is so minute that it went unnoticed, such a deterministic dynamics can therefore explain a random or stochastic process. Most remarkably, there exists today a quantitative theory for such properties as dynamical randomness and sensitivity to initial conditions. The historical milestones are the following: Dynamical randomness has been first characterized by the information theory of Shannon around 1948 [11]. The quantitative characterization of dynamical randomness was established in the work of Kolmogorov and Sinai in 1959 with the concept of entropy per unit time [12, 13]. This latter characterizes the disorder of the trajectories along the time axis, while the thermodynamic entropy is a measure of disorder in space at a given time; otherwise, both entropies have similar interpretations. On the other hand, the concept of Lyapunov exponents was introduced to describe quantitatively the sensitivity to initial conditions [14], which Lorenz discovered in the context of meteorology in 1963 [15]. The Lyapunov exponents are defined as the rates of exponential separation between a reference trajectory and a perturbed one:

$$\lambda_i = \lim_{t \to \infty} \frac{1}{t} \ln \frac{\| \delta \mathbf{\Gamma}_i(t) \|}{\| \delta \mathbf{\Gamma}_i(0) \|} \qquad (46)$$

where $\delta \mathbf{\Gamma}_i(t)$ denotes the infinitesimal perturbation at time t on the reference trajectory. There exist as many Lyapunov exponents as there are directions $i = 1, 2, \ldots, 2f = 6N$ in the phase space. In 1977, Pesin proved that dynamical randomness finds its origin in the sensitivity to initial conditions [16]. Thereafter, fundamental connections between these new concepts from dynamical systems theory and nonequilibrium statistical mechanics have been discovered since 1990 [1].

It is essential to understand that the aforementioned dynamical randomness is quantitatively comparable to the one seen in Brownian motion or other stochastic processes of nonequilibrium statistical mechanics. Indeed, the dynamical

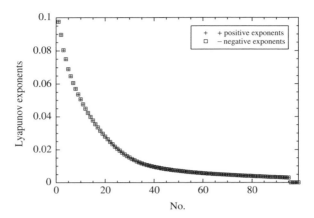

Figure 3. Spectrum of Lyapunov exponents of a dynamical system of 33 hard spheres of unit diameter and mass at unit temperature and density 0.001. The positive Lyapunov exponents are superposed to minus the negative ones showing that the Lyapunov exponents come in pairs $\{\lambda_i, -\lambda_i\}$ as expected in Hamiltonian systems. Eight Lyapunov exponents vanish because the system has four conserved quantities, namely, energy and the three components of momentum and because of the pairing rule. The total number of Lyapunov exponents is equal to $6 \times 33 = 198$.

randomness of the stochastic processes can also be characterized by a positive entropy per unit time. Since the mid-1990s, we can compute the Kolmogorov–Sinai entropy per unit time as well as the spectrum of Lyapunov exponents is the many-particle systems of statistical mechanics such as the hard-sphere gas [17–19]. In a dilute gas, a typical Lyapunov exponent is of the order of the inverse of the intercollisional time and the Kolmogorov–Sinai entropy per unit time of the order of the Avogadro number multiplied by a typical Lyapunov exponent, showing that many-particle systems have a huge dynamical randomness [1].

Figure 3 depicts the spectrum of Lyapunov exponents in a hard-sphere system. The area below the positive Lyapunov exponent gives the value of the Kolmogorov–Sinai entropy per unit time. The positive Lyapunov exponents show that the typical trajectories are dynamically unstable. There are as many phase-space directions in which a perturbation can amplify as there are positive Lyapunov exponents. All these unstable directions are mapped onto corresponding stable directions by the time-reversal symmetry. However, the unstable phase-space directions are physically distinct from the stable ones. Therefore, systems with positive Lyapunov exponents are especially propitious for the spontaneous breaking of the time-reversal symmetry, as shown below.

The Lyapunov exponents and the Kolmogorov–Sinai entropy per unit time concern the short time scale of the kinetics of collisions taking place in the fluid. The longer time scales of the hydrodynamics are instead characterized by the decay of the statistical averages or the time correlation functions of the observables. Here, we consider a statistical ensemble of trajectories described by a probability density $p(\mathbf{\Gamma})$, which is known to evolve in time according to the famous Liouville equation of nonequilibrium statistical mechanics [4]:

$$\partial_t p = \hat{L} p \tag{47}$$

The Liouvillan operator is defined in terms of the Poisson bracket with the Hamiltonian:

$$\hat{L} \equiv \{H, \ \} = \sum_{a=1}^{N} \left(\frac{\partial H}{\partial \boldsymbol{r}_a} \cdot \frac{\partial}{\partial \boldsymbol{p}_a} - \frac{\partial H}{\partial \boldsymbol{p}_a} \cdot \frac{\partial}{\partial \boldsymbol{r}_a} \right) \tag{48}$$

The time evolution of the probability density is induced by Hamiltonian dynamics so that it has its properties—in particular, the time-reversal symmetry. However, the solutions of Liouville's equation can also break this symmetry as it is the case for Newton's equations. This is the case if each trajectory (43) has a different probability weight than its time reversal (44) and that both are physically distinct (45).

The idea of Pollicott–Ruelle resonances relies on this mechanism of spontaneous breaking of the time-reversal symmetry [20, 21]. The Pollicott–Ruelle resonances are generalized eigenvalues s_j of Liouvillian operator associated with decaying eigenstates which are singular in the stable phase-space directions but smooth in the unstable ones:

$$\hat{L} \Psi_j = s_j \Psi_j \tag{49}$$

These are the classical analogues of quantum scattering resonances except that these latter ones are associated with the wave eigenfunctions of the energy operator, although the eigenstates of the Liouvillian operator are probability densities or density matrices in quantum mechanics. Nevertheless, the mathematical method to determine the Pollicott–Ruelle resonances is similar, and they can be obtained as poles of the resolvent of the Liouvillian operator

$$\frac{1}{s - \hat{L}} \tag{50}$$

at real or complex values of the variable s which is a complex frequency. The poles can thus be identified by analytic continuation of the resolvent toward complex frequencies. This idea has already been proposed in the early 1960s in the context of nonequilibrium statistical mechanics [4]. In the mid-1980s, Pollicott and Ruelle provides rigorous and systematic tools in order to determine these resonances in fully chaotic systems [20, 21]. The periodic-orbit theory was

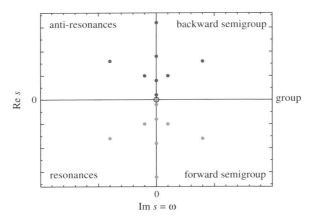

Figure 4. Complex plane of the variable s. The vertical axis $\mathrm{Re}\, s$ is the axis of the rates or complex frequencies. The horizontal axis $\mathrm{Im}\, s$ is the axis of real frequencies ω. The resonances are the poles in the lower half-plane contributing to the forward semigroup. The antiresonances are the poles in the upper half-plane contributing to the backward semigroup. The resonances are mapped onto the antiresonances by time reversal. Complex singularities such as branch cuts are also possible but not depicted here. The spectrum contributing to the unitary group of time evolution is found on the axis $\mathrm{Re}\, s = 0$.

developed which clearly showed that these resonances are intrinsic to the dynamics of the system [1, 22]. The knowledge of these resonances allows us to decompose the time evolution of the statistical averages of the observables

$$\langle A \rangle_t = \langle A | \exp(\hat{L}t) | p_0 \rangle = \int A(\Phi^t \Gamma_0)\, p_0(\Gamma_0)\, d\Gamma_0 = \int A(\Gamma)\, p_t(\Gamma)\, d\Gamma \quad (51)$$

into decaying exponential functions. The resonances obtained by analytic continuation toward negative values of $\mathrm{Re}\, s$ are associated with exponential decays for positive times (see Fig. 4). The corresponding expansion defines the forward semigroup:

$$\langle A \rangle_t = \langle A | \exp(\hat{L}t) | p_0 \rangle \simeq \sum_j \langle A | \Psi_j \rangle\, \exp(s_j t)\, \langle \tilde{\Psi}_j | p_0 \rangle + \cdots \quad (52)$$

which is only valid for $t > 0$. The dots denote the contributions beside the simple exponentials due to the resonances. These extra contributions may include Jordan-block structures if a resonance has a multiplicity m_j higher than unity. In this case, the exponential decay is modified by a power-law dependence on time as $t^{m_j-1} e^{s_j t}$ [1]. It is important to notice that the expansion (52) is obtained without assuming that the Liouvillian operator is anti-Hermitian, so the

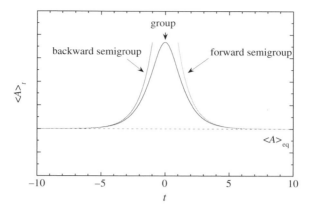

Figure 5. Time evolution of the statistical average (51) according to the expansion (52) of the forward semigroup valid for $t > 0$ and the expansion (53) of the backward semigroup valid for $t < 0$. See color insert.

right-hand eigenstate Ψ_j will in general differ from the left-hand eigenstate $\tilde{\Psi}_j$. As a corollary, these eigenstates do not belong to a Hilbert space of square integrable functions; instead they can be singular distributions. In particular, the right-hand eigenstates associated with nonvanishing eigenvalue $s_j \neq 0$ are smooth in the unstable phase-space directions but singular in the stable ones. On long time scales, the statistical average (52) tends toward the slowest decay with the eigenvalue for which $|\text{Re }s_j|$ is minimum (see Fig. 5). If this eigenvalue is $s_0 = 0$, the statistical average converges toward a stationary state which defines the state of thermodynamic equilibrium. Here, we recover the fundamental property of mixing introduced by Gibbs in 1902 according to which the statistical averages and the time correlation functions weakly converge toward their equilibrium value [10].

On the other hand, the antiresonances obtained by analytic continuation toward positive values of $\text{Re }s$ are associated with exponential decays for negative times. The corresponding expansion defines the backward semigroup:

$$\langle A \rangle_t = \langle A | \exp(\hat{L}t) | p_0 \rangle \simeq \sum_j \langle A | \Psi_j \circ \Theta \rangle \, \exp(-s_j t) \, \langle \tilde{\Psi}_j \circ \Theta | p_0 \rangle + \cdots \quad (53)$$

valid for $t < 0$ (see Fig. 5).

We notice that it is the analytic continuation which has the effect of breaking the time-reversal symmetry. If we contented ourselves with the continuous spectrum of eigenvalues with $\text{Re }s = 0$, we would obtain the unitary group of time evolution valid for positive and negative times. The unitary spectral decomposition is as valid as the spectral decompositions of the forward or backward semigroups. However, only the semigroups provide us with well-defined relaxation rates that are intrinsic to the system. The continuous spectrum of the unitary group does not display characteristic times.

V. MICROSCOPIC CONSTRUCTION OF THE DIFFUSIVE MODES

A. The Diffusive Modes as Liouvillian Eigenstates

When applied to spatially extended dynamical systems, the Pollicott–Ruelle resonances give the dispersion relations of the hydrodynamic and kinetic modes of relaxation toward the equilibrium state. This can be illustrated in models of deterministic diffusion such as the multibaker map, the hard-disk Lorentz gas, or the Yukawa-potential Lorentz gas [1, 23]. These systems are spatially periodic. Their time evolution Frobenius-Perron operator

$$\hat{P}^t = \exp(\hat{L}t) \tag{54}$$

is invariant under a discrete group of spatial translations $\hat{T}^a$:

$$[\hat{P}^t, \hat{T}^a] = 0 \tag{55}$$

Since these operators commute, they admit common eigenstates:

$$\hat{P}^t \Psi_k = \exp(s_k t) \Psi_k \tag{56}$$

$$\hat{T}^a \Psi_k = \exp(i\mathbf{k} \cdot \mathbf{a}) \Psi_k \tag{57}$$

The generalized eigenvalue s_k is a Pollicott–Ruelle resonance associated with the eigenstate Ψ_k. The hydrodynamic modes can be identified as the eigenstates associated with eigenvalues s_k vanishing with the wavenumber $\mathbf{k}$.

Here, we consider the diffusive processes in which independent particles are transported in a lattice where they perform a random walk. According to a formula by Van Hove [24], the dispersion relation of diffusion is given by

$$s_k = \lim_{t \to \infty} \frac{1}{t} \ln \langle \exp[i\mathbf{k} \cdot (\mathbf{r}_t - \mathbf{r}_0)] \rangle = -\mathcal{D}k^2 + O(k^4) \tag{58}$$

where $\mathbf{r}_t$ is the position of the particle in the lattice. The diffusion coefficient $\mathcal{D}$ is obtained by expanding in powers of the wavenumber and is given by Einstein and Green–Kubo formulas [25, 26]. The dispersion relation of diffusion (58) is nothing other than a Liouvillian eigenvalue and should therefore be found among the Pollicott–Ruelle resonances of the dynamical system. The corresponding

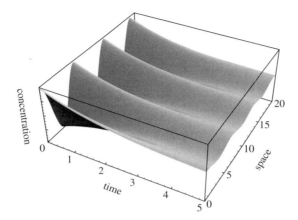

Figure 6. Schematic representation of the relaxation of a diffusive mode in space and time toward the uniform equilibrium state.

diffusive mode has the wavelength $2\pi/\|\boldsymbol{k}\|$ and decays exponentially at the rate $-s_{\boldsymbol{k}}$ as depicted in Fig. 6.

At the microscopic level of description, the hydrodynamic mode of diffusion is defined as the Liouvillian eigenstate:

$$\hat{L}\Psi_{\boldsymbol{k}} = s_{\boldsymbol{k}}\,\Psi_{\boldsymbol{k}} \tag{59}$$

Since this latter is expected to be a singular distribution, its density $\Psi_{\boldsymbol{k}}$ cannot be depicted as a function. Accordingly, we consider its cumulative function by integrating over some curve in phase space. This curve is followed by changing the parameter θ:

$$F_{\boldsymbol{k}}(\theta) = \lim_{t\to\infty} \frac{\int_0^\theta d\theta'\,\exp[i\boldsymbol{k}\cdot(\boldsymbol{r}_t - \boldsymbol{r}_0)_{\theta'}]}{\int_0^{2\pi} d\theta'\,\exp[i\boldsymbol{k}\cdot(\boldsymbol{r}_t - \boldsymbol{r}_0)_{\theta'}]} \tag{60}$$

This function is normalized to take the unit value for $\theta = 2\pi$. For vanishing wavenumber, the cumulative function is equal to $F_{\boldsymbol{k}}(\theta) = \theta/(2\pi)$, which is the cumulative function of the microcanonical uniform distribution in phase space. For nonvanishing wavenumbers, the cumulative function becomes complex. These cumulative functions typically form fractal curves in the complex plane $(\mathrm{Re}\,F_{\boldsymbol{k}}, \mathrm{Im}\,F_{\boldsymbol{k}})$. Their Hausdorff dimension D_{H} can be calculated as follows. We can decompose the phase space into cells labeled by ω and represent the trajectories by the sequence $\boldsymbol{\omega} = \omega_0\omega_1\omega_2\ldots\omega_{n-1}$ of cells visited at regular time interval $0, \tau, 2\tau, \ldots, (n-1)\tau$. The integral over the phase-space curve in Eq. (60) can be discretized into a sum over the paths $\boldsymbol{\omega}$. The weight of each path $\boldsymbol{\omega}$ is inversely proportional to the stretching factor $\Lambda(\omega)$ by which perturbations are amplified due to the dynamical instability in the phase space, so we get

$$F_k(\theta) = \lim_{n\to\infty} \sum_\omega \frac{1}{|\Lambda(\omega)|} \, e^{ik\cdot(r_{n\tau}-r_0)_\omega} \, e^{-s_k n\tau} = \lim_{n\to\infty} \sum_\omega \Delta F_k(\omega) \qquad (61)$$

where we have used the fact that the denominator in Eq. (60) behaves as $\exp(s_k t)$. This provides the approximation of the cumulative function as a polygonal curve formed by a sequence of complex vectors. The Hausdorff dimension of this curve is obtained by the condition

$$\sum_\omega |\Delta F_k(\omega)|^{D_H} \sim 1 \qquad \text{for} \quad n \to \infty \qquad (62)$$

This condition can be rewritten in terms of Ruelle's function defined as the generating function of the Lyapunov exponents and their statistical moments:

$$P(\beta) \equiv \lim_{n\tau\to\infty} \frac{1}{n\tau} \ln \sum_\omega \frac{1}{|\Lambda(\omega)|^\beta} \qquad (63)$$

Combining Eqs. (61), (62), and (63), we finally obtain the formula

$$P(D_H) = D_H \, \text{Re} \, s_k \qquad (64)$$

which gives the Hausdorff dimension in terms of the dispersion relation of diffusion and Ruelle's function [23].

Since Ruelle's function vanishes if its argument takes the unit value $P(1) = 0$, the Hausdorff dimension can be expanded in powers of the wavenumber as

$$D_H(k) = 1 + \frac{\mathcal{D}}{\lambda} \, k^2 + \mathcal{O}(k^4) \qquad (65)$$

so the diffusion coefficient can be obtained from the Hausdorff dimension and the Lyapunov exponent by the formula

$$\mathcal{D} = \lambda \lim_{k\to 0} \frac{D_H(k) - 1}{k^2} \qquad (66)$$

This formula has been verified for the following dynamical systems sustaining deterministic diffusion [23].

B.   Multibaker Model of Diffusion

One of the simplest models of deterministic diffusion is the multibaker map, which is a generalization of the well-known baker map into a spatially periodic system [1, 27, 28]. The map is two dimensional and rules the motion of a particle which can jump from square to square in a random walk. The equations of the map are given by

$$\phi(l, x, y) = \begin{cases} \left(l - 1, 2x, \frac{y}{2}\right), & 0 \leq x \leq \frac{1}{2} \\ \left(l + 1, 2x - 1, \frac{y+1}{2}\right), & \frac{1}{2} < x \leq 1 \end{cases} \tag{67}$$

where (x, y) are the coordinates of the particle inside a square, while $l \in \mathbb{Z}$ is an integer specifying in which square the particle is currently located. This map acts as a baker map, but, instead of mapping the two stretched halves into themselves, they are moved to the next-neighboring squares as shown in Fig. 7.

The multibaker map preserves the vertical and horizontal directions, which correspond respectively to the stable and unstable directions. Accordingly, the diffusive modes of the forward semigroup are horizontally smooth but vertically singular. Both directions decouple, and it is possible to write down iterative equations for the cumulative functions of the diffusive modes, which are known as de Rham functions [1, 29]

$$F_k(y) = \begin{cases} \alpha F_k(2y), & 0 \leq y \leq \frac{1}{2} \\ (1 - \alpha) F_k(2y - 1) + \alpha, & \frac{1}{2} < y \leq 1 \end{cases} \tag{68}$$

with

$$\alpha = \frac{\exp(ik)}{2 \cos k} \tag{69}$$

For each value of the wavenumber k, the de Rham functions depict fractal curves as seen in Fig. 8. The fractal dimension of these fractal curves can be calculated

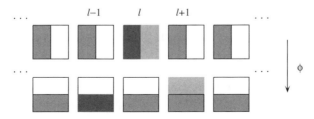

Figure 7.   Schematic representation of the multibaker map ϕ acting on an infinite sequence of squares.

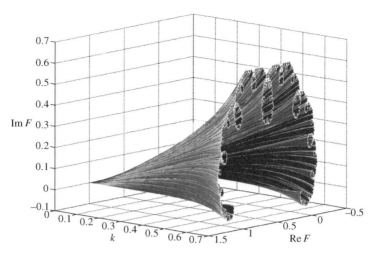

Figure 8. The diffusive modes of the multibaker map represented by their cumulative function depicted in the complex plane $(\operatorname{Re} F_k, \operatorname{Im} F_k)$ versus the wavenumber k.

using Eq. (64). Since the multibaker map is equivalent to a random walk in a one-dimensional lattice with probabilities $\frac{1}{2}$ to jump to the left- or right-hand sides, the dispersion relation of diffusion is given by

$$s_k = \ln \cos k = -\frac{1}{2}k^2 + O(k^4) \tag{70}$$

so the diffusion coefficient is equal to $\mathcal{D} = \frac{1}{2}$. Since the dynamics is uniformly expanding by a factor 2 in the multibaker map, Ruelle's function (63) has the form

$$P(\beta) = (1 - \beta)\ln 2 \tag{71}$$

whereupon the Hausdorff dimension of the diffusive mode takes the value

$$D_{\mathrm{H}} = \frac{\ln 2}{\ln 2 \cos k} \tag{72}$$

according to Eq. (64) [30].

C. Periodic Hard-Disk Lorentz Gas

The periodic hard-disk Lorentz gas is a two-dimensional billiard in which a point particle undergoes elastic collisions on hard disks which are fixed in the plane in the form of a spatially periodic lattice. Bunimovich and Sinai have proved that

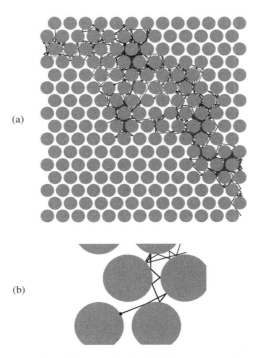

Figure 9. Two trajectories of the periodic hard-disk Lorentz gas. They start from the same position but have velocities that differ by one part in a million. (a) Both trajectories depicted on large spatial scales. (b) Initial segments of both trajectories showing the sensitivity to initial conditions.

the motion is diffusive if the horizon seen by the particles is always finite [31]. This is the case for a hexagonal lattice under the condition that the disks are large enough to avoid the possibility of straight trajectories moving across the whole lattice without collision. The dynamics of this system is ruled by the free-particle Hamiltonian:

$$H = \frac{p^2}{2m} \tag{73}$$

supplemented by the rules of elastic collisions on the disks. Because of the defocusing character of the collisions on the disks, the motion is chaotic. Two trajectories issued from slightly different initial conditions are depicted in Fig. 9. The dynamics is very sensitive to the initial conditions because the trajectories separate after a few collisions as seen in Fig. 9b. On long times, the trajectories perform random walks on the lattice (see Fig. 9a).

The cumulative functions of the diffusive modes can be constructed by using Eq. (60). The trajectories start from the border of a disk with an initial position

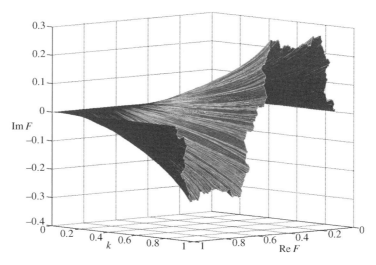

Figure 10. The diffusive modes of the periodic hard-disk Lorentz gas represented by their cumulative function depicted in the complex plane $(\operatorname{Re} F_k, \operatorname{Im} F_k)$ versus the wavenumber $\boldsymbol{k}$.

at an angle θ with respect to the horizontal and an initial velocity normal to the disk. The result is depicted in Fig. 10, where we see the fractal character of these curves developing as the wavenumber $\boldsymbol{k}$ increases. The Hausdorff dimension satisfies Eq. (65) as shown elsewhere [23].

D. Periodic Yukawa-Potential Lorentz Gas

This other Lorentz gas is similar to the previous one except that the hard disks are replaced by Yukawa potentials centered here at the vertices of a square lattice. The Hamiltonian of this system is given by

$$H = \frac{\boldsymbol{p}^2}{2m} - \sum_i \frac{\exp(-ar_i)}{r_i} \tag{74}$$

where a is the inverse screening length. Knauf [32] has proved that this system is chaotic and diffusive if the energy of the moving particles is large enough. The sensitivity to initial conditions is illustrated in Fig. 11, which depicts two trajectories starting from very close initial conditions. The particles undergo a random walk on long time scales.

The cumulative functions of the diffusive modes can here also be constructed by using Eq. (60) with trajectories integrated with a numerical algorithm based on the rescaling of time at the singular collisions. The initial position is taken on a small circle around a scattering center at an angle θ with respect to the

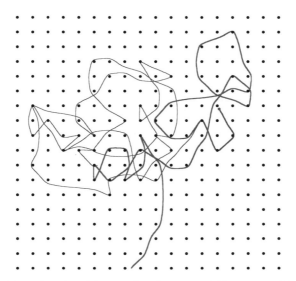

Figure 11.   Two trajectories of the periodic Yukawa-potential Lorentz gas. They start from the same position but have velocities that differ by one part in a million.

horizontal direction and the initial velocity is normal and pointing to the exterior of this circle. The results are shown in Fig. 12 for two nonvanishing wavenumbers. The Hausdorff dimension of these fractal curves also satisfies Eq. (65) as shown elsewhere [23].

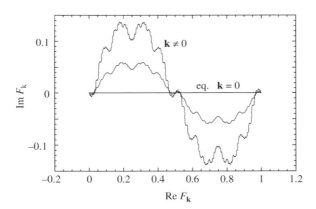

Figure 12.   The diffusive modes of the periodic Yukawa-potential Lorentz gas represented by their cumulative function depicted in the complex plane $(\operatorname{Re} F_k, \operatorname{Im} F_k)$ for two different nonvanishing wavenumbers k. The horizontal straight line is the curve corresponding to the vanishing wavenumber $k = 0$ at which the mode reduces to the invariant microcanonical equilibrium state.

Beside the diffusive modes, the chemical modes can also be constructed in models of reaction-diffusion processes [33–36].

E. Remarks

The hydrodynamic modes are rigorously constructed at the microscopic level of description in the phase space of the Hamiltonian dynamics by suitably weighting each trajectory in order to obtain eigenstates of the microscopic Liouvillian operator. Since these modes are exponentially decaying at positive times, they break the time-reversal symmetry of Liouville's equation in complete compatibility with the fundamental microscopic Hamiltonian dynamics. Time-reversal symmetry continues to be satisfied in the sense that the backward semigroup holds for negative times while the forward semigroup holds for positive times. Liouville's theorem is also satisfied which allows us to give a physical meaning to the probability distribution and to define a corresponding entropy. The transport coefficients are exactly defined so that the characteristic times of relaxation are well defined thanks to the spontaneous breaking of the time-reversal symmetry.

VI. *AB INITIO* DERIVATION OF ENTROPY PRODUCTION

The singular character of the diffusive modes allows their exponential relaxation at the rate given by the dispersion relation of diffusion. Their explicit construction can be used to perform an *ab initio* derivation of entropy production directly from the microscopic Hamiltonian dynamics [8, 9].

The phase-space region $\mathcal{M}_l$ corresponding to the lattice vector l is partitioned into cells $\mathcal{A}$. The probability that the system is found in the cell $\mathcal{A}$ at time t is given by

$$P_t(\mathcal{A}) = \int_{\mathcal{A}} p_t(\boldsymbol{\Gamma})\, d\boldsymbol{\Gamma} \tag{75}$$

in terms of the probability density $p_t(\boldsymbol{\Gamma})$ which evolves in time according to Liouville's equation. The knowledge of the Pollicott–Ruelle resonances s_j of the forward semigroup allows us to specify the approach to the equilibrium state as $t \to +\infty$:

$$P_t(\mathcal{A}) = P_{\mathrm{eq}}(\mathcal{A}) + \sum_j C_j \exp(s_j t) + \cdots \tag{76}$$

where the coefficients C_j are calculated using the eigenstates associated with the resonances.

The coarse-grained entropy is defined in terms of these probabilities as

$$S_t(\mathcal{M}_l|\{\mathcal{A}\}) = -k_{\mathrm{B}} \sum_{\mathcal{A}} P_t(\mathcal{A}) \ln P_t(\mathcal{A}) \tag{77}$$

where k_B is Boltzmann's constant. As the consequence of Gibbs' mixing property and the decomposition (76), the coarse-grained entropy converges toward its equilibrium value S_{eq} at long times. We notice that the rates of convergence are given by the Pollicott–Ruelle resonances and are thus intrinsic to the system. At long times, the phase-space probability density becomes more and more inhomogeneous and acquires the singular character of the diffusive modes that control the long-time evolution. Therefore, the approach of the entropy toward its equilibrium value is determined by the diffusion coefficient. The time variation of the coarse-grained entropy over a time interval τ

$$\Delta^\tau S = S_t(\mathcal{M}_l|\{\mathcal{A}\}) - S_{t-\tau}(\mathcal{M}_l|\{\mathcal{A}\}) \tag{78}$$

can be separated into the *entropy flow*

$$\Delta_e^\tau S \equiv S_{t-\tau}(\Phi^{-\tau}\mathcal{M}_l|\{\mathcal{A}\}) - S_{t-\tau}(\mathcal{M}_l|\{\mathcal{A}\}) \tag{79}$$

and the *entropy production*

$$\Delta_i^\tau S \equiv \Delta^\tau S - \Delta_e^\tau S = S_t(\mathcal{M}_l|\{\mathcal{A}\}) - S_t(\mathcal{M}_l|\{\Phi^\tau \mathcal{A}\}) \tag{80}$$

This latter can be calculated using the construction of the diffusive modes described here above, as shown elsewhere [1, 8, 9]. Finally, we obtain the fundamental result that the entropy production takes the value expected from nonequilibrium thermodynamics

$$\Delta_i^\tau S \simeq \tau k_B \, \mathcal{D} \, \frac{(\nabla n)^2}{n} \tag{81}$$

where $n = P_t(\mathcal{M}_l)$ is the particle density [8, 9]. Remarkably, this result holds for partitions into arbitrarily fine cells and is therefore a general result independent of the particular coarse graining which is adopted. This result is a consequence of the singular character of the diffusive modes defined as Liouvillian eigenstates. It can indeed be shown that the entropy production would vanish if these modes had a density given by a function instead of a singular distribution [1, 8, 9]. This result shows that the singular character of the nonequilibrium states plays a fundamental role in giving a positive entropy production. The thermodynamic value of the entropy production is thus coming from the microscopic Hamiltonian dynamics itself.

VII.   ESCAPE-RATE THEORY

The spontaneous breaking of time-reversal symmetry also manifests itself in the escape-rate theory, which consists in putting the system out of equilibrium by allowing the escape of trajectories from some phase-space region [1, 37–40]. If this region is adequately chosen the escape rate can be directly related to a transport coefficient. This escape rate is the leading Pollicott–Ruelle resonance of the dynamics in such open systems. Dynamical systems theory has shown that this escape rate is given by the difference between the sum of positive Lyapunov exponents and the Kolmogorov–Sinai entropy per unit time [14]. Accordingly, the escape-rate theory allows us to establish relationships between the transport coefficients and the characteristic quantities of the microscopic dynamics.

A. Helfand Moments and Transport Coefficients

It is well known that each transport coefficients is given by a Green–Kubo formula or, equivalently, by an Einstein formula:

$$\alpha = \int_0^\infty \langle j_\alpha(t) j_\alpha(0) \rangle \, dt = \lim_{t \to \infty} \frac{1}{2t} \langle (G_\alpha(t) - G_\alpha(0))^2 \rangle \tag{82}$$

in terms of the Helfand moment defined as the time integral of the current [41]:

$$G_\alpha(t) = G_\alpha(0) + \int_0^t j_\alpha(t') \, dt' \tag{83}$$

The Helfand moment is the center of mass, energy or momentum of the moving particles, depending on whether the transport property is diffusion, heat conductivity, or viscosity. The Helfand moments associated with the different transport properties are given in Table III. Einstein formula shows that the Helfand moment undergoes a diffusive random walk, which suggests to set up a

TABLE III
Helfand Moments of Different Transport Properties

Irreversible Property (α)	Helfand Moment G_α
Shear viscosity (η)	$\frac{1}{\sqrt{V k_B T}} \sum_{a=1}^N x_a \, p_{ay}$
Other viscosity ($\psi = \zeta + \frac{4}{3}\eta$)	$\frac{1}{\sqrt{V k_B T}} \sum_{a=1}^N x_a \, p_{ax}$
Heat conduction (κ)	$\frac{1}{\sqrt{V k_B T^2}} \sum_{a=1}^N x_a \, (E_a - \langle E_a \rangle)$
Electric conductivity (σ)	$\frac{1}{\sqrt{V k_B T}} \sum_{a=1}^N e Z_a \, x_a$
Diffusion ($\mathcal{D}$)	x_a
Reaction (L)	$\frac{1}{\sqrt{V}} (N^{(r)} - \langle N^{(r)} \rangle)$

V is the volume of the system and T the temperature. The particles have positions $r_a = (x_a, y_a, z_a)$, momenta $p_a = (p_{ax}, p_{ay}, p_{az})$ $(a = 1, 2, \ldots, N)$, energy $E_a = \frac{p_a^2}{2m} + \frac{1}{2}\sum_{b(\neq a)} u(r_{ab})$, and electric charge $e Z_a$. $N^{(r)}$ is the number of reactive events during a chemical reaction.

first-passage problem beyond a certain threshold for the Helfand moment. This threshold corresponds to the boundary of some region in the phase space of all the particles. Most of the trajectories escape from this phase-space region, but there might exist a zero-probability set of trajectories remaining inside this region. Typically, this set is a fractal composed of unstable trajectories.

The diffusive random walk of the Helfand moment is ruled by a diffusion equation. If the phase-space region is defined by requiring $|G_\alpha(t)| < \chi/2$, the escape rate can be computed as the leading eigenvalue of the diffusion equation with these absorbing boundary conditions for the Helfand moment [37, 39]:

$$\gamma \simeq \alpha \left(\frac{\pi}{\chi}\right)^2 \qquad \text{for} \quad \chi \to \infty \tag{84}$$

hence the proportionality between the escape rate γ and the transport coefficient α.

B.   Escape-Rate Formula in Dynamical Systems Theory

A natural invariant probability measure can be constructed on the set of trajectories which are remaining in the phase-space region delimited by the threshold on the Helfand moment. It is invariant under the microscopic Hamiltonian dynamics. This invariant probability measure can be built from the eigenstate of the Liouvillian operator associated with the escape rate, which is the leading Pollicott–Ruelle resonance. The phase-space region is partitioned into cells labeled by ω. The trajectories visiting the sequence of cells $\omega = \omega_0\omega_1\omega_2\ldots\omega_{n-1}$ at the successive times $t = 0, \tau, 2\tau, \ldots, (n-1)\tau$ form a path or history. A perturbation on the trajectories of this path is amplified by the stretching factor $\Lambda(\omega) = \Lambda(\omega_0\omega_1\omega_2\ldots\omega_{n-1})$. The sum of positive Lyapunov exponents of these trajectories is then given by

$$\sum_{\lambda_i > 0} \lambda_i = \lim_{n\to\infty} \frac{1}{n\tau} \ln|\Lambda(\omega_0\omega_1\omega_2\ldots\omega_{n-1})| \tag{85}$$

The classical mechanics is naturally weighting the paths according to their instability. The higher the instability, the smaller the probability weight according to

$$P(\omega) = \frac{|\Lambda(\omega)|^{-1}}{\sum_\omega |\Lambda(\omega)|^{-1}} \tag{86}$$

in the limit $n \to \infty$ [1]. This probability is normalized to unity by the denominator which decays at the escape rate:

$$\gamma = \lim_{n\tau\to\infty} -\frac{1}{n\tau} \ln \sum_\omega \frac{1}{|\Lambda(\omega)|} \tag{87}$$

so that

$$\frac{P(\omega)}{|\Lambda(\omega)|^{-1}} \sim \exp(\gamma t) \tag{88}$$

On the other hand, the dynamical randomness is characterized by the Kolmogorov–Sinai entropy per unit time:

$$h_{KS} = \mathrm{Sup}_{\mathcal{P}} \lim_{n \to \infty} -\frac{1}{n\tau} \sum_{\omega} P(\omega) \ln P(\omega) \tag{89}$$

where the supremum is taken over all the possible partitions $\mathcal{P}$ of the phase-space region containing the nonescaping trajectories into cells ω [1, 14]. The Kolmogorov–Sinai entropy is the mean decay rate of the probability (86):

$$P(\omega) \sim \exp(-h_{KS} t) \tag{90}$$

A similar relation holds for the stretching factors:

$$|\Lambda(\omega)| \sim \exp\left(\sum_{\lambda_i > 0} \lambda_i t\right) \tag{91}$$

Introducing these relations in Eq. (88) for the invariant probability, we obtain the escape-rate formula

$$\gamma = \sum_{\lambda_i > 0} \lambda_i - h_{KS} \tag{92}$$

according to which the escape rate is the difference between the sum of positive Lyapunov exponents and the Kolmogorov–Sinai entropy per unit time [1, 14]. In a closed system without escape, the escape rate vanishes $\gamma = 0$, and we recover Pesin relation $h_{KS} = \sum_{\lambda_i > 0} \lambda_i$ showing that dynamical randomness finds its origin in the sensitivity to initial conditions [16]. However, in nonequilibrium systems with escape, there is a disbalance between dynamical randomness and the dynamical instability due to the escape of trajectories as schematically depicted in Fig. 13. Out of equilibrium, the system has less dynamical randomness than possible by the dynamical instability.

In systems with two degrees of freedom such as the two-dimensional Lorentz gases, there is a single positive Lyapunov exponent λ and the partial Hausdorff dimension of the set of nonescaping trajectories can be estimated by the ratio of the Kolmogorov–Sinai entropy to the Lyapunov exponent [1, 38]

$$d_H \simeq \frac{h_{KS}}{\lambda} \tag{93}$$

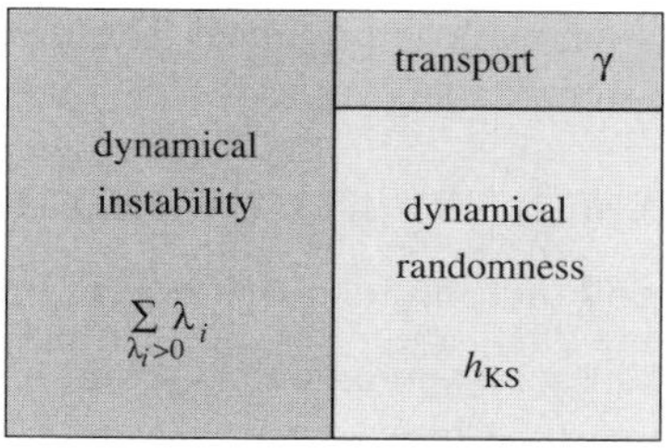

Figure 13. Diagram showing how dynamical instability characterized by the sum of positive Lyapunov exponents $\sum_{\lambda_i > 0} \lambda_i$ contributes to dynamical randomness characterized by the Kolmogorov–Sinai entropy per unit time h_{KS} and to the escape γ due to transport according to the chaos-transport formula (95).

so that the escape rate can be directly related to the fractal dimension and the Lyapunov exponent:

$$\gamma \simeq \lambda(1 - d_H) \tag{94}$$

C. The Chaos-Transport Formula

If we combine the escape-rate formula (92) with the result (84) that the escape rate is proportional to the transport coefficient, we obtain the following large-deviation relationships between the transport coefficients and the characteristic quantities of chaos [37, 39]:

$$\alpha = \lim_{\chi, V \to \infty} \left(\frac{\chi}{\pi}\right)^2 \left(\sum_{\lambda_i > 0} \lambda_i - h_{KS} \right)_\chi \tag{95}$$

This formula has already been applied to diffusion in periodic and random Lorentz gases [38, 42], reaction-diffusion [34], and viscosity [43, 44].

For diffusion in the open two-dimensional periodic Lorentz gas with parallel absorbing walls separated by the distance L, Eq. (95) shows that the diffusion coefficient is given by [38]

$$\mathcal{D} = \lim_{L \to \infty} \left(\frac{L}{\pi}\right)^2 \lambda \left(1 - d_H\right)\Big|_L \tag{96}$$

which is a formula similar to Eq. (66) given that the wavenumber is here equal to $k = \pi/L$. A difference is that the Hausdorff dimension of the fractal curves in the complex plane are larger than unity while the partial Hausdorff dimension of the nonescaping trajectories is smaller than unity. These chaos-transport formulas apply to Hamiltonian systems satisfying Liouville's theorem.

VIII. TIME ASYMMETRY IN DYNAMICAL RANDOMNESS

A. Randomness of Fluctuations in Nonequilibrium Steady States

In this section, we consider a system in a nonequilibrium steady state, such as a conductor between two particle reservoirs at different chemical potentials (see Fig. 14). The state ω of the system at a given time can be represented by the numbers $\{N_i\}_{i=1}^L$ of particles in the different cells $\{X_i\}_{i=1}^L$ composing the conductor [2]. These numbers randomly change with time according to the motion of the particles along the conductor. In a nonequilibrium steady state, the conductor is crossed by a mean current from the reservoir at the highest chemical potential to the other one. The path of the system is the sequence of states $\omega = \omega_0\omega_1\omega_2\ldots\omega_{n-1}$ at the successive times $t = 0, \tau, 2\tau, \ldots, (n-1)\tau$. In a stationary state, a probability $P(\omega) = P(\omega_0\omega_1\omega_2\ldots\omega_{n-1})$ is assigned to each path. This invariant probability distribution describes the fluctuations in the numbers of particles in the cells of the conductor. In a nonequilibrium steady state, we should expect that the particles enter into the conductor at the two concentrations fixed by the chemical potentials of both reservoirs. However, they exit the conductor from either the left- or the right-hand side after having been mixed by the dynamics internal to the conductor, so that the outgoing particles are statistically correlated on fine phase-space scales. Reversing the time would require that the particles are injected with these fine statistical correlations in order that they exit unmixed at both reservoirs. Although the time-reversal steady state is possible, it is highly improbable and clearly distinct from the actual steady state. This reasoning shows that the probability of a typical path $\omega = \omega_0\omega_1\omega_2\ldots\omega_{n-1}$ should be different from the probability of its time reversal $\Theta(\omega) = \omega^R = \omega_{n-1}\ldots\omega_2\omega_1\omega_0$. This is possible because the trajectories of typical paths are distinct from their time reversal as pointed out in Section III. The nonequilibrium boundary conditions explicitly break the time-reversal symmetry.

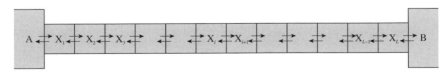

Figure 14. Schematic representation of the diffusion process of particles in a conductor composed of L cells $\{X_i\}_{i=1}^L$ of volume ΔV between two particle reservoirs, A and B.

The dynamical randomness of the nonequilibrium process can be characterized by the decay of the path probabilities as defined by the entropy per unit time [12–14]:

$$
\begin{aligned}
h &\equiv \lim_{n\to\infty} -\frac{1}{n\tau} \sum_{\omega} P(\omega) \ln P(\omega) \\
&= \lim_{n\to\infty} -\frac{1}{n\tau} \sum_{\omega_0\omega_1\omega_2\ldots\omega_{n-1}} P(\omega_0\omega_1\omega_2\ldots\omega_{n-1}) \ln P(\omega_0\omega_1\omega_2\ldots\omega_{n-1})
\end{aligned}
\tag{97}
$$

The Kolmogorov–Sinai entropy per unit time is defined in Eq. (89) as the supremum of h over all the possible partitions $\mathcal{P}$. Since we expect that the probability of the nonequilibrium steady state is not time-reversal symmetric, the probability of the time-reversed paths should decay at a different rate, which can be called a time-reversed entropy per unit time [3]

$$
\begin{aligned}
h^{R} &\equiv \lim_{n\to\infty} -\frac{1}{n\tau} \sum_{\omega} P(\omega) \ln P(\omega^{R}) \\
&= \lim_{n\to\infty} -\frac{1}{n\tau} \sum_{\omega_0\omega_1\omega_2\ldots\omega_{n-1}} P(\omega_0\omega_1\omega_2\ldots\omega_{n-1}) \ln P(\omega_{n-1}\ldots\omega_2\omega_1\omega_0)
\end{aligned}
\tag{98}
$$

For almost all paths with respect to the probability measure P, we would have

$$
P(\omega) = P(\omega_0\omega_1\omega_2\ldots\omega_{n-1}) \sim \exp(-hn\tau) \tag{99}
$$

$$
P(\omega^{R}) = P(\omega_{n-1}\ldots\omega_2\omega_1\omega_0) \sim \exp(-h^{R}n\tau) \tag{100}
$$

The first line is known as the Shannon–McMillan–Breiman theorem [1], the second is its extension for stationary states which are not time-reversal symmetric. The entropy per unit time h characterizes the dynamical randomness of the process. The faster the decay of the path probabilities, the larger the proliferation of these paths as time increases. Therefore, the larger the entropy per unit time h, the higher the temporal disorder of the time evolution. The time-reversed entropy per unit time h^{R} characterizes the decay of the time reversals of the typical paths in a similar way, and it thus characterizes the dynamical randomness of the backward paths.

B. Entropy Production

The most remarkable result is that the difference between both entropies per unit time (98) and (97) gives the entropy production of the nonequilibrium steady state:

$$
\frac{1}{k_{B}} \frac{d_{i}S}{dt} = h^{R} - h \geq 0 \tag{101}
$$

in the limit $\tau \to 0$ [3]. An equivalent statement is that the ratio of the probabilities (99) and (100) grows as the entropy production

$$\frac{P(\omega)}{P(\omega^R)} = \frac{P(\omega_0\omega_1\omega_2\ldots\omega_{n-1})}{P(\omega_{n-1}\ldots\omega_2\omega_1\omega_0)} \sim \exp[n\tau(h^R - h)] = \exp\left(\frac{n\tau}{k_B}\frac{d_iS}{dt}\right) \quad (102)$$

For an isothermal process, the logarithm of the ratio of the probabilities thus gives the work dissipated along the path:

$$\ln\frac{P(\omega)}{P(\omega^R)} \simeq \frac{W_{\text{diss}}(\omega)}{k_B T} \quad (103)$$

as $n \to \infty$ while $\tau \to 0$.

The formula (101) gives a non-negative entropy production in agreement with the second law of thermodynamics. Indeed, the non-negativity of the right-hand member is guaranteed by the fact that the difference between Eqs. (98) and (97) is a relative entropy that is known to be non-negative:

$$\frac{1}{k_B}\frac{d_iS}{dt} = \lim_{n\to\infty}\frac{1}{n\tau}\sum_\omega P(\omega)\ln\frac{P(\omega)}{P(\omega^R)} \geq 0 \quad (104)$$

Consequently, we have the general property that

$$h^R \geq h \quad (105)$$

The equality $h^R = h$ holds if and only if the probabilities of the paths are equal to the probabilities of their time reversals:

$$P(\omega) = P(\omega^R) \qquad \text{for all} \quad \omega \quad (106)$$

This is the condition of detailed balance which holds in the equilibrium state where entropy production vanishes.

The property (105) means that, in a nonequilibrium steady state, the probabilities of the typical paths decay more slowly than the probabilities of their time reversals (see Fig. 15). In this sense, the temporal disorder is smaller for typical paths than for their time reversal. We thus have the following

Principle of Temporal Ordering. *In nonequilibrium steady states, the typical paths are more ordered in time than their corresponding time reversals.*

This principle and the formula (101) show that entropy production results from a time asymmetry in the dynamical randomness in nonequilibrium steady states.

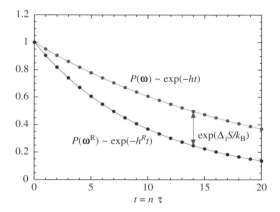

Figure 15. Diagram showing the probability of a path and the probability of the corresponding time reversal as a function of the number n of time intervals τ, illustrating the formula (101). $\Delta_i S$ denotes the entropy production during the whole duration $n\tau$. See color insert.

It is remarkable that this ordering phenomenon appears if we consider the evolution of the process as a movie, looking for regularities in the sequence of pictures. The temporal ordering is compatible with Boltzmann's interpretation of the second law according to which spatial disorder increases with time. In a sense, the temporal ordering is possible at the expense of spatial disorder.

We notice the similarity of the structures of Eq. (101) with the chaos-transport relationship (95). Indeed, both formulas are large-deviation dynamical relationships giving an irreversible property from the difference between two quantities characterizing the dynamical randomness or instability of the microscopic dynamics (see Fig. 16).

Formula (101) can be proved in different contexts.

1. Proof for Continuous-Time Jump Processes

These stochastic processes are ruled by master equations of Pauli type:

$$\frac{d}{dt}P_t(\omega) = \sum_{\rho,\omega'(\neq\omega)} [P_t(\omega')W_\rho(\omega'|\omega) - P_t(\omega)W_{-\rho}(\omega|\omega')] \qquad (107)$$

for the probability $P_t(\omega)$ to find the system in the state ω by the time t [5]. $W_\rho(\omega|\omega')$ denotes the rate of the transition $\omega \to \omega'$ for the elementary process ρ. A transition rate is associated with several possible elementary processes $\rho = \pm 1, \pm 2, \ldots, \pm r$. The probabilities are constant in a steady state: $dP(\omega)/dt = 0$.

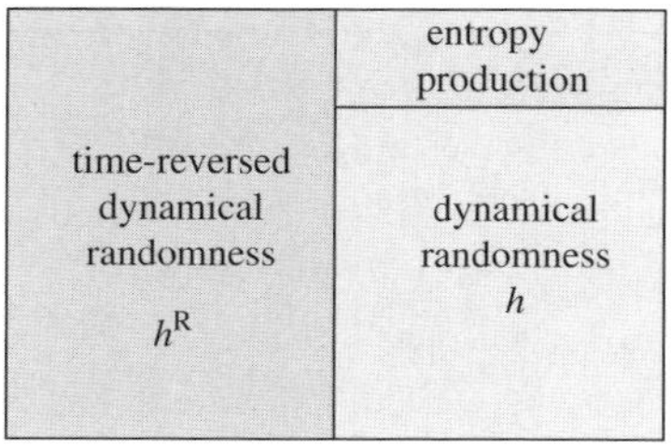

	entropy production
time-reversed dynamical randomness h^R	dynamical randomness h

Figure 16. Diagram showing how the dynamical randomness of the backward paths characterized by the time-reversed entropy per unit time h^R contributes to dynamical randomness characterized by the entropy per unit time h and to the entropy production $(1/k_B)d_iS/dt$ according to the formula (101). We notice the similarity of the structure with Fig. 13.

The dynamical randomness of such stochastic processes can be characterized by the τ-entropy per unit time introduced more than 10 years ago [45],

$$h(\tau) = \left(\ln\frac{e}{\tau}\right) \sum_{\rho,\omega,\omega'} P(\omega)\, W_\rho(\omega|\omega') - \sum_{\rho,\omega,\omega'} P(\omega)\, W_\rho(\omega|\omega')\, \ln W_\rho(\omega|\omega') + O(\tau)$$

(108)

and the time-reversed τ-entropy per unit time [3],

$$h^R(\tau) = \left(\ln\frac{e}{\tau}\right) \sum_{\rho,\omega,\omega'} P(\omega)\, W_\rho(\omega|\omega') - \sum_{\rho,\omega,\omega'} P(\omega)\, W_\rho(\omega|\omega')\, \ln W_{-\rho}(\omega'|\omega) + O(\tau)$$

(109)

We notice that the expressions (108) and (109) differ by the transition rate in the logarithm. Their difference is equal to the known expression for the entropy production of these stochastic processes [46, 47]:

$$h^R(\tau) - h(\tau) = \frac{1}{2} \sum_{\rho,\omega,\omega'} \left[P(\omega)\, W_\rho(\omega|\omega') - P(\omega')\, W_{-\rho}(\omega'|\omega) \right]$$
$$\times \ln \frac{P(\omega)\, W_\rho(\omega|\omega')}{P(\omega')\, W_{-\rho}(\omega'|\omega)} + O(\tau) \simeq \frac{1}{k_B} \frac{d_iS}{dt}$$

(110)

in the limit $\tau \to 0$. This proves the relationship (101) [3].

2. *Proof for Thermostated Dynamical Systems*

Thermostated dynamical systems are deterministic systems with non-Hamiltonian forces modeling the dissipation of energy toward a thermostat [48]. The non-Hamiltonian forces are chosen in such a way that the equations of motion are time-reversal symmetric. In these systems, the stochastic fluctuations of the reaction of the thermostat back onto the system are neglected. The nonequilibrium steady states are described by SRB invariant probability measures which spontaneously break the time-reversal symmetry.

The phase space is partitioned into cells ω of diameter δ. In the limit of an arbitrarily fine partition, the entropy per unit time tends to the Kolmogorov–Sinai entropy per unit time which is equal to the sum of positive Lyapunov exponents by Pesin theorem [16]:

$$\lim_{\delta \to 0} h = \sum_{\lambda_i > 0} \lambda_i = h_{KS} \tag{111}$$

In the same limit, the time-reversed entropy per unit time is equal to minus the sum of negative Lyapunov exponents:

$$\lim_{\delta \to 0} h^R = -\sum_{\lambda_i < 0} \lambda_i \tag{112}$$

The difference between both entropies per unit time is minus the sum of all the Lyapunov exponents which is the rate of contraction of the phase-space volumes under the effects of the nonHamiltonian forces:

$$\lim_{\delta \to 0} \left(h^R - h \right) = -\sum_i \lambda_i = \frac{1}{k_B} \frac{d_i S}{dt} \tag{113}$$

This contraction rate has been identified in these models as the rate of entropy production [49], which proves the formula (101) in this case as well.

3.   Proof with the Escape-Rate Theory

The formula (101) can also be proved with the escape-rate theory. We consider the escape of particles by difffusion from a large reservoir, as depicted in Fig. 17. The density of particles is uniform inside the reservoir and linear in the slab where diffusion takes place. The density decreases from the uniform value N/V of the reservoir down to zero at the exit where the particles escape. The width of the diffusive slab is equal to L so that the gradient is given by $\nabla n = -N/(VL)$ and the particle current density $J = -\mathcal{D}\nabla n = \mathcal{D}N/(VL)$. Accordingly, the number of particles in the reservoir decreases at the rate

$$\frac{dN}{dt} = -\int J \cdot d\Sigma = \int \mathcal{D}\nabla n \cdot d\Sigma = -\mathcal{D}\frac{A}{VL}N \tag{114}$$

Therefore, the number of particles decreases exponentially as

$$N(t) = N(0)\exp(-\gamma t) \tag{115}$$

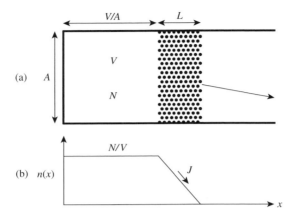

Figure 17. (a) Escape of N particles from a reservoir of volume V and area A by diffusion through a slab of width L and area A. (b) The profile of concentration $n(x)$ which is uniform in the reservoir and linear in the diffusive slab.

at the escape rate

$$\gamma = \mathcal{D} \frac{A}{VL} \tag{116}$$

In the limit of an arbitrarily large reservoir $N, V \to \infty$ with $n = N/V$ constant, the escape rate vanishes and a nonequilibrium steady state establishes itself in the diffusive slab.

The entropy production can be calculated by using the thermodynamic expression (81) as

$$\frac{1}{k_\mathrm{B}} \frac{d_i S}{dt} = \int \mathcal{D} \frac{(\nabla n)^2}{n} d^3 r = \mathcal{D} \frac{A}{VL} N = \gamma N \tag{117}$$

which is proportional to the escape rate of individual particles.

According to dynamical systems theory, the escape rate is given by the difference (92) between the sum of positive Lyapunov exponents and the Kolmogorov–Sinai entropy. Since the dynamics is Hamiltonian and satisfies Liouville's theorem, the sum of positive Lyapunov exponents is equal to minus the sum of negative ones:

$$\gamma = \sum_{\lambda_i > 0} \lambda_i - h_{\mathrm{KS}} = - \sum_{\lambda_i < 0} \lambda_i - h_{\mathrm{KS}} \tag{118}$$

Now, the entropies per unit time in Eq. (101) concern the dynamics of the N particles. Here, the particles are independent of each other so that the entropies per unit time of the N-particle system is equal to N times the entropies per unit time of the one-particle system. On the other hand, we have seen with Eq. (112) that minus the sum of negative Lyapunov exponents should correspond to the time-reversed entropy per unit time so that we can here also prove Eq. (101):

$$\frac{1}{k_B}\frac{d_i S}{dt} = N\left(-\sum_{\lambda_i < 0} \lambda_i - h_{KS}\right)_1 = (h^R - h)_N \tag{119}$$

in the infinite-reservoir limit N, $V \to \infty$ with $n = N/V$ constant with a nonequilibrium steady state in the diffusive slab.

C.   Markov Chains and Information Theoretic Aspects

The entropies per unit time as well as the thermodynamic entropy production entering in the formula (101) can be interpreted in terms of the numbers of paths satisfying different conditions. In this regard, important connections exist between information theory and the second law of thermodynamics.

In order to show these aspects, let us consider discrete-time Markov chains. The matrix of transition probabilities is denoted $P_{\omega\omega'}$, which is the conditional probability that the system is in the state ω' at time $n + 1$ if it was in the state ω at time n. The time is counted in units of the time interval τ. The matrix of transition probabilities satisfies

$$\sum_{\omega'} P_{\omega\omega'} = 1 \tag{120}$$

In a steady state, the probabilities p_ω to find the system in given states are obtained by solving the stationary equation

$$\sum_{\omega} p_\omega P_{\omega\omega'} = p_{\omega'} \tag{121}$$

The dynamical randomness of this Markov chain is characterized by the Kolmogorov–Sinai entropy per unit time:

$$h = -\sum_{\omega,\omega'} p_\omega P_{\omega\omega'} \ln P_{\omega\omega'} \tag{122}$$

On the other hand, the time-reversed entropy per unit time is here given by

$$h^R = -\sum_{\omega,\omega'} p_\omega P_{\omega\omega'} \ln P_{\omega'\omega} \tag{123}$$

We notice that the only difference between both dynamical entropies is the exchange of ω and ω' in the transition probabilities appearing in the logarithm. According to Eq. (101), the thermodynamic entropy production of this process would be equal to

$$\Delta_i S = h^R - h \tag{124}$$

in units of the time interval τ and Boltzmann's constant k_B.

For Markov chains with two states $\{0, 1\}$, it turns out that we always have the equality $h^R = h$ so that they are not appropriate to model nonequilibrium processes.

Next, we can consider a Markov chain with three states $\{1, 2, 3\}$ with the following matrix of transition probabilities:

$$P = \begin{pmatrix} \frac{a}{2} & 1-a & \frac{a}{2} \\ \frac{a}{2} & \frac{a}{2} & 1-a \\ 1-a & \frac{a}{2} & \frac{a}{2} \end{pmatrix} \tag{125}$$

where $0 \le a \le 1$ is a parameter. The entropies per unit time of this Markov chain are given by

$$h = -a \ln \frac{a}{2} - (1-a) \ln(1-a) \tag{126}$$

$$h^R = -\left(1 - \frac{a}{2}\right) \ln \frac{a}{2} - \frac{a}{2} \ln(1-a) \tag{127}$$

whereupon the entropy production is

$$\Delta_i S = h^R - h = \left(1 - \frac{3a}{2}\right) \ln \frac{2(1-a)}{a} \ge 0 \tag{128}$$

These quantities are depicted in Fig. 18. The entropy production vanishes at equilibrium $a = 2/3$ where $h = h^R = \ln 3$. The entropy production is infinite and the process fully irreversible at $a = 1$ where $h = \ln 2$ and $a = 0$ where $h = 0$, in which case the process is perfectly cyclic $\ldots 123123123123 \ldots$

The number of typical paths generated by the random process increases as $\exp(hn)$. In this regard, the Kolmogorov–Sinai entropy per unit time is the rate of production of information by the random process. On the other hand, the time-reversed entropy per unit time is the rate of production of information by the time reversals of the typical paths. The thermodynamic entropy production is the difference between these two rates of information production. With the formula (101), we can recover a result by Landauer [50] and Bennett [51] that erasing information in the memory of a computer is an irreversible process of

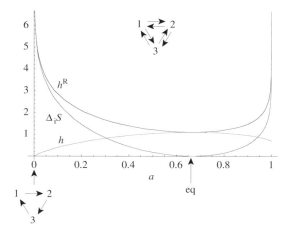

Figure 18. The dynamical entropies (126) and (127) as well as the entropy production (128) for the three-state Markov chain defined by the matrix (125) of transition probabilities versus the parameter a. The equilibrium corresponds to the value $a = 2/3$. The process is perfectly cyclic at $a = 0$ where the path is $\ldots 123123123123\ldots$ and the Kolmogorov–Sinai entropy h vanishes as a consequence. See color insert.

entropy production equal to $k_B \ln 2$. Indeed, erasing the bits of a sequence of zeros and ones is a process without surprise so that its entropy per unit time vanishes $h = 0$. On the other hand, the time-reversed process lets the zeros and ones appear with surprises so that the time-reversed entropy per unit time is equal to $h^R = \ln 2$. The difference between them gives the entropy production $\Delta_i S = k_B (h^R - h) = k_B \ln 2$ [3]. The above reasoning confirms the solution of the paradox of Maxwell's demon given by Landauer and Bennett [50, 51], according to whom this demon produces entropy while erasing information so that its behavior is in agreement with the second law of thermodynamics.

D. Fluctuation Theorem for the Currents

Further large-deviation dynamical relationships are the so-called fluctuation theorems, which concern the probability than some observable such as the work performed on the system would take positive or negative values under the effect of the nonequilibrium fluctuations. Since the early work of the fluctuation theorem in the context of thermostated systems [52–54], stochastic [55–59] as well as Hamiltonian [60] versions have been derived. A fluctuation theorem has also been derived for nonequilibrium chemical reactions [62]. A closely related result is the nonequilibrium work theorem [61] which can also be derived from the microscopic Hamiltonian dynamics.

Beside the work performed on the system, interesting quantities are the currents crossing a system in a nonequilibrium steady state. Recently, we have been able to derive a fluctuation theorem for the currents in the framework of the stochastic master equation using Schnakenberg graph analysis [63–66]. According to Schnakenberg [46], a graph can be associated with some stochastic process: The states of the process are the vertices of the graph and the transitions the bonds. This analysis is very useful because it allows us to identify the affinities—that is, the thermodynamic forces imposed on the system by the differences of temperature or chemical potentials between the external reservoirs of heat or molecules. These affinities are given, for instance, by free energy differences $A_\alpha = \Delta F_\alpha / T$ and are the causes of the currents crossing the system. At the microscopic level, these currents fluctuate and depend on the path followed by the system: $j_\alpha(t; \omega)$. These currents can be the electric current in a conductor, the number of molecules of some reactant consumed in a chemical reaction, the rotation velocity of a rotary molecular motor, or local currents in a fluid element, as listed in Table II. A system can sustain several independent currents. The mean current over a time interval t is defined by

$$J_\alpha(\omega) = \frac{1}{t} \int_0^t j_\alpha(t'; \omega) \, dt' \qquad (129)$$

which is a fluctuating variable. The currents are odd under time reversal:

$$J_\alpha(\omega^R) = -J_\alpha(\omega) \qquad (130)$$

In a nonequilibrium steady state, the entropy production is the sum of the affinities multiplied by the mean currents:

$$\left. \frac{d_i S}{dt} \right|_{neq} = \sum_\alpha A_\alpha \langle J_\alpha \rangle > 0 \qquad (131)$$

At the mesoscopic level, the currents fluctuate and the previous result (102) suggests that

$$\frac{P(\omega)}{P(\omega^R)} = \frac{P(\omega_0 \omega_1 \omega_2 \ldots \omega_{n-1})}{P(\omega_{n-1} \ldots \omega_2 \omega_1 \omega_0)} \simeq \exp\left(\frac{t}{k_B} \sum_\alpha A_\alpha J_\alpha(\omega) \right) \qquad (132)$$

where $t = n\tau$. Using this assumption, a schematic proof for the fluctuation theorem follows:

$$P(\{J_\alpha = \xi_\alpha\}) \simeq \sum_\omega P(\omega) \, \delta[J_\alpha(\omega) - \xi_\alpha] \qquad (133)$$

$$= \sum_\omega P(\omega^R) \, \delta[J_\alpha(\omega^R) - \xi_\alpha] \qquad (134)$$

$$\simeq \sum_\omega P(\omega) \, e^{-\frac{t}{k_B} \sum_\alpha A_\alpha J_\alpha(\omega)} \, \delta[-J_\alpha(\omega) - \xi_\alpha] \qquad (135)$$

$$= e^{\frac{t}{k_B} \sum_\alpha A_\alpha \xi_\alpha} \sum_\omega P(\omega) \, \delta[J_\alpha(\omega) + \xi_\alpha] \qquad (136)$$

$$\simeq e^{\frac{t}{k_B} \sum_\alpha A_\alpha \xi_\alpha} P(\{J_\alpha = -\xi_\alpha\}) \qquad (137)$$

where $\delta(\cdot)$ denotes the product of Dirac delta distributions over all the currents of interest in the system. The first line is the definition of the probability that each fluctuating current J_α takes the value ξ_α. The second line is obtained because the sum over all the paths ω is equivalent to the sum over all the time-reserved paths ω^R. Now, the probability $P(\omega^R)$ of the time-reversed paths can be expressed by using Eq. (132) and we can use the odd parity (130) of the currents. In the third line, the delta shows that each current J_α now takes the negative value $-\xi_\alpha$, so that the exponential can exit the sum after specifying these values and the result is obtained. This leads to the *fluctuation theorem for the currents*

$$\frac{P(\{J_\alpha = \xi_\alpha\})}{P(\{J_\alpha = -\xi_\alpha\})} \simeq \exp\left(\frac{t}{k_B} \sum_\alpha A_\alpha \xi_\alpha\right) \qquad (138)$$

for $t \to \infty$. The rigorous proof of this fluctuation theorem is given elsewhere [63–65].

Different expressions can be given to the fluctuation theorem. If we introduce the decay rates of the probabilities that the currents take some values as

$$P(\{J_\alpha = \xi_\alpha\}) \sim \exp[-H(\{\xi_\alpha\})t] \qquad (139)$$

the fluctuation theorem writes [64, 65]

$$H(\{-\xi_\alpha\}) - H(\{\xi_\alpha\}) = \frac{1}{k_B} \sum_\alpha A_\alpha \xi_\alpha \qquad (140)$$

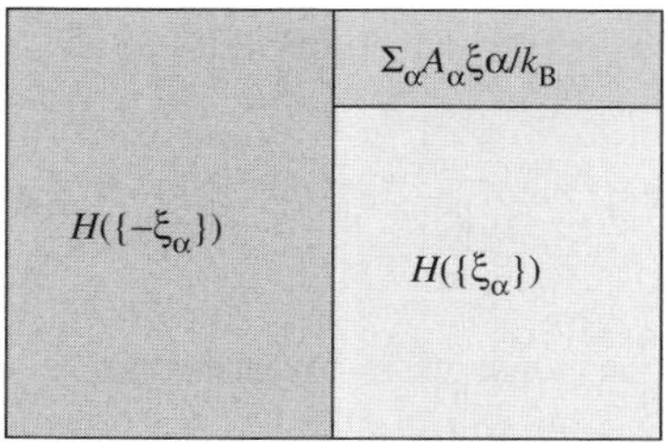

Figure 19. Diagram illustrating the fluctuation theorem for the currents (140) showing how the decay rate at negative values of the currents equates the decay rate at positive value plus the irreversible value. Compare with Figs. 13 and 16.

This form shows the deep similarity with the chaos-transport formula (95) and Eq. (101).

On the other hand, we can introduce the generating function of the mean currents and their statistical moments as

$$Q(\{\lambda_\alpha\}) = \lim_{t \to \infty} -\frac{1}{t} \ln \left\langle \exp\left[-\sum_\alpha \lambda_\alpha G_\alpha(t)\right]\right\rangle \qquad (141)$$

with the Helfand moments $G_\alpha(t) = \int_0^t j_\alpha(t'; \omega)dt' = t J_\alpha(\omega)$ associated with the fluctuating currents. This generating function is the Legendre transform of the decay function defined in Eq. (139). The fluctuation theorem for the currents is expressed in terms of this generating function as [63]

$$Q(\{A_\alpha - \lambda_\alpha\}) = Q(\{\lambda_\alpha\}) \qquad (142)$$

We notice that the generating and decay functions characterize the nonequilibrium process in the steady state and, consequently, have a general dependence on the affinities which play the role of nonequilibrium parameters.

E. Onsager Reciprocity Relations and Generalizations to Nonlinear Response

In nonequilibrium steady states, the mean currents crossing the system depend on the nonequilibrium constraints given by the affinities or thermodynamic forces which vanish at equilibrium. Accordingly, the mean currents can be expanded in powers of the affinities around the equilibrium state. Many nonequilibrium processes are in the linear regime studied since Onsager classical work [7]. However, chemical reactions are known to involve the nonlinear regime. This is also the case for nanosystems such as the molecular motors as recently shown [66]. In the nonlinear regime, the mean currents depend on powers of the affinities so that it is necessary to consider the full Taylor expansion of the currents on the affinities:

$$\langle J_\alpha \rangle = \sum_\beta L_{\alpha\beta} A_\beta + \frac{1}{2}\sum_{\beta,\gamma} M_{\alpha\beta\gamma} A_\beta A_\gamma + \frac{1}{6}\sum_{\beta,\gamma,\delta} N_{\alpha\beta\gamma\delta} A_\beta A_\gamma A_\delta + \cdots \qquad (143)$$

The coefficients $L_{\alpha\beta}$ are Onsager's linear-response coefficients. The other coefficients $M_{\alpha\beta\gamma}$, $N_{\alpha\beta\gamma\delta}$, and so on, characterize the nonlinear response of the system.

A direct consequence of the fluctuation theorem for the currents (138) is that these coefficients obey remarkable relations [63–65], the first of which are

Onsager reciprocity relations as well as the Green–Kubo and Einstein formulas for these coefficients:

$$L_{\alpha\beta} = L_{\beta\alpha} = \frac{1}{2} \int_{-\infty}^{+\infty} \langle [j_\alpha(t) - \langle j_\alpha \rangle][j_\beta(0) - \langle j_\beta \rangle] \rangle_{eq} \, dt = \lim_{t \to \infty} \frac{1}{2t} \langle \Delta G_\alpha(t) \Delta G_\beta(t) \rangle_{eq}$$

$$(144)$$

with $\Delta G_\alpha = G_\alpha - \langle G_\alpha \rangle$. Next, the third-order response coefficients can be expressed as

$$M_{\alpha\beta\gamma} = \tilde{R}_{\alpha\beta,\gamma} + \tilde{R}_{\alpha\gamma,\beta} \tag{145}$$

in terms of the sensitivity of the linear-response coefficients to nonequilibrium perturbations:

$$\tilde{R}_{\alpha\beta,\gamma} \equiv \frac{\partial}{\partial A_\gamma} \lim_{t \to \infty} \frac{1}{2t} \langle \Delta G_\alpha(t) \Delta G_\beta(t) \rangle_{neq} \Big|_{\{A_\epsilon = 0\}} \tag{146}$$

Furthermore, the fourth-order response coefficients are given by

$$N_{\alpha\beta\gamma\delta} = \tilde{T}_{\alpha\beta,\gamma\delta} + \tilde{T}_{\alpha\gamma,\beta\delta} + \tilde{T}_{\alpha\delta,\beta\gamma} - \tilde{S}_{\alpha\beta\gamma,\delta} \tag{147}$$

in terms of

$$\tilde{T}_{\alpha\beta,\gamma\delta} = \frac{\partial}{\partial A_\gamma} \frac{\partial}{\partial A_\delta} \lim_{t \to \infty} \frac{1}{2t} \langle \Delta G_\alpha(t) \Delta G_\beta(t) \rangle_{neq} \Big|_{\{A_\epsilon = 0\}} \tag{148}$$

which are similar to the coefficients (146), and

$$\tilde{S}_{\alpha\beta\gamma,\delta} = \frac{\partial}{\partial A_\delta} \lim_{t \to \infty} \frac{1}{2t} \langle \Delta G_\alpha(t) \Delta G_\beta(t) \Delta G_\gamma(t) \rangle_{neq} \Big|_{\{A_\epsilon = 0\}} \tag{149}$$

which can be proved to be totally symmetric under all the permutations of the four indices [64, 65]. Similar relations also hold at higher orders. These results show that the fluctuation theorem, which is a consequence of microreversibility, plays a central role in nonequilibrium statistical mechanics because previous results and generalizations can be deduced thereof.

IX.   CONCLUSIONS AND PERSPECTIVES

In this chapter, we have described recent advances in nonequilibrium statistical mechanics and have shown that they constitute a breakthrough which opens very new perspectives in our understanding of nonequilibrium processes and the second law of thermodynamics.

First of all, the time asymmetry of nonequilibrium processes finally finds an explanation at the microscopic level of description. It is nowadays possible to show that the modes of relaxation toward the equilibrium as well as the nonequilibrium steady states involve a spontaneous breaking of the time-reversal symmetry of the microscopic Hamiltonian dynamics at the level of the statistical description. Relaxation modes such as the diffusive modes can be constructed as eigenstates of the fundamental Liouvillian operator ruling the time evolution of the probability density in phase space. These eigenstates are associated with the Pollicott–Ruelle resonances. They are singular distributions that are smooth in the unstable phase-space directions but singular in the stable ones. This reflects the breaking of the time-reversal symmetry at the origin of these eigenmodes. Furthermore, their singular character can be demonstrated to imply the positive entropy production expected from nonequilibrium thermo-dynamics.

On the other hand, the nonequilibrium steady states are constructed by weighting each phase-space trajectory with a probability which is different for their time reversals. As a consequence, the invariant probability distribution describing the nonequilibrium steady state at the microscopic phase-space level explicitly breaks the time-reversal symmetry.

This time asymmetry has manifestations which can be directly tested in experiments. In particular, the transport coefficients can be obtained by considering the escape of trajectories from suitably chosen phase-space regions. The escape rate is the leading Pollicott–Ruelle resonance of these open systems and is related to the properties of the microscopic motion such as the Lyapunov exponents, the Kolmogorov–Sinai entropy per unit time, and the fractal dimensions characterizing the singularities of the nonequilibrium states. In this way, relationships have been established between the irreversible properties of transport and the characteristic quantities of the dynamical randomness and instability of the underlying microscopic dynamics. These escape-rate formulas are large-deviation relationships that include the fluctuation theorems as well as the formula (101) giving the entropy production in nonequilibrium steady states. These different relationships are given in Table IV, which shows their

TABLE IV
Large-Deviation Dynamical Relationships of Nonequilibrium Statistical Mechanics

Transport coefficient	$\alpha = \lim\limits_{\chi, V \to \infty} \left(\frac{\chi}{\pi}\right)^2 \left(\sum\limits_{\lambda_i > 0} \lambda_i - h_{KS}\right)_\chi$ (95)	$\dfrac{P(\omega_0 \omega_1 \omega_2 \ldots \omega_{n-1})}{\lvert \Lambda(\omega_0 \omega_1 \omega_2 \ldots \omega_{n-1}) \rvert^{-1}} \sim \exp\left[\alpha\left(\frac{\pi}{\chi}\right)^2 t\right]$ (88)
Entropy production	$\dfrac{1}{k_B} \dfrac{d_i S}{dt} = h^R - h$ (101)	$\dfrac{P(\omega_0 \omega_1 \omega_2 \ldots \omega_{n-1})}{P(\omega_{n-1} \ldots \omega_2 \omega_1 \omega_0)} \sim \exp\left(\frac{t}{k_B} \frac{d_i S}{dt}\right)$ (102)
Fluctuating currents	$\dfrac{1}{k_B} \sum\limits_\alpha A_\alpha \xi_\alpha = H(\{-\xi_\alpha\}) - H(\{\xi_\alpha\})$ (140)	$\dfrac{P(\{J_\alpha = \xi_\alpha\})}{P(\{J_\alpha = -\xi_\alpha\})} \sim \exp\left(\frac{t}{k_B} \sum\limits_\alpha A_\alpha \xi_\alpha\right)$ (138)

similarities. In one version, they relate an irreversible property to the difference between rates characterizing the microscopic fluctuations. In another version, it is the ratio of two path properties which turns out to grow at a given irreversible rate. In each case, the counter term is the decay rate of some path probability, such as the Kolmogorov–Sinai entropy per unit time in the escape-rate formula or $H(\{\xi_\alpha\})$ in the fluctuation theorem.

It is most remarkable that the entropy production in a nonequilibrium steady state is directly related to the time asymmetry in the dynamical randomness of nonequilibrium fluctuations. The entropy production turns out to be the difference in the amounts of temporal disorder between the backward and forward paths or histories. In nonequilibrium steady states, the temporal disorder h^R of the time reversals is larger than the temporal disorder h of the paths themselves. This is expressed by the *principle of temporal ordering*, according to which the typical paths are more ordered than their corresponding time reversals in nonequilibrium steady states. This principle is proved with nonequilibrium statistical mechanics and is a corollary of the second law of thermodynamics. Temporal ordering is possible out of equilibrium because of the increase of spatial disorder. There is thus no contradiction with Boltzmann's interpretation of the second law. Contrary to Boltzmann's interpretation, which deals with disorder in space at a fixed time, the principle of temporal ordering is concerned by order or disorder along the time axis, in the sequence of pictures of the nonequilibrium process filmed as a movie. The emphasis of the dynamical aspects is a recent trend that finds its roots in Shannon's information theory and modern dynamical systems theory. This can explain why we had to wait the last decade before these dynamical aspects of the second law were discovered.

We here have a principle showing that the second law of thermodynamics plays a constructive role in nonequilibrium systems and thus does not have the need to exceed some threshold in the nonequilibrium constraints as was the case for the Glansdorff–Prigogine dissipative structures [67]. The principle of temporal ordering is valid as soon as the system is out of equilibrium and holds arbitrarily far from equilibrium.

These results have significant consequences in the nanosciences and biology. In nanosciences, the fluctuation theorem as well as the nonequilibrium work theorem provide a framework to understand how the nanoscale fluctuations can be taken into account, and they have recently contributed to determine thermodynamic quantities of such nanosystems as single biomolecules [68, 69]. Furthermore, these new relationships are valid arbitrarily far from equilibrium and are not limited to the linear regime. This goes along with the observation that the molecular motors typically work in the nonlinear regime [66]. Here it should be pointed out that applications in nanosciences often deal with nonequilibrium nanosystems. In this regard, the aforementioned results are decisive contributions to the statistical thermodynamics of nonequilibrium nanosystems. The nanosciences are deeply changing the traditional view according to which the world of atoms and molecules is separated by Avogadro's number from the macroscopic world. On the one hand, it should be noticed that Avogadro's number is an artifact of the metric system conventions, and recent work shows that the macroscopic properties of thermodynamics or chemical kinetics already hold with a few hundred or thousand atoms or molecules—that is, on the scale of nanometers. Collective effects such as nonequilibrium oscillations can already emerge on this small scale [70]. On the other hand, the traditional view cannot explain why the biological systems have a multiscale architecture with molecules assembling into supramolecular structures forming organels, cells, multicellular organisms, and so on. Today, the nanosciences are discovering the existence of structures on the nanoscale in both the inorganic and organic worlds, which clearly shows that structures can exist on all scales by the interaction of atoms and molecules, contrary to the traditional view.

The biological systems are characterized by two main properties: their metabolism and their capacity to self-reproduce. The first property refers to nonequilibrium thermodynamics. Until recently, this connection between biology and thermodynamics have been limited to the macroscopic aspects. The new advances here reported open new avenues toward nonequilibrium nanosystems of biological significance such as pulled biomolecules and molecular motors. Thanks to these new advances, the thermodynamics of biological systems can nowadays apply down to the nanoscale and make the connection with molecular biology. In this perspective, the aforementioned principle of temporal ordering shows at a fundamental level how a spontaneous generation of information is possible out of equilibrium and how random fluctuations can be converted into information [71]. This question remains a puzzle from the viewpoint of equilibrium statistical mechanics because of the balance between the forward and backward fluctuations. However, in nonequilibrium systems, the principle of temporal ordering shows that the random fluctuations should be biased, allowing the emergence of dynamical order. If this dynamical order can be memorized by the system—for instance, during a process of copolymerization—we could understand that information can be generated. If furthermore this information can be restored back to the nonequilibrium dynamics of the system, the self-reproduction of biological systems could be explained. From the viewpoint given by the principle of out-of-equilibrium temporal ordering, the biological systems would be physico-chemical systems with a built-in thermodynamic arrow of time. In this sense, the principle of temporal ordering could provide an explanation of Monod's teleonomy [72] on the basis of nonequilibrium statistical mechanics and the second law of thermodynamics. In this regard, the new advances contribute to understand the origins of biological systems in the physicochemical laws.

Acknowledgments

The author thanks Professor G. Nicolis for support and encouragement in this research. This research was financially supported by the "Communauté française de Belgique" (contract "Actions de Recherche Concertées" No. 04/09-312).

References

1. P. Gaspard, *Chaos, Scattering, and Statistical Mechanics*, Cambridge University Press, Cambridge, UK, 1998.

2. P. Gaspard, New. *J. Phys.* **7**, 77 (2005).

3. P. Gaspard, *J. Stat. Phys.* **117**, 599 (2004).

4. R. Balescu, *Equilibrium and Nonequilibrium Statistical Mechanics*, John Wiley & Sons, New York, 1975.

5. G. Nicolis and I. Prigogine, *Self-Organization in Nonequilibrium Systems*, John Wiley & Sons, New York, 1977.

6. T. De Donder and P. Van Rysselberghe, *Affinity*, Stanford University Press, Menlo Park, CA, 1936.

7. L. Onsager, *Phys. Rev.* **37**, 405 (1931).

8. T. Gilbert, J. R. Dorfman, and P. Gaspard, *Phys. Rev. Lett.* **85**, 1606 (2000).

9. J. R. Dorfman, P. Gaspard, and T. Gilbert, *Phys. Rev. E* **66**, 026110 (2002).

10. J. W. Gibbs, *Elementary Principles in Statistical Mechanics* Yale University Press, New Haven, 1902; reprinted by Dover Publ. Co., New York, 1960.

11. C. Shannon, *Bell System Tech. J.* **27**, 379, 623 (1948).

12. A. N. Kolmogorov, *Dokl. Akad. Nauk SSSR* **124**, 754 (1959).

13. Ya. G. Sinai, *Dokl. Akad. Nauk SSSR* **124**, 768 (1959).

14. J.-P. Eckmann and D. Ruelle, *Rev. Mod. Phys.* **57**, 617 (1985).

15. E. Lorenz, *J. Atmos. Sci.* **20**, 130 (1963).

16. Ya. B. Pesin, *Russian Math. Surveys* **32**, 55 (1977).

17. Ch. Dellago, H. A. Posch, and W. G. Hoover, *Phys. Rev. E* **53**, 1485 (1996).

18. H. van Beijeren, J. R. Dorfman, H. A. Posch, and Ch. Dellago, *Phys. Rev. E* **56**, 5272 (1997).

19. P. Gaspard and H. van Beijeren, *J. Stat. Phys.* **109**, 671 (2002).

20. M. Pollicott, *Invent. Math.* **81**, 413 (1985); *Invent. Math.* **85**, 147 (1986).

21. D. Ruelle, *Phys. Rev. Lett.* **56**, 405 (1986); *J. Stat. Phys.* **44**, 281 (1986).

22. P. Cvitanović and B. Eckhardt, *J. Phys. A: Math. Gen.* **24**, L237 (1991).

23. P. Gaspard, I. Claus, T. Gilbert, and J. R. Dorfman, *Phys. Rev. Lett.* **86**, 1506 (2001).

24. L. Van Hove, *Phys. Rev.* **95**, 249 (1954).

25. M. S. Green, *J. Chem. Phys.* **20** 1281 (1952); **22**, 398 (1954).

26. R. Kubo, *J. Phys. Soc. Jpn.* **12** 570 (1957).

27. P. Gaspard, *J. Stat. Phys.* **68**, 673 (1992).

28. S. Tasaki and P. Gaspard, *J. Stat. Phys.* **81**, 935 (1995).

29. S. Tasaki and P. Gaspard, *Bussei Kenkyu Cond. Matt.* **66**, 23 (1996).

30. T. Gilbert, J. R. Dorfman, and P. Gaspard, *Nonlinearity* **14**, 339 (2001).

31. L. A. Bunimovich, and Ya. G. Sinai, *Commun. Math. Phys.* **78**, 247, 479 (1980).

32. A. Knauf, *Commun. Math. Phys.* **110**, 89 (1987); *Ann. Phys. (N.Y.)* **191**, 205 (1989).

33. I. Claus and P. Gaspard, *J. Stat. Phys.* **101**, 161 (2000).

34. I. Claus and P. Gaspard, *Phys. Rev. E* **63**, 036227 (2001).

35. P. Gaspard and I. Claus, *Philos. Trans. Roy. Soc. Lond. A* **360**, 303 (2002).

36. I. Claus and P. Gaspard, *Physica D* **168–169**, 266 (2002).

37. P. Gaspard and G. Nicolis, *Phys. Rev. Lett.* **65**, 1693 (1990).

38. P. Gaspard and F. Baras, *Phys. Rev. E* **51**, 5332 (1995).

39. J. R. Dorfman and P. Gaspard, *Phys. Rev. E* **51**, 28 (1995).

40. P. Gaspard and J. R. Dorfman, *Phys. Rev. E* **52**, 3525 (1995).

41. E. Helfand, *Phys. Rev.* **119**, 1 (1960).

42. H. van Beijeren, A. Latz, and J. R. Dorfman, *Phys. Rev. E* **63**, 016312 (2000).

43. S. Viscardy and P. Gaspard, *Phys. Rev. E* **68**, 041204 (2003).

44. S. Viscardy and P. Gaspard, *Phys. Rev. E* **68**, 041205 (2003).

45. P. Gaspard and X.-J. Wang, *Phys. Rep.* **235**, 291 (1993).

46. J. Schnakenberg, *Rev. Mod. Phys.* **48**, 571 (1976).

47. Luo Jiu-li, C. Van den Broeck, and G. Nicolis, *Z. Phys. B - Condensed Matter* **56**, 165 (1984).

48. D. J. Evans and G. P. Morriss, *Statistical Mechanics of Nonequilibrium Liquids*, Academic Press, London, 1990.

49. L. Andrey, *Phys. Lett. A* **11**, 45 (1985).

50. R. Landauer, *IBM J. Res. Dev.* **5**, 183 (1961).

51. C. H. Bennett, *Int. J. Theor. Phys.* **21**, 905 (1982).

52. D. J. Evans, E. G. D. Cohen, and G. P. Morriss, *Phys. Rev. Lett.* **71**, 2401 (1993).

53. D. J. Evans and D. J. Searles, *Phys. Rev. E* **50**, 1645 (1994).

54. G. Gallavotti and E. G. D. Cohen, *Phys. Rev. Lett.* **74**, 2694 (1995).

55. J. Kurchan, *J. Phys. A: Math. Gen.* **31**, 3719 (1998).

56. G. E. Crooks, *Phys. Rev. E* **60**, 2721 (1999).

57. J. L. Lebowitz and H. Spohn, *J. Stat. Phys.* **95**, 333 (1999).

58. C. Maes, *J. Stat. Phys.* **95**, 367 (1999).

59. U. Seifert, *Europhys. Lett.* **70**, 36 (2005).

60. P. Gaspard, *Physica A* **369**, 201 (2006).

61. C. Jarzynski, *Phys. Rev. Lett.* **78**, 2690 (1997).

62. P. Gaspard, *J. Chem. Phys.* **120**, 8898 (2004).

63. D. Andrieux and P. Gaspard, *J. Chem. Phys.* **121**, 6167 (2004).

64. D. Andrieux and P. Gaspard, preprint cont-mat/0512254 (2005).

65. D. Andrieux and P. Gaspard, *J. Stat. Mech.: Th. Exp.*, P01011 (2006).

66. D. Andrieux and P. Gaspard, *Phys. Rev. E* **74**, 011906 (2006).

67. P. Glansdorff and I. Prigogine, *Thermodynamics of Structure, Stability, and Fluctuations*, Wiley-Interscience, New York, 1971.

68. C. Bustamante, J. Liphardt, and F. Ritort, *Phys. Today* **58**, 43 (2005).

69. D. Collin, F. Ritort, C. Jarzynski, S. B. Smith, I. Tinoco, Jr., and C. Bustamante, *Nature* **437**, 231 (2005).

70. P. Gaspard, *J. Chem. Phys.* **117**, 8905 (2002).

71. M. Eigen, *Steps Towards Life: A Perspective on Evolution*, Oxford University Press, Oxford, 1992.

72. J. Monod, *Chance and Necessity*, Knopf, New York, 1971.

QUANTUM AND CLASSICAL DYNAMICS
OF NONINTEGRABLE SYSTEMS[*]

ILYA PRIGOGINE,[†] TOMIO PETROSKY, AND GONZOLO ORDONEZ[‡]

*Center for Studies in Statistical Mechanics and Complex Systems,
The University of Texas, Austin, Texas 78712, USA; and International Solvay
Institutes for Physics and Chemistry, CP231, B-1050 Brussels, Belgium*

CONTENTS

I. INTRODUCTION

Classical dynamics and orthodox quantum mechanics are constructed along the model of integrable systems in the sense of Poincaré. Our aim is to construct dynamics for nonintegrable systems. As far as we know, this is a new attempt, which has its roots in the early work of the Brussels School [1–9]. The main result is that we have to replace the unitary transformation U by a nonunitary

---

[*]This chapter is the final work contributed by Professor Ilya Prigogine and was originally prepared to report a summary of recent developments of his research with his colleagues in the *Journal (PNAS) of the National Academy of Science*, where Professor Prigogine was a member. We are pleased to realize its publication in this Proceedings of the symposium in his memory.
[†]Deceased.
[‡]Present Address: Butler University, Indianapolis, Indiana 46208.

Special Volume in Memory of Ilya Prigogine: Advances in Chemical Physics, Volume 135,
edited by Stuart A. Rice

("star-unitary") transformation Λ. The construction has been done for examples going from relatively simple systems such as the Friedrichs model to more complicated systems such as the N-body problem and interacting fields. The requirements on Λ are the following:

1. Λ is invertible.
2. The Λ transformation is obtained by analytic continuation of resonance denominators in the unitary transformation U. When there is no resonance singularity, Λ reduces to U.
3. Λ preserves the measure of the phase space (or the trace in quantum mechanics).
4. Λ maps real variables to real variables (Hermitian operators to Hermitian operators in quantum mechanics).
5. Λ is analytic with respect to the coupling constant λ at $\lambda = 0$.
6. Λ leads to closed Markovian kinetic equations.

An important point is that $\Lambda\Lambda^{-1} = 1$. So we can go from the initial representation to the Λ representation and come back by Λ^{-1}. We have verified this property. The transformation to the Λ representation leads to a number of new properties hidden in the initial representation. The equations for density matrices become irreducible to classical trajectories or wave amplitudes. For nonintegrable systems, amplitudes have to be replaced by probabilities.

It is remarkable that nonintegrability introduces deep changes. One obtains really an extended formulation of classical and quantum mechanics. One of the most fascinating aspects is that we can now introduce an entropy on a purely microscopic basis. This entropy shows that the basic processes of microscopic physics such as quantum transitions or interactions of fields lead to irreversible processes, including entropy. Entropy dominates the physics of nonintegrable systems, which are the great majority of systems we observe. This is contary to the view that it is only from coarse graining that one obtains entropy for microscopic systems. We come back to the Boltzmann picture of an evolving universe without the contradiction of Boltzmann's presentation due to the use of classical mechanics.

This chapter gives a brief summary of our new approach without showing detailed proofs of the formulae presented here. However, when we use the formulae, we indicate the original papers where one can find their proofs. Here we concentrate on the qualitative aspects.

II. UNITARY TRANSFORMATION FOR INTEGRABLE SYSTEMS

We first consider quantum systems with Hamiltonian

$$H = H_0 + \lambda V \tag{1}$$

where H_0 is the unperturbed Hamiltonian describing free motion, and λV is the interaction with a dimensionless coupling constant λ. A simple example is the one-dimensional Friedrichs model of a harmonic oscillator (particle) coupled to field modes,

$$H = \hbar\omega_1 a_1^\dagger a_1 + \sum_k \hbar\omega_k a_k^\dagger a_k + \lambda \sum_k V_k(a_1^\dagger a_k + a_k^\dagger a_1) \qquad (2)$$

with $\omega_1 > 0$, $\omega_k = |k| > 0$, and

$$[a_\alpha, a_\beta^\dagger] = \delta_{\alpha,\beta} \qquad (3)$$

The operator with dagger implies Hermitian conjugate of the operator, as usual. We put the system is in a one-dimensional box of size L with periodic boundaries. As a result, the wave numbers are discrete. We have $k = 2\pi n/L$ with n integer. The spectrum of frequencies ω_k is discrete as well.

The Hamiltonian is bilinear with respect to the creation and annihilation operators. If $\omega_1 \neq \omega_k$ for all k (i.e., there is no resonance), then we can diagonalize the Hamiltonian as

$$H = \hbar\bar{\omega}_1 A_1^\dagger A_1 + \sum_k \hbar\bar{\omega}_k A_k^\dagger A_k \qquad (4)$$

The new creation and annihilation operators are related to the original operators by a unitary transformation $U^{-1} = U^\dagger$:

$$A_r = U^\dagger a_r, \qquad A_r^\dagger = U^\dagger a_r^\dagger \qquad (5)$$

for $r = 1$ or k. $U^\dagger$ is a factorizable superoperator of the form $(a \times b)$ defined as $(a \times b)c \equiv acb$, where a, b, and c are linear operators acting on wave functions. We have

$$U^\dagger = u^\dagger \times u \qquad (6)$$

where u is a unitary operator.

The $U^\dagger$ operator gives a representation where the interactions are eliminated. The change of representation is obtained by the transformation of density operators as

$$\rho \Rightarrow U^{-1}\rho = U^\dagger \rho \qquad (7)$$

and transformation of observables as

$$A \Rightarrow U^\dagger A \tag{8}$$

This change of representation leaves expectation values unchanged:

$$\text{Tr}(A^\dagger \rho) = \text{Tr}(A^\dagger U U^{-1} \rho) = \text{Tr}((U^\dagger A)^\dagger U^\dagger \rho) \tag{9}$$

Because there is no resonance between the particle and the field, $U^\dagger$ is expandable in a power series in λ, around $\lambda = 0$. This means the system is integrable in the sense of Poincaré.

The superoperator $U^\dagger$ has the distributive property:

$$A = BC \Rightarrow U^\dagger A = (U^\dagger B)(U^\dagger C) \tag{10}$$

for arbitrary operators A, B, C. This implies that commutation relations are preserved:

$$A = [B, C] \Rightarrow U^\dagger A = [U^\dagger B, U^\dagger C] \tag{11}$$

Let us now consider the Liouville equation:

$$i \frac{\partial \rho}{\partial t} = L_H \rho \tag{12}$$

where $L_H = [H,]$ is the Liouville–von Neumann operator. For $\rho = |\psi\rangle\langle\psi|$ this is reducible to a pair of equations:

$$i \frac{\partial |\psi\rangle}{\partial t} = H|\psi\rangle, \qquad -i \frac{\partial \langle\psi|}{\partial t} = \langle\psi|H \tag{13}$$

Writing $\bar\rho = U\rho$, we get

$$i \frac{\partial \bar\rho}{\partial t} = \bar{L}_0 \bar\rho \tag{14}$$

where $\bar{L}_0 = U L_H U^\dagger = [\bar{H}_0,]$ with $\bar{H}_0 = u H u^\dagger$ has the same form as the unperturbed Hamiltonian, but with renormalized frequencies. The transformed Liouville equation [Eq. (14)] is thus reduced to a pair of transformed Schrödinger equations

$$i \frac{\partial |\bar\psi\rangle}{\partial t} = \bar{H}_0 |\bar\psi\rangle, \qquad -i \frac{\partial \langle\bar\psi|}{\partial t} = \langle\bar\psi|\bar{H}_0 \tag{15}$$

where

$$|\bar{\psi}\rangle = u|\psi\rangle \tag{16}$$

The transformed Schrödinger equations describe free motion among each degree of freedom.

In short, the distributivity of the transformation $U^\dagger$ implies that $U^\dagger$ retains the reducibility of the Liouville equation into a pair of Schrödinger equations. Furthermore, this transformation retains the time-reversal invariance of these equations, since the free-motion equations [Eqs. (15)] are time-reversal invariant.

III. NONUNITARY TRANSFORMATION FOR NONINTEGRABLE SYSTEMS

The situation changes drastically for nonintegrable systems. As we will see, the transformation $U^\dagger = U^{-1}$ is replaced by a nonunitary transformation $\Lambda^\dagger \neq \Lambda^{-1}$.

Let us consider the large volume limit $L \to \infty$. In this limit we have a continuous spectrum for the field modes k, and we replace summations over modes by integrations

$$\frac{2\pi}{L}\sum_k \Rightarrow \int dk \tag{17}$$

As a result, we have Poincaré's resonance singularity at $\omega_k = \omega_1$ for $\omega_1 > 0$ in the series expansion of $U^\dagger$ in λ. The Friedrichs model discussed above may become nonintegrable.

As shown in previous papers [10–13], the regularization of the divergences in the series expansion leads to the new pair of transformations $\Lambda^\dagger$ and Λ^{-1}. The regularization involves the analytic continuation of real frequencies appearing in $U^\dagger$ into the complex complex plane. $\Lambda^\dagger$ and Λ^{-1} are mutually related to each other by complex conjugation of the complex frequencies [in the present case, z_1 in Eq. (21) below].

As discussed below, $\Lambda^\dagger$ (as well as Λ^{-1}) is nondistributive with respect to multiplication:

$$A = BC \Rightarrow \Lambda^\dagger A \neq (\Lambda^\dagger B)(\Lambda^\dagger C) \tag{18}$$

As a consequence, commutation relations are not preserved:

$$A = [B, C] \Rightarrow \Lambda^\dagger A \neq [\Lambda^\dagger B, \Lambda^\dagger C] \tag{19}$$

Still, we have the following relation for canonical variables x and p:

$$[x, p] = i\hbar \Rightarrow \Lambda^\dagger[x, p] = i\hbar \tag{20}$$

due to the preservation of the unit operator $\Lambda^\dagger 1 = 1$, which is a direct consequence of the analyticity of $\Lambda^\dagger$ required in the introduction [requirement (5)].

Let us briefly summarize the derivation of $\Lambda^\dagger$ (see Ref. 14). To construct it, we use the known fact that for the nonintegrable case, Friedrichs' Hamiltonian is diagonalizable in the complex energy plane as [10, 15, 16]

$$H = \hbar z_1 A_1^\dagger \tilde{A}_1 + \sum_k \hbar \omega_k A_k^\dagger \tilde{A}_k \tag{21}$$

where $z_1 = \tilde{\omega}_1 - i\gamma$ is the complex pole of the resolvent operator of the Hamiltonian H, giving the renormalized frequency $\tilde{\omega}_1$ and the inverse half-lifetime $\gamma > 0$. The new modes are linear combinations of the original modes appearing in Eq. (2). For example,

$$\tilde{A}_1 = \tilde{c}_{11} a_1 + \sum_k \tilde{c}_{1k} a_k \tag{22}$$

See Refs. 10 and 15 for explicit forms of $\tilde{c}_{ij}$ and corresponding linear transformations for $\tilde{A}_k$, $A_1^\dagger$, and $A_k^\dagger$.

The particle operators (also called Gamow modes) $\tilde{A}_1$, $A_1^\dagger$ have a strictly exponential time dependence,

$$e^{iL_H t}\tilde{A}_1^\dagger = e^{iz_1^* t}\tilde{A}_1^\dagger, \quad e^{iL_H t}A_1^\dagger = e^{iz_1 t}A_1^\dagger \tag{23}$$

$$e^{iL_H t}\tilde{A}_1 = e^{-iz_1 t}\tilde{A}_1, \quad e^{iL_H t}A_1^\dagger = e^{-iz_1^* t}A_1^\dagger \tag{24}$$

The distinction between the operators $A_1^\dagger$ and $\tilde{A}_1$ comes from the fact that the first is associated with the pole z_1, while the second is associated with z_1^*.

These operators satisfy remarkable commutation relations (see Refs. 10 and 15):

$$[\tilde{A}_1, \tilde{A}_1^\dagger] = 0 \tag{25}$$

$$[\tilde{A}_1, A_1^\dagger] = 1 \tag{26}$$

Note that there is a discontinuity at $\lambda = 0$ in Eq. (25) due to Poincaré resonances, because for $\lambda = 0$ the Gamow operators reduce to the bare operators a_1, $a_1^\dagger$, whose commutator is 1 and not 0.

We construct the explicit forms of $\Lambda^\dagger$ and Λ^{-1} by their action on monomials of a_r and $a_r^\dagger$. We first consider the simplest case that they act on a single a_r or $a_r^\dagger$ operators. From the requirements on Λ stated in the Introduction we obtain the relations for $\Lambda^\dagger$,

$$\tilde{A}_r = \Lambda^\dagger a_r, \quad \tilde{A}_r^\dagger = \Lambda^\dagger a_r^\dagger \tag{27}$$

and the relations for Λ^{-1},

$$A_r = \Lambda^{-1} a_r, \qquad A_r^\dagger = \Lambda^{-1} a_r^\dagger \tag{28}$$

(see Ref. 14). These relations are consistent with the perturbative method to construct $\Lambda^\dagger$ or Λ^{-1} based on kinetic operators [11]. Since the explicit forms of the linear transformations for $\tilde{A}_r$ and A_r are known as mentioned at Eq. (22), we know the forms of $\Lambda^\dagger$ and Λ^{-1} in the case where they act on a single a_r or $a_r^\dagger$.

Next we construct $\Lambda^\dagger$ by their action on the product $a_1^\dagger a_1$. A first idea would be to keep the distributive relation $\Lambda^\dagger a_1^\dagger a_1 = (\Lambda^\dagger a_1^\dagger)(\Lambda^\dagger a_1) = \tilde{A}_1^\dagger \tilde{A}_1$ analogous to Eq. (10). But the product $\tilde{A}_1^\dagger \tilde{A}_1$ contains terms nonanalytic at $\lambda = 0$ [see discussion below Eq. (26)] due to Poincaré resonances.

Thus distributivity is incompatible with analyticity of $\Lambda^\dagger$ at $\lambda = 0$. Hence, if we want to keep analyticity as listed in the Introduction, we have to give up the distributivity.

To obtain the transformed product $\Lambda^\dagger a_1^\dagger a_1$, we start with the expression for the integrable case, $U^\dagger a_1^\dagger a_1$ and we analitycally continue the denominators in $U^\dagger a_1^\dagger a_1$ to the complex energy plane for the nonintegrable case in such a way that $\Lambda^\dagger a_1^\dagger a_1$ is analytic at $\lambda = 0$. This leads to the following form (see Ref. 14):

$$\Lambda^\dagger a_1^\dagger a_1 = \tilde{A}_1^\dagger \tilde{A}_1 + Y \tag{29}$$

or

$$\Lambda^\dagger a_1^\dagger a_1 - (\Lambda^\dagger a_1^\dagger)(\Lambda^\dagger a_1) = Y \tag{30}$$

where

$$Y = \sum_k b_k a_k^\dagger a_k \tag{31}$$

The quantity b_k has a physical meaning as the spectral density of photons emitted by the dressed excited particle, and its explicit form is rather complicated [11]. However, for weak coupling case it reduces to a simple form:

$$b_k \approx \frac{2\pi}{L} \delta(\omega_k - \omega_1) \tag{32}$$

For the integrable case (with no resonance), b_k vanishes.

One can see the physical meaning of the operator Y for the case where the field is in a thermal equilibrium state. Indeed, by taking the ensemble average with a Bose–Einstein distribution of field modes at temperature T, we get

$$\langle Y \rangle = \sum_k b_k (\exp[\omega_k/(k_B T)] - 1)^{-1}$$

$$\approx (\exp[\omega_1/(k_B T)] - 1)^{-1} \tag{33}$$

This shows that in this case, $\langle Y \rangle$ is the average number of excitations of the oscillator in the thermal equilibrium state.

One can extend the above procedure of the construction of $\Lambda^\dagger$ to the case where it acts on arbitrary monomial of $a_1^\dagger$ and a_1. The resultant expression is given by (see Refs. 14)

$$e^{iL_H t} \Lambda^\dagger a_1^{\dagger m} a_1^n = e^{i(mz_1^* - nz_1)t} \sum_{l=0}^{\min(m,n)} \frac{m!n!}{(m-l)!(n-l)!l!} \Lambda^\dagger a_1^{\dagger m-l} a_1^{n-l} Y^l (e^{2\gamma t} - 1)^l \tag{34}$$

where $\gamma = -\mathrm{Im}(z_1)$, as before.

From the above expressions presented in this section, we can already conclude several interesting properties of the transformation $\Lambda^\dagger$. First, Eqs. (27) and (28) show that it is nonunitary transformation, $\Lambda^\dagger \neq \Lambda^{-1}$. Actually, one can show that this transformation satisfies a more general symmetric property called *star unitarity* (see Refs. 5, 10, 13, and 14):

$$\Lambda^{-1} = \Lambda^* \tag{35}$$

Star conjugation is essentially Hermitian conjugation followed by a complex conjugation of the complex energies (z_1 in the present case).

Second, expectation values are preserved as

$$\mathrm{Tr}(A^\dagger \rho) = \mathrm{Tr}(A^\dagger \Lambda \Lambda^{-1} \rho) = \mathrm{Tr}((\Lambda^\dagger A)^\dagger \Lambda^{-1} \rho) \tag{36}$$

This property follows from the existence of the inverse operator Λ^{-1} such that $\Lambda \Lambda^{-1} = 1$.

Third, Eq. (31) shows that $\Lambda^\dagger$ is nondistributive, and b_k determines fluctuations. Since there is a fluctuation, we can expect that the time evolution in Eq. (34) may be related to a stochastic process. Indeed, one can show that the time evolution (34) is identical to the time evolution generated by the set of Langevin equations for the stochastic operators $\alpha_1^\dagger(t)$, $\alpha_k^\dagger(t)$ (see Ref. 14):

$$\frac{d}{dt}\alpha_1^\dagger(t) = -iz_1\alpha_1^\dagger(t) + \sum_k \xi_k^*(t)\alpha_k^\dagger(t) \tag{37}$$

$$\frac{d}{dt}\alpha_k^\dagger(t) = -i\omega_k\alpha_k^\dagger(t) + \xi_k^*(t)\alpha_1^\dagger(t) \tag{38}$$

where $\xi_k^*(t)$ is a Gaussian white-noise source with noise average

$$\overline{\xi_k^*(t)\xi_{k'}(t')} = 2\gamma b_k \delta_{kk'} \delta(t - t') \tag{39}$$

and

$$[\alpha_r(t), \alpha_s^\dagger(t)] = \delta_{rs} \tag{40}$$

The equivalence between the stochastic variables and the $\Lambda^\dagger$-transformed variables is given by the relation

$$\overline{\alpha_1^\dagger(t)^m \alpha_1(t)^n} = \langle e^{iL_H t} \Lambda^\dagger a_1^{\dagger m} a_1^n \rangle \tag{41}$$

IV.   EXACT QUANTUM MASTER EQUATION

In the previous sections we have considered the transformation of observables. In this section we consider the transformation of the density matrix. The transformed Liouville equation for $\tilde{\rho}(t) = \Lambda\rho(t)$ is

$$i\frac{\partial}{\partial t}\tilde{\rho}(t) = \tilde{\theta}\tilde{\rho}(t) \tag{42}$$

where $\tilde{\theta} = \Lambda L_H \Lambda^{-1}$ is a *collision operator*. Collision operators are the central objects in kinetic theory or nonequilibrium statistical mechanics. As will be shown below, instead of giving factorized equations describing free motion [see Eq. (15)] the transformation gives an irreversible equation, which is irreducible to a pair of Schrödinger equations.

We will consider the expectation value of observables $G_1(a_1^\dagger, a_1)$, depending on polynomials of the creation and annihilation operators $a_1^\dagger$, a_1. From Eq. (42) we obtain

$$i\frac{\partial}{\partial t}\text{Tr}[G_1\tilde{\rho}(t)] = \text{Tr}[G_1\tilde{\theta}\tilde{\rho}(t)] \tag{43}$$

The connection to the $\Lambda^\dagger$ transformation is obtained through

$$\text{Tr}[G_1\tilde{\rho}(t)] = \text{Tr}[(\Lambda^\dagger G_1)^\dagger\rho(t)] \tag{44}$$

We can always write G_1 in normal order of $a_1^\dagger$, a_1. Therefore G_1 can be written as a superposition of the monomials $a_1^{\dagger m} a_1^n$, with $m, n \geq 0$ integers. Hence, we will consider the equation

$$i\frac{\partial}{\partial t}\text{Tr}[a_1^{\dagger m} a_1^n \tilde{\rho}(t)] = \text{Tr}[a_1^{\dagger m} a_1^n \tilde{\theta}\tilde{\rho}(t)] \tag{45}$$

Using Eq. (34), one can obtain (with $n_r = a_r^\dagger a_r$) [14]

$$
i\frac{\partial}{\partial t}\mathrm{Tr}[a_1^{\dagger m}a_1^n\tilde{\rho}(t)] = -\mathrm{Tr}\left[\left\{(mz_1^* - nz_1)a_1^{\dagger m}a_1^n - 2i\gamma mn\sum_k b_k n_k a_1^{\dagger m-1}a_1^{n-1}\right\}\tilde{\rho}(t)\right]
$$
(46)

This leads to

$$
\begin{aligned}
i\frac{\partial}{\partial t}\mathrm{Tr}[G_1\tilde{\rho}] = \mathrm{Tr}\Big(G_1\Big\{&\tilde{\omega}_1[a_1^\dagger a_1, \tilde{\rho}]\\
&+ i\gamma\sum_k b_k(n_k + 1)(2a_1\tilde{\rho}a_1^\dagger - a_1^\dagger a_1\tilde{\rho} - \tilde{\rho}a_1^\dagger a_1)\\
&+ i\gamma\sum_k b_k n_k(2a_1^\dagger\tilde{\rho}a_1 - a_1 a_1^\dagger\tilde{\rho} - \tilde{\rho}a_1 a_1^\dagger)\Big\}\Big)
\end{aligned}
$$
(47)

which is exact in all orders of λ (see Ref. 17). Since G_1 is an arbitrary polynomial of $a_1^\dagger$ and a_1, one can equate $i\partial\tilde{\rho}/\partial t$ to the expression inside the curly bracket on the right-hand side of Eq. (47). Of course, the equivalence holds in a week sense under the trace. This equation for $\tilde{\rho}(t)$ is our exact master equation, which is a Markovian kinetic equation.

In contrast to the integrable case, the master equation contains the dissipative terms proportional to γ, in addition to the free-motion term. These terms break time symmetry and introduce diffusion, which causes the collapse of wave functions.

For the weak coupling case with Eq. (32), our master equation reduces to the well-known quantum master equation, obtained through the $\lambda^2 t$ approximation, widely used in quantum optics. This equation describes, among other things, quantum decoherence due to Brownian motion. Hence, we have derived an *exact* quantum master equation for the transformed density operator $\tilde{\rho}$ that describes exact decoherence. Furthermore, our master equation cannot keep the "purity" of the transformed density matrix. Indeed, one can show that if $\tilde{\rho}(t)$ is factorized into a product of transformed wave functions at $t = 0$, it will not be factorized into their product for $t > 0$. This is consistent the nondistributivity of the nonunitary transformation (18).

One can derive another interesting equation from Eq. (47). Indeed, for $m = n = 1$ one can obtain

$$
\frac{\partial}{\partial t}\mathrm{Tr}[n_1\tilde{\rho}(t)] = -2\gamma\left(\mathrm{Tr}[n_1\tilde{\rho}(t)] - \sum_k b_k\mathrm{Tr}[n_k\tilde{\rho}(t)]\right)
$$
(48)

In the weak coupling limit with Eq. (32), this reduces to the well-known kinetic equation for the average number of the excited particle obtained by the $\lambda^2 t$ approximation.

V. STOCHASTIC SCHRÖDINGER EQUATION

As mentioned in the previous section, the exact master equation cannot be reduced to the Schrödinger equation for a transformed wave function that is a deterministic equation in time. However, it can reduce to a stochastic Schrödinger equation [18].

In order to see this, we consider Eqs. (37) and (38). It is easy to see that this set of equations can be written in the form of the "Heisenberg" equations

$$\frac{d}{dt}\alpha_r^\dagger(t) = -i[\tilde{H}(t), \alpha_r^\dagger(t)] \tag{49}$$

where $\tilde{H}(t)$ is a stochastic dissipative Hamiltonian

$$
\begin{aligned}
\tilde{H}(t) = {} & z_1\alpha_1^\dagger(t)\alpha_1(t) + \sum_k \omega_k\alpha_k^\dagger(t)\alpha_k(t) \\
& + \sum_k \left(\xi_k^*(t)\alpha_k^\dagger(t)\alpha_1(t) + \xi_k(t)\alpha_1^\dagger(t)\alpha_k(t) \right)
\end{aligned} \tag{50}
$$

Hence we obtain the stochastic Schrödinger equation

$$i\frac{\partial|\psi\rangle}{\partial t} = \tilde{H}(t)|\psi\rangle \tag{51}$$

Previously, stochastic Schrödinger equations for a quantum Brownian motion have been derived only for the particle component through approximated equations, such as the master equation obtained by the Markovian approximation [18]. In contrast, our stochastic Schrödinger equation is exact. Moreover, our stochastic equation includes both the particle and the field components, so it does not rely on integrating out the field bath modes.

VI. FLUCTUATIONS IN POSITION AND MOMENTUM

Resonances are nonlocal in time and space. We expect that for systems that have resonances there are new types of fluctuations associated with the nondistributivity of $\Lambda^\dagger$. To see this, consider the position and momentum operators defined by

$$
\begin{aligned}
x_1 &\equiv \sqrt{\frac{\hbar}{2m_1\omega_1}}(a_1 + a_1^\dagger) \\
p_1 &\equiv -i\sqrt{\frac{\hbar m_1\omega_1}{2}}(a_1 - a_1^\dagger)
\end{aligned} \tag{52}
$$

They obey the commutation relation $[x_1, p_1] = i\hbar$. We consider the transformed position and momentum operators given by

$$
\begin{aligned}
\tilde{x}_1 &= \Lambda^\dagger x_1 \\
\tilde{p}_1 &= \Lambda^\dagger p_1
\end{aligned}
\tag{53}
$$

Let us define the "fluctuations" of the transformed operators as

$$
\begin{aligned}
(\Delta x_1)^2 &\equiv \Lambda^\dagger x_1^2 - (\Lambda^\dagger x_1)^2 \\
(\Delta p_1)^2 &\equiv \Lambda^\dagger p_1^2 - (\Lambda^\dagger p_1)^2
\end{aligned}
\tag{54}
$$

Using Eq. (25), we obtain

$$
(\Delta x_1)^2 = \frac{\hbar}{m_1 \omega_1} \left(\frac{1}{2} + \sum_k b_k a_k^\dagger a_k \right)
\tag{55}
$$

$$
(\Delta p_1)^2 = \hbar m_1 \omega_1 \left(\frac{1}{2} + \sum_k b_k a_k^\dagger a_k \right)
\tag{56}
$$

This leads to the "uncertainty relation" for our fluctuations as

$$
\Delta x_1 \Delta p_1 = \hbar \left(\frac{1}{2} + \sum_k b_k a_k^\dagger a_k \right)
\tag{57}
$$

This relation is due to the nondistributive property of the $\Lambda^\dagger$ transformation. For the integrable case (no resonances), there is no uncertainty relation of this type, since we have $\Lambda^\dagger \Rightarrow U^\dagger$ and $\Delta x_1 = \Delta p_1 = 0$. Although the relation (57) is reminiscent of Heinseberg's uncertainty relation, the relation (57) has a different origin. It is due to resonances allowing the transfer of fluctuations from the field to the particle. The term $1/2$ is due to vacuum fluctuations of the field modes, and the term involving b_k is due to the nonzero temperature distribution of field modes. Both terms are rooted in the resonance singularity of the nonintegrable case.

VII.   STOCHASTIC CLASSICAL WAVE EQUATION

So far we have considered quantum systems. Similar considerations can be applied to classical systems and, in particular, to the classical Friedrichs model [13, 19]. In the Λ representation we have fluctuations. We can define transformed action variables like in the quantum case,

$$
\tilde{J}_r = \Lambda^\dagger J_r, \qquad J_r = a_r^* a_r
\tag{58}
$$

where a_r^* is the complex conjugate of the normal mode a_r, which is the classical correspondent of the quantum annihilation operator, with commutators replaced by Poisson brackets. Note that a_r has the dimension of the square root of action variable. As one can expect, the transformed actions $\tilde{J}_k$ for the field present fluctuations due to the resonance. Indeed, similarly to the above argument, we can write a stochastic wave equation for the classical field [19].

To see this, we introduce the field

$$\phi(x,t) = \sum_k \left(\frac{1}{2L\omega_k}\right)(\alpha_k(t)e^{ikx} + \text{c.c.}) \tag{59}$$

as well as the stochastic fields

$$\Phi(x,t) = \sum_k \left(\frac{1}{2L\omega_k}\right)(\xi_k(t)e^{ikx} + \text{c.c.})$$

$$\Psi(x,t) = \sum_k \left(\frac{\omega_k}{2L}\right)(\xi_k(t)e^{ikx} + \text{c.c.}) \tag{60}$$

with the same Gaussian white noise as in Eq. (39) (with a suitable unit of the action variable.) Then, by a straightforward calculation from Eq. (38), one can obtain the stochastic "Maxwell–Lorentz" equation

$$\left(\frac{\partial^2}{\partial x^2} - \frac{\partial^2}{\partial t^2}\right)\phi(x,t) = -j(x,t) \tag{61}$$

with the stochastic flow

$$j(x,t) = \text{Re}[\alpha_1(t)]\Psi(x,t) - \frac{\partial}{\partial t}[\text{Im}[\alpha_1(t)]\Phi(x,t)] \tag{62}$$

VIII.   INVERTIBILITY OF Λ

An essential property of Λ is the existence of the inverse transformation Λ^{-1}. This allows us to go back and forth between Hamiltonian dynamics and Markovian dynamics. In other words, Λ maps deterministic reversible dynamics to irreversible stochastic dynamics.

In Ref. 13, we have proved that the Λ transformation constructed is invertible for the classical model discussed in the previous section. Here, using the same system discussed in the previous section, we demonstrate the invertiblity of our transformation by a numerical calculation of the time evolution of the action variable $J_1(t)$ for an initial condition where all the field actions are zero [20]. Due to radiation damping, $J_1(t)$ follows an approximately exponential decay. However, there are deviations from exponential in the exact evolution both at short and long time scales as compared with the relaxation time scale. In Fig. 1, we present numerical results.

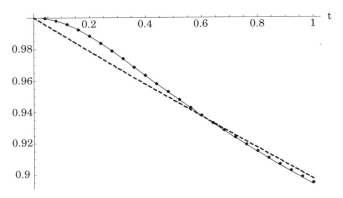

Figure 1. Numerical result of the time evolution of the bare action $J_1(t)$ (solid line), and the transformed action $\hat{J}_1(t)$ (dashed line) for times much smaller than the relaxation time. The bold dots are the result of applying Λ to $\hat{J}_1(t)$ at several points in time—that is, the result of inverse transformation applied on points on the transformed dashed line.

The solid line in the figure represents $J_1(t)$, which is given by

$$J_1(t) = \int J_1 e^{-iL_H t} \rho(0)\, d\Gamma \tag{63}$$

where $d\Gamma$ is the infinitesimal phase volume in phase space and the integration is taken over all the action-angle variables J_r and β_r, and with the initial condition

$$\rho(0) = \delta(J_1 - J_{10})\delta(\beta_1 - \beta_{10}) \prod_k \delta(J_k)\delta(\beta_k - \beta_{k0}) \tag{64}$$

This solid line is obtained by numerical integration of Hamilton's equations of motion for the original variables.

The dashed line represents

$$\begin{aligned}
\hat{J}_1(t) &= \int d\Gamma J_1 e^{-iL_H t}\Lambda^{-1}\rho(0) \\
&= \int d\Gamma (\Lambda^{-1}\rho(0))(e^{iL_H t}J_1)
\end{aligned} \tag{65}$$

The dashed line was drawn by using the numerical values of $\exp[iL_H t]J_1$ obtained from the numerical integration of the equations of motion for the original variables and then applying the theoretical form of $\Lambda^{-1}\rho(0)$ using explicit form of the Λ transformation presented in Ref. 13. Note that $\hat{J}_1(t)$ shown in the figure has a strictly exponential decay.

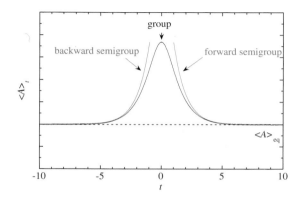

Figure 5. Time evolution of the statistical average (51) according to the expansion (52) of the forward semigroup valid for $t > 0$ and the expansion (35) of the backward semigroup valid for $t < 0$.

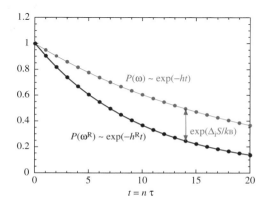

Figure 15. Diagram showing the probability of a path and the probability of the corresponding time reversal as a function of the number n of time intervals τ, illustrating the formula (101). $\Delta_i S$ denotes the entropy production during the whole duration $n\tau$.

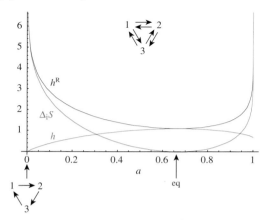

Figure 18. The dynamical entropies (126) and (127) as well as the entropy production (128) for the three-state Markov chain defined by the matrix (125) of transition probabilities versus the parameter a. The equilibrium corresponds to the value $a = 2/3$. The process is perfectly cyclic at $a = 0$ where the path is ... 123123123123 ... and the Kolmogorov–Sinai entropy h vanishes as a consequence.

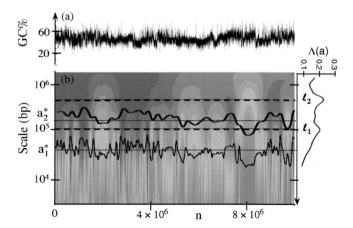

Figure 2. Space-scale representation of the GC content of a 10-Mbp-long fragment of human chromosome 22 when using a Gaussian smoothing filter $g^{(0)}(x)$ [Eq. (6)]. (a) GC content fluctuations computed in adjacent 1 kbp intervals. (b) Color coding of the convolution product $W_{g(0)}[GC](n, a) = (GC * g^{(0)}(\cdot/a))(n)$ using 256 colors from black (0) to red (max); superimposed are shown the smoothed GC profiles obtained at scales $a_1^* = 40$ kbp and $a_2^* = 160$ kbp. On the right-hand side we see vertically the scale (frequency^{-1}) spectrum $\Lambda(a)$ [Eq. (9)] computed with the complex Morlet wavelet [Eq. (8)] over the entire chromosome 22. The horizontal dashed lines in the color picture correspond to the two main characteristic oscillations length $l_1 = 100$ kbp and $l_2 = 400$ kbp.

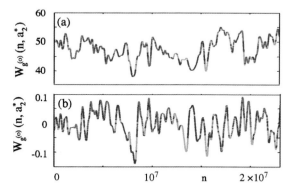

Figure 3. Compositional oscillations observed in the human chromosome 22 fragment (23 Mbp, NT_011520.8) after low-pass filtering at scale $a_2^* = 160$ kbp (see Fig. 2). (a) GC content. (b) Total skew $S = S_{TA} + S_{GC}$ [Eq. (2)]. The red (blue) portions of the profiles correspond to the location of sense (antisense) genes that have the same (opposite) orientation than the sequence. The location of the immunoglobulin locus is shown in pink.

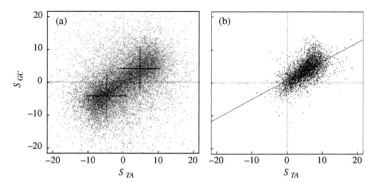

Figure 5. (a) S_{TA} and S_{GC} skews in human introns [36]: Each point corresponds to one of the 14,854 intron-containing genes; repeated elements are removed from the analysis (Section 2.1); red points correspond to sense genes (7508) with the same orientation as the Watson strand; blue points correspond to antisense genes (7346) with opposite orientation; black crosses represent the standard deviations of the distributions. (b) Correlation between S_{TA} and S_{GC} skews determined on the coding strand from intronic regions without repeats: Each point corresponds to a gene for which the total length of intronic regions is $l > 25$ kbp (7797 genes); Pearson's correlation coefficient r equals 0.61 (the slope of the regression line is 0.58).

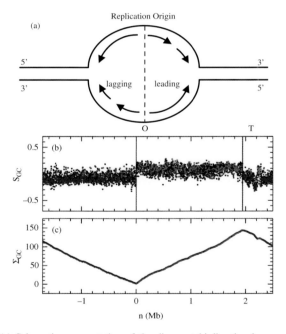

Figure 8. (a) Schematic representation of the divergent bi-directional progression of the two replication forks from the replication origin. (b) S_{GC} calculated in 1-kbp windows along the genomic sequence of *Bacillus subtilis*. (c) Cumulated skew Σ_{GC}. The vertical lines correspond respectively to the replication origin (O) and termination (T) positions. In (b) and (c), red (blue) points correspond to sense (antisense) genes that have the same (opposite) orientation than the sequence.

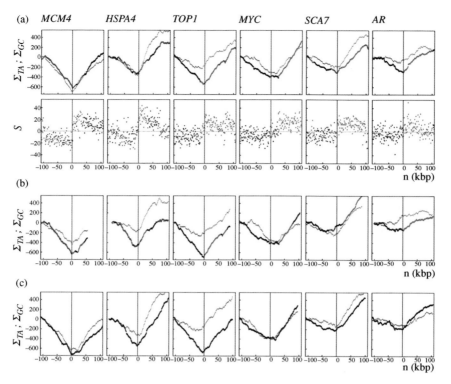

Figure 9. TA and GC skew profiles around experimentally determined human replication origins [38]. (a) The skew profiles were determined in 1-kbp windows in regions surrounding (±100 kbp without repeats) experimentally determined human replication origins (see Section II.A). (Upper) TA and GC cumulated skew profiles Σ_{TA} (thick line) and Σ_{GC} (thin line). (Lower) Skew S calculated in the same regions. The ΔS amplitude associated with these origins, calculated as the difference of the skews measured in 20-kbp windows on both sides of the origins, are: MCM4 (31%), HSPA4 (29%), TOP1 (18%), MYC (14%), SCA7 (38%), and AR (14%). (b) Cumulated skew profiles calculated in the six regions of the mouse genome homologous to the human regions analyzed in (a). (c) Cumulated skew profiles in the six regions of the dog genome homologous to human regions analyzed in (a). The abscissa (n) represents the distance (in kbp) of a sequence window to the corresponding origin; the ordinate represents the values of S given in percent. The colors have the following meaning: red, sense genes (coding strand identical to the Watson strand); blue, antisense genes (coding strand opposite to the Watson strand); black, intergenic regions. In (c), genes are not represented.

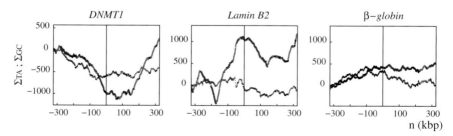

Figure 10. Cumulated skew profiles calculated around the origin of replication DNMT1, Lamin B2, and β-globin in the human genome: Σ_{TA} (thick line) and Σ_{GC} (thin line). The colors have the same meaning as in Fig. 9.

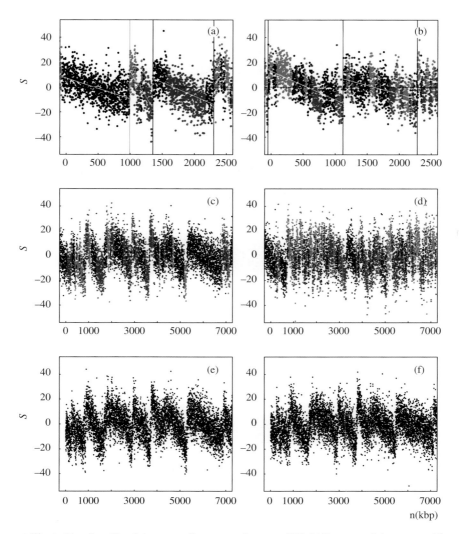

Figure 11. *S* profiles along mammalian genome fragments [38]. (a) Fragment of chromosome 20 including the TOP1 origin (red vertical line). (b and c) Chromosome 4 and chromosome 9 fragments, respectively, with low GC content (36%). (d) Chromosome 22 fragment with larger GC content (48%). In (a) and (b), vertical lines correspond to selected putative origins (see Section VI.A); yellow lines are linear fits of the *S* values between successive putative origins. Black, intergenic regions; red, sense genes; blue, antisense genes. Note the fully intergenic regions upstream of TOP1 in (a) and from positions 5290–6850 kbp in (c). (e) Fragment of mouse chromosome 4 homologous to the human fragment shown in (c). (f) Fragment of dog chromosome 5 syntenic to the human fragment shown in (c). In (e) and (f), genes are not represented.

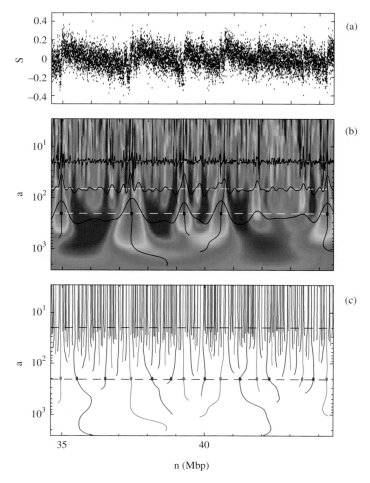

Figure 14. (a) Skew profiles of a fragment of human chromosome 12. (b) WT of S using $g^{(1)}$; $W_{g^{(1)}}[S](n, a)$ is coded from black (min) to red (max); three cuts of the WT at constant scale $a = a^* = 200\,\text{kbp}$, 70 kbp and 20 kbp are superimposed together with five maxima lines identified as pointing to upward jumps in the skew profile. (c) WT skeleton defined by the maxima lines in blue (respectively, red) when corresponding to positive (respectively, negative) values of the WT. At the scale $a^* = 200\,\text{kbp}$, one thus identify 7 upward (blue dots) and 8 downward (red dots) jumps. The black dots in (b) correspond to the five WTMM of largest amplitude that have been identified as putative replication origins; it is clear that the associated maxima lines point to the five major upward jumps in the skew profile in the limit $a \to 0^+$.

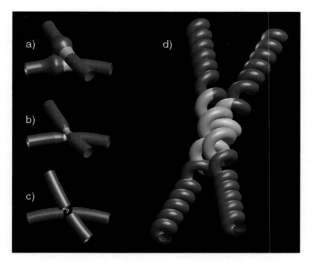

Figure 19. Examples of possible local defects along the fiber. (a) Local swelling or attachment of an external agent (e.g., RNA polymerases in the model of Cook [22, 23]); (b) local shrinking; (c) any form of fiber denaturation inducing a depletive potential well, according to the position along the fiber and the entry–exit angle; (d) as an example of (c), the fiber seen as a compact helix (condensed nucleosomal array) with local partial decondensation illustrating a situation where the excluded volume gain is quite important and the entry-exit angle is fixed.

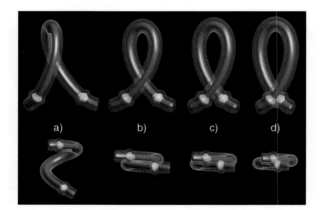

Figure 20. Steps involved in loop formation. (a) Free evolution of the tube in depletive environment; (b) formation of an unstable loop at around 3.4 l_p; (c) gliding of the loop governed by the positions of the two contact points along the fiber and the entry–exit angle; (d) trapping of the loop by local defects. The translucent green surface represents the excluded volume for the fluid of hard spheres; in (b,c,d) one sees that some of the excluded volume is reduced from the overlap resulting from formation of the loop.

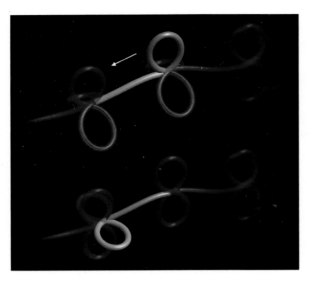

Figure 22. Illustration of the fiber defects clustering dynamics. The number of leaves per rosette fluctuates from one rosette to the next; this is due to both statistical fluctuations and variations in the local environment. From one cell cycle (up) to the next (down), leaves can be exchanged between neighboring rosettes.

Figure 23. Illustration of the spontaneous emergence of rosette-like folding of the chromatin fiber in the crowded environment of the cell nucleus.

The dots on the solid line are the numerical result of the inverse transformation. This is obtained by applying the explicit theoretical form of Λ on the numerical values of $\hat{J}_1(t)$ at several points of time. The figure shows that by applying Λ to $\hat{J}_1(t)$, we indeed go back to $J_1(t)$. This numerically demonstrates that there is no loss of information by the nonunitary transformation.

IX. ENTROPY

As discussed in Refs. 5, 10, and 12, we can introduce the microscopic "entropy" operator as

$$M(t) = e^{iL_H t} \Lambda^\dagger \Lambda^{-iL_H t} \tag{66}$$

For N-particle systems, we define reduced operators depending on a finite number of particles. For the Friedrichs model considered in Section II we define the reduced operator associated to the particle by

$$M_1(t) = e^{iL_H t} \Lambda^\dagger |a_1^\dagger a_1\rangle\rangle\langle\langle a_1^\dagger a_1 | \Lambda^{-iL_H t} \tag{67}$$

where we use the notation $\langle\langle A|B\rangle\rangle = \mathrm{Tr}(A^\dagger B)$. The expectation value $\mathcal{H}(t) \equiv \langle\langle \rho|M_1|\rho\rangle\rangle$ is a Lyapounov function ($\mathcal{H}$-function) that decreases monotonically for all times. For example, for $\rho = a_1^\dagger |0\rangle\langle 0|a_1$ we have

$$\mathcal{H}(t) = e^{-4\gamma t}\mathcal{H}(0) \tag{68}$$

An important feature of the $\mathcal{H}$-function is that its monotonic behavior in time is independent of probabilistic arguments.

This function is somewhat analogous to Boltzmann's $\mathcal{H}$-function; however, an important difference is that our $\mathcal{H}$-function includes correlations. This has already been discussed in simple examples in Refs. 5, 10, 12, and 21.

Boltzmann's $\mathcal{H}$-function is not monotonic after we perform a velocity inversion of every particle—that is, if we perform time inversion. In contrast, our $\mathcal{H}$-function is always monotonic as long as the system is isolated. When a velocity inversion is performed, the $\mathcal{H}$-function jumps discontinuously due to the flow of entropy from outside. After this, the $\mathcal{H}$-function continues its monotonic decrease [10]. Our $\mathcal{H}$-function breaks time symmetry, because $\Lambda^\dagger$ itself breaks time symmetry.

Velocity inversion is a special case of velocity rotation of every particle, which corresponds to acceleration. Thus one can expect that velocity rotations and acceleration will also produce a change of the $\mathcal{H}$-function or a change of entropy. The effects of velocity rotations on the $\mathcal{H}$-function have been considered in Ref. 22.

X. CONCLUDING REMARKS

Nonintegrable systems form the core of our universe. To construct dynamics for these systems, we have developed a method in which we replace the unitary transformation U by the star-unitary transformation Λ. It is appealing that the new representation of these systems bring in a number of new features such as dissipation and fluctuations that were hidden in the original representation in terms of classical trajectories or wave amplitudes. These features change the physics. Distribution functions or density matrices become irreducible to classical trajectories or wave amplitudes, and wave equations become stochastic. Moreover, one can introduce an entropy on a purely microscopic basis. Therefore for nonintegrable systems, physics is not determined by trajectories or amplitudes but by probabilities. Nevertheless, it is remarkable that the Λ transformation is invertible. We do not lose any information by the Λ transformation. We have demonstrated these properties of Λ without using any approximations in the Friedrichs model. This is a striking difference from the traditional approach for irreversible processes based on approximations such as coarse graining that resort to the limitation of human's controllability of nature.

The fluctuations are the consequence of nondistributivity of the Λ transformation. We need a new mathematical framework (i.e., nondistributive algebra) to analyze nonintegrable systems. This fact reminds us that whenever we found new aspects in physics, we needed new mathematical frameworks, such as calculus for Newton mechanics, noncommutative algebra for quantum mechanics, and the Riemann geometry for relativity.

We believe that our approach is only the starting point for new investigations on physics of nonintegrable systems. The horizon seems widely open for new results.

References

1. I. Prigogine, Dissipative processes, quantum states and field theory, XIV Conseil de Physique Solvay, *Fundamental Problems in Elementary Particle Physics*, Interscience Publishers, New York, 1968, p. 155.

2. I. Prigogine, F. Henin, and C. George, *Proc. Natl. Acad. Sci.* **59**, 7 (1968).

3. F. Henin, *Physica* **39**, 599 (1968).

4. P. Mandel, *Physica* **54**, 241 (1971).

5. I. Prigogine, C. George, F. Henin, and L. Rosenfeld, *Chem. Scripta* **4**, 5 (1973).

6. M. de Haan and F. Henin, *Physica* **67**, 197 (1973); M. de Haan, *Bull. Acad. Roy. Belg., Cl. Sci.* **63**, 69 (1977); *Bull. Acad. Roy. Belg., Cl. Sci.*, **63**, 317 (1977); *Bull. Acad. Roy. Belg., Cl. Sci.* **63**, 605(E) (1977).

7. C. George, F. Henin, F. Mayne, and I. Prigogine, *Hadr. J.* **7**, 520 (1978).

8. T. Petrosky and I. Prigogine, *Physica A* **147**, 439 (1988).

9. I. Prigogine and T. Petrosky, *Physica A* **147**, 461 (1988).

10. T. Petrosky, I. Prigogine, and S. Tasaki, *Physica A* **173**, 175 (1991).

11. G. Ordonez, T. Petrosky, and I. Prigogine, *Phys. Rev. A* **63**, 052106 (2001).

12. T. Petrosky, G. Ordonez, and I. Prigogine, *Phys. Rev. A* **64**, 062101 (2001).

13. T. Petrosky, G. Ordonez, and I. Prigogine, *Phys. Rev. A* **68**, 022107 (2003).

14. S. Kim and G. Ordonez, *Phys. Rev. E* **67**, 056117 (2003).

15. S. Tasaki, *Phys. Rep.* **219**, 110 (1992).

16. I. E. Antoniou, M. Gadella, E. Karpov, I. Prigogine, and G. Pronko, *Chaos, Solitons and Fractals* **12**, 2757 (2001).

17. B. A. Tay and G. Ordonez, *Phys. Rev. E* **73**, 016120 (2006).

18. U. Weiss, *Quantum Dissipative Systems* World Scientific, Singapore, 1983, Chapter 2.

19. T. Petrosky, *Int. J. Quantum Chem.* **98**, 103 (2004).

20. The numerical simulations were performed by Dr. Chu-Ong Ting.

21. T. Petrosky, and I. Prigogine, *Adv. Chem. Phys.* **99**, 1 (1997).

22. I. Prigogine and G. Ordonez, *Chaos and Complexity Lett.* **1**, 1 (2005).

STATISTICAL MECHANICS OF A
GRAVITATIONAL PLASMA

YVES POMEAU*

*Laboratoire de Physique Statistique de l'Ecole Normale Supérieure, 75231
Paris Cedex 05, France*

CONTENTS

I. INTRODUCTION

In a very influential paper, Prigogine and Severne [5] examined the problem of
building nonequilibrium statistical mechanics for a gas of point masses
interacting gravitationally. Following the schema of derivation of kinetic
equations that had been so successful in plasma physics, they derived a kinetic
equation for the short-time evolution of a gas filling homogeneously the whole
3D space with particles immobile and uncorrelated at the beginning. Such an
endeavor meets major difficulties, partly because of the lack of any small
parameter in the theory. However, it is possible to look at real astrophysical
situations dominated by gravitational interactions and where a good expansion
parameter exists, the inverse number of stars in the self-gravitating cluster.
This is the situation of "globular clusters," which we shall consider below.

Present Address: Department of Mathematics, The University of Arizona, Tucson, Arizona 85721,
USA.

Special Volume in Memory of Ilya Prigogine: Advances in Chemical Physics, Volume 135,
edited by Stuart A. Rice

In our Galaxy, as well as in others, populations of stars aggregate from place to place in the so-called "globular clusters." There the local number density of stars is far bigger than in the rest of the Galaxy except perhaps in its core.

References to this observation can be found in Ref. 2. To analyze those clusters, a thermodynamical approach has been tried, assuming the existence of some sort of thermal equilibrium in the gas of stars. Kinetic theory has been used also, by assuming each star to be a point mass interacting with the others via the gravitational attraction. Each approach has its drawbacks. Any standard thermodynamical approach of this problem rests on shaky ground because no thermal equilibrium exists for point masses interacting via Newton's law. Indeed almost the same thing could be said of a *classical* plasma of electrons and ions. But there short distance divergences are taken care of by the *quantum* effects, although the Debye screening cancels the long-range effects of the electrostatic interactions: Positive and negative charges cancel their effect at large scales in a globally neutral plasma. Neither phenomena have a counterpart for gravitating macroscopic masses. As far as we know, Newton's constant is the same everywhere with the same sign. Therefore we assume that gravitation is always attracting. The star–star interaction is certainly changed when the interstar distance becomes of the order of their diameter, but this is irrelevant for globular clusters: Even in their dense core, the distance between nearby stars is far bigger than their diameter.

As often in the study of astrophysical phenomena, a hard question is the one of the initial data: Nobody knows the kind of initial condition that leads to the globular clusters one observes now. In particular, it is not clear if they have anything to do with the so-called Jean's instability. Because of this instability, density fluctuations of a homogeneous gravitational gas without velocity tend to grow exponentially in the linear approximation. During the growth of this instability the initial potential energy is converted into kinetic energy, the problem studied by Prigogine and Severne [5]. It is unknown if, in the past, globular clusters started like swarms of stars with little or no kinetic energy in their relative motion and collapsed afterwards. Similar questions could be raised concerning the idea of "fast relaxation" according to which, at the beginning, globular clusters were in a non-equilibrium state that relaxed "quickly"— namely, over a time scale of the order of the time of growth of the Jean's instability. Therefore a careful (perhaps too careful) way of approaching this question is to assume that a globular cluster has the kind of structure we can observe now and to try to see what can be predicted for the evolution of such a self-gravitating system. This assumes somehow that, on the "short" time scale—that is, the period of the orbit of a star in the mean field of the others (about the time of growth of the Jean's instability)—the cluster is in Vlasov–Newton equilibrium (see Section II for a definition). Therefore its velocity/position statistical distribution is one out of the many stationary solutions of the

Vlasov–Newton equations for a self-gravitating system. This is not sufficient for finding the velocity/position probability distribution, given that the Vlasov–Newton equilibria depend on arbitrary functions. Furthermore, one knows that over the lifetime of the globular clusters (believed to be very old at the scale of the age of the Universe), a fair number of close collisions has taken place (in the range of 10 collisions per star) that should have changed very significantly the velocity distribution. "Close collision" or "collision" here must be understood in the sense of classical kinetic theory. By definition, this is a collision such that the two stars meet at a distance short enough for their trajectories to be scattered significantly. This is not to say that the two stars bump on each other by meeting at distances of order of their physical diameter (a "catastrophic" event). The collision in the first sense are far more frequent than the "catastrophic" ones and will be the one we shall refer to by "close collision" or "collision."

These collisions take place very rarely compared to the orbital period of a star in the field of the others (whence the possibility of a rational expansion). Typically, a star makes a collision with another one about once every N orbits in the mean field, N number of stars in the cluster. Therefore, it remains to write and solve the equations for this slow evolution of the Vlasov–Newton equilibria under the effect of the rare collisions. It is possible to write the kinetic theory for this collisional relaxation. The relevant equations are very complicated because of the constraint of conservation of energy and angular momentum [1, 3, 5], and it is hard to know what these equations predict. Moreover, it is not clear altogether that the binary collisions make the dominant process of relaxation. To understand this point, let us start from the Boltzmann kinetic equation with gravitational interaction and pair collisions only. As is well known, the kinetic operator diverges logarithmically because of the grazing binary collisions. In a plasma, this divergence is taken care of by the Debye screening at large distances that cancel the charge fluctuations. However, gravitational interactions are not screened at large distances. Therefore, it does not make any sense to put some Debye-like cutoff in the diverging logarithm. One has to go back to the full kinetic theory. This was done by Luciani and Pellat [1], who showed that the dominant process of relaxation is due to the resonant interaction between stars with periods of their orbital motion rationally related. This makes this system belong to what Prigogine called "the large Poincaré sytems," where relaxation is by the resonance between commensurate frequencies. Given the uncertainty in the true mechanism of relaxation, it makes sense to try to build a theory for the irreversible evolution of the self-gravitating cluster by retaining the dominant features of the system and staying as simple as possible without introducing any meaningless temperature and without trying to solve the very complex equations of kinetic theory. The theory presented below gives up the hope to know both the velocity and position distribution of the stars. The idea is to write a diffusion equation for the number density of stars, preserving both their total number (that is easy) and the total energy (less easy). The physical mechanism of diffusion is assumed to come from the collisions between individual stars, although it could be also the resonances between the frequencies of the orbits of the stars. To make things simpler, the binary collisions will be assumed to give the scaling laws for the relaxation.

We shall be precise regarding the way the various assumptions are made to derive our model. Let $\rho(\mathbf{r}, t)$ be the number density of a gas of stars, called later on the "gravitational plasma" following Prigogine and Severne [5]. All the stars have the same mass m, and ρ is normalized in such a way that

$$\int d^3\mathbf{r}\, \rho(\mathbf{r}, t) = N \tag{1}$$

In the equation above, N is the total number of stars, a constant, large but finite, and $\mathbf{r}$ is the position in the 3D space, with boldface representing vectors. For a given number density the potential energy of the system is

$$\mathcal{E}_{pot} = -\frac{Gm^2}{2} \int d^3\mathbf{r} \int d^3\mathbf{r}' \, \frac{\rho(\mathbf{r})\rho(\mathbf{r}')}{|\mathbf{r} - \mathbf{r}'|} \tag{2}$$

In the Eq. (2), G is the (positive) Newton's constant. Below we shall consider first the mean-field problem—that is, the one relevant for the short-time relaxation of the distribution. As it will appear, this is insufficient to yield an unique or even a restricted set of equilibria. Therefore, we shall use a more elaborate scheme to describe some sort of relaxation by irreversible process.

II.   VLASOV–NEWTON EQUILIBRIA: A SHORT SUMMARY

Before we present our idea, we give a short summary of the mean-field theory. The core concept there is the velocity ($\mathbf{v}$)–position ($\mathbf{r}$) probability distribution of stars, $f(\mathbf{r}, \mathbf{v}, t)$, a positive and integrable function that is *a priori* time-dependent. The number density $\rho(\mathbf{r}, t)$ is the integral over velocities of $f(\mathbf{r}, \mathbf{v}, t)$:

$$\int d^3\mathbf{v} f(\mathbf{r}, \mathbf{v}, t) = \rho(\mathbf{r}, t) \tag{3}$$

The Vlasov–Newton equation of motion of f is the closed equation:

$$\frac{\partial f(\mathbf{r}, \mathbf{v}, t)}{\partial t} + \mathbf{v}.\frac{\partial f(\mathbf{r}, \mathbf{v}, t)}{\partial \mathbf{r}} + \mathbf{\Gamma}(\mathbf{r}, t).\frac{\partial f(\mathbf{r}, \mathbf{v}, t)}{\partial \mathbf{v}} = 0 \tag{4}$$

In this equation, $\Gamma(\mathbf{r}, t)$ is the acceleration of a mass at position $(\mathbf{r})$ due to attraction by the other stars, in the mean-field approximation:

$$\Gamma(\mathbf{r}, t) = +Gm \int d^3\mathbf{r}' \, \rho(\mathbf{r}', t) \, \frac{(\mathbf{r}' - \mathbf{r})}{|\mathbf{r}' - \mathbf{r}|^3} \tag{5}$$

The Vlasov–Newton equation has many steady solutions describing a self-gravitating cluster. This is easy to show in the spherically symmetric case (the situation we shall restrict in this work, except for a few remarks at the end of this section). If one assumes a given $\Gamma(\mathbf{r})$ in the steady state, the general steady solution of Eq. (4) is a somewhat arbitrary function of the constants of the motion of a single mass in this given external field, namely a function $f(E, L^2)$ where mE is the total energy of a star in a potential $\Phi(r)$ such that $\Gamma(\mathbf{r}) = -(\mathbf{r}/r) \, [d\Phi(r)/dr]$ and where $L^2 = v^2 r^2 - (\mathbf{r} \cdot \mathbf{v})^2$ is the square of the angular momentum divided by the mass. If one assumes that the distribution depends on E only—that is, of $E = \frac{1}{2} v^2 + \Phi(r)$— one obtains an integral equation for Φ by writing Poisson's equation, the mass density being related to f by Eq. (3), adapted to the present case:

$$\frac{1}{r^2} \frac{d}{dr} \left(r^2 \frac{d\Phi}{dr} \right) = 16\pi^2 Gm \int_{\Phi}^{+\infty} dE f(E) \sqrt{2(E - \Phi)} \tag{6}$$

Again, this is an equation for the unknown function $\Phi(r)$, $f(E)$ being given. Because one has introduced functions depending on energy, one needs to define a reference value for this energy. We choose to have $\Phi \to 0$ as r tends to infinity, so that Φ is negative or zero. This implies in particular that the right-hand side of equation (6) becomes equal to zero as soon as Φ becomes positive. This requires that $f(E)$ is zero for E positive (if one puts $\Phi = 0$ on the right side, one gets an integral that can vanish only if $f(E) = 0$ for E positive, $f(.)$ being positive or zero in general). Therefore the integal on the right-hand side can be written as $\int_{\Phi}^{0} dE f(E) \sqrt{2(E - \Phi)}$.

It is of interest also to notice that the solution of Eq. (6) minimizes the following Euler–Lagrange functional with respect to variations of $\Phi(r)$:

$$\mathcal{L}[\Phi] = \int_0^\infty dr r^2 \left[\frac{1}{2} \left(\frac{d\Phi}{dr} \right)^2 - v(\Phi) \right] \tag{7}$$

In Eq. (7), $v(\Phi)$ is given by

$$v(\Phi) = \frac{32\pi^2 Gm}{3} \int_{\Phi}^{0} dE f(E) \, (2(E - \Phi))^{3/2}.$$

Although the sign of the $v(\Phi)$ term is opposite to the one of the term with a derivative, namely $\frac{1}{2}(d\Phi/dr)^2$, a true lower bound of the functional likely exists, because a large negative Φ does not make it decrease to minus infinity the functional, since the positive term is quadratic in Φ although the negative term is, for large negative Φ, of order $-|\Phi|^{3/2}$ and thus not dominant. A more tricky question is the uniqueness of the minimum of $\mathcal{L}[\Phi]$. Since this functional is bounded from below, we could always choose as a relevant solution the one with the lowest value of this functional.

Suppose Eq. (6) has a solution with the given asymptotic conditions, which holds true in a wide range of cases [2]; then one associates to a given $f(E)$ a solution of the steady state of the Vlasov–Newton equation. There are various restrictions on possible functions $f(E)$: It must be positive or zero and such that the total mass is finite. Of course, as we said, this is not enough to tell what function $f(E)$ is to be chosen. Moreover, knowing $\Phi(r)$, it is possible in principle to find the function $f(E)$ from Eq. (6) by writing the left-hand side as a function of Φ (instead of r). Then there remains to invert an Abel transform to get back $f(E)$. We shall comment now on the impossibility of applying the usual methods of equilibrium statistical mechanics to the present problem (that is, the determination of $f(E)$ from a principle of maximization of entropy for instance).

Another important issue is the stability of the steady solution(s) of the Vlasov–Newton equilibria. This is discussed in Ref. 2. For the case of spherically symmetric equilibria, the situation seems to be well understood. The case of nonspherical equilibria is far more complex, and to our opinion it is not well understood yet. Nonspherical equilibria are found when the globular cluster has some angular momentum. Then the steady distribution depends on the energy E as for the spherical case and on the component of the individual angular momentum of each star, namely the vector product $(r_x v_y - r_y v_x)$, r_x being the x-Cartesian component of the position, and so on, all this assuming that the conserved angular momentum is along the z-direction. The algebra for the condition derived from Poisson's equation becomes rather heavy, because the number density and the potential Φ now depend on $(r_x^2 + r_y^2)$ and on r_z^2, although they depend on the radial distance only for the spherical case. The motion of a single star in the axisymetric potential is not integrable anymore. Therefore one may expect coexistence of random and of quasiperiodic trajectories, depending on how close they are to the bottom of the minimum of the self-consistent potential. In particular, the chaotic trajectories are exponentially unstable, and one may believe that this instability induces an instability of the steady solution too. It could well be that steady smooth solutions of the Vlasov–Newton equations for this case are irrelevant and have to be replaced by solutions including a constraint that chaotic individual trajectories have no weight in the steady state, a constraint not compatible with a smooth probability distribution. Perhaps this has to do with some of the large-scale structures seen in Galaxies.

III. IRRELEVANCE OF CLASSICAL EQUILIBRIUM STATISTICAL MECHANICS FOR THE GRAVITATIONAL PROBLEM

This irrelevance can be seen in a number of ways. At the more fundamental level, one can try to use the standard Gibbs–Boltzmann approach. This meets several (we believe unsurmountable) difficulties. A Gibbs–Boltzmann statistical ensemble cannot be used because of divergences in the partition functions, the normalizing factor in front of the various statistical weights which depend on the ensemble one looks at. According to the principles of equilibrium statistical mechanics, the most "fundamental" ensemble is the microcanonical one at imposed total energy and number of particles. As argued in Ref. 4 the normalization factor of this microcanonical ensemble diverges as soon as it is more than three point masses in a box (notice that equilibrium statistical mechanics does not require us to take the thermodynamic limit, $N \to \infty$). The divergence is interesting in itself: The strongest divergence is related to the unlimited growth of the available volume in the momentum part of the phase-space where one pair of stars gets very close with a very negative potential energy. This divergence is the strongest with the smallest possible number of stars in a deeply bound state. Of course, in a discrete system, this smallest possible number is two. It does not make sense to discuss the case of the canonical ensemble (that is, to assume that the system is at a given temperature): Such ensembles are defined for systems exchanging energy with a thermostat. There is nothing like a thermostat around globular clusters. Moreover, the Boltzmann–Gibbs canonical distribution $e^{-U/k_B T}$ does not make any sense, U being the total energy, k_B the Boltzmann factor, and T the absolute temperature. The normalization factor (that is, the integral of the Boltzmann factor over space) diverges massively because of the negative contribution to the interaction energy coming from the interaction potential with very close pairs: It diverges like the integral over $\mathbf{r}_i$ and $\mathbf{r}_j$ of $e^{Gm^2/k_B T|\mathbf{r}_i - \mathbf{r}_j|}$, $\mathbf{r}_i$ position of star i. Because this yields a positive contribution to $(-U/k_B T)$, the short-distance divergence of the integral defining the normalization is dramatic. It has been argued that a short-distance cutoff can take care of this divergence, which is formally true. However, this is irrelevant for globular clusters. Once the partition function is made convergent thanks to this cutoff, the main statistical weight is for configurations that would make the partition function diverge without cutoff—that is, for stars accumulating at interparticle distances of the order of this arbitrary cutoff. This would be consistent with a hard-core repulsive interaction at the cutoff distance, all things without counterpart in the real situation of globular clusters.

One could try also a less direct application of the idea of equilibrium statistical mechanics. A macroscopic thermodynamical approach, without recourse to the underlying statistical mechanics, does not seem possible because it requires a concept of temperature that is meaningless for this problem. Another approach is to try to plug into Eq. (6) a form of the energy distribution $f(E)$ more or less inspired from similar expressions for thermodynamic systems (although the need of a function that is exactly zero for positive energies makes it hard to justify a function $f(E)$ looking vaguely as its counterparts in standard statistical mechanics). Moreover, as we have shown, this function, at least for the mean-field model, can be chosen rather freely without hurting any basic principle.

Indeed the right way to derive an expression of $f(E)$ compatible with the irreversible process occurring during the collisions is to write the corresponding kinetic equation and to solve it in one way or another (here and below we mean by "collision" the process that bring irreversibe changes in $f(E)$, be it due to two-body collisions in the usual sense or to a more complex interaction between resonant trajectories). This is a very arduous task because even a simplified form of the kinetic equations with collisions is very complex. Below we present an alternative approach, half-way between the full kinetic theory and a more macroscopic approach. It allows us to draw conclusions without recourse to an ill–defined temperature. In particular, it permits us to discuss rather straightfor-wardly questions regarding the existence of stationary solutions and finite time singularity in the evolution problem.

An interesting issue is the one of irreversibility. The full kinetic theory including two-body collisions surely obeys a kind of H-theorem, showing the growth of a form of entropy. However, one cannot conclude from this that there should be ultimately relaxation toward thermodynamic equilibrium. This is primarily because there is no such thing as a thermal equilibrium for this problem. Furthermore, at least with binary collisions, a crucial effect is missing—that is, the formation of Keplerian pairs with a negative energy heating the rest of the gas by transforming this potential energy into kinetic energy, a process favored by the increase of volume in phase space. This kind of process is completely lacking if one keeps two-body collisions only and if one starts from single particles. Indeed Keplerian pairs are formed by triple close collisions. But these collisions are even rarer than the binary collisions, by a factor of $1/N$. Practically, if there has been 10 close collisions per star during the lifetime of the cluster, no triple collision per star took place on the same duration. Therefore the time evolution we are looking at is such that the system cannot relax to its most probable state, but to some intermediate state, without significant influence of the triple collisions and of the transfer of energy from Keplerian pairs to kinetic motion. It seems reasonable anyway to believe that once the triple collisions have enough relevance (and so enough pairs are created), the cluster will evaporate, with (a) all the interaction energy stored into Keplerian pairs and (b) a kinetic energy far above the escape energy in the cluster. This ultimate stage of the evolution will be reached after a time of order N^2 times the orbital period, although we investigate time scales of order of the orbital period times N, much shorter than the evaporation time. Therefore, in this intermediate range, very little information can be obtained from considerations of volume of phase space, which is dominated by Keplerian pairs and not yet formed at this stage of the evolution.

IV. DERIVATION OF AN EINSTEIN–SMOLUKOWSKI-LIKE EQUATION

In this section we describe another approach to the problem of relaxation of a cluster of stars in gravitational interaction. We borrow ideas of chemical physics, particularly the one of relaxation of density toward equilibrium by collisions in a given external potential, as described by the Einstein–Smolukowski [6] equations. Besides obvious differences of scales between molecular process and globular clusters, other important differences are also relevant: First, the potential of a self-gravitating cluster is not an external potential but is instead created by the distribution of stars in space itself. There is some analogy there with the Poisson–Boltzmann theory of electrolytes describing the screening of a charge by a Debye cloud in thermal equilibrium. But for the globular cluster there is no thermostat and the concept of temperature is meaningless. Therefore, one has to adapt the general concept of diffusion in an external potential to this special case. This is done below.

Because the total number of stars is constant, the number density satisfies an equation of the form

$$\frac{\partial \rho}{\partial t} + \nabla . \mathbf{j} = 0 \tag{8}$$

We face now the standard closure problem: To keep going, one should relate in some way or another the current $\mathbf{j}$ to the number density. The usual relationship is Fick's law:

$$\mathbf{j} = -D \nabla \rho \tag{9}$$

This equation with D positive represents the tendency of stars to spread under the effect of collisions. Such a tendency surely exists in the case of globular clusters. Each star in the mean field of the others remains on an orbit that is bounded and even integrable if the cluster is spherically symmetric. Collisions bring the star to another orbit and thus yield some diffusion in the position space. This process is constrained by the fact that the total energy of the cluster is constant. This constraint is not met by the simple Fick's law (9). The other difficulty there is somewhat hidden; namely, what is conserved is the total energy, kinetic plus potential. Fortunately, it does not make any difference here to consider separately the conservation of kinetic plus potential energy, or the conservation of potential energy only. At the dominant order in the small parameter $1/N$, each star moves in the mean field of the others and so its trajectory satisfies the virial theorem, making proportional to each other the kinetic and potential energy with a constant ratio (specifically the kinetic energy on average is $-1/2$ the potential energy for gravitational interactions, a classical result [9]). Therefore, the energy of the system is well represented by the potential energy, itself a function(al) of the density. This prevents us from defining a temperature in the usual sense, because there is no such thing and because all parameters of the gravitating gas can be scaled with the potential energy.

Let us define the potential energy per unit mass, a space-dependent quantity:

$$\Phi(\mathbf{r}) = -G m \int d^3 \mathbf{r}' \, \frac{\rho(\mathbf{r}')}{|\mathbf{r} - \mathbf{r}'|} \tag{10}$$

Later on we shall call Φ "the energy density." The integral over space of $\Phi(\mathbf{r})$ times $\frac{m}{2}\rho(\mathbf{r})$ gives the total potential energy [Eq. (2)]:

$$\mathcal{E}_{\text{pot}} = \frac{m}{2} \int d^3 \mathbf{r} \, \Phi(\mathbf{r}) \, \rho(\mathbf{r})$$

The constraint that this potential energy (actually the full energy, potential and kinetic, by the virial theorem) is conserved can be written as a constraint that the potential energy, a functional of $\rho(\mathbf{r})$ because of Eq. (10), does not change when the number density itself evolves according to Eq. (8). Putting the expression of $\partial\rho/\partial t$ given in this equation into the time derivative of the potential energy, one obtains (after one integration by part) that the time derivative of this potential energy vanishes if

$$\int d^3 \mathbf{r} \, \mathbf{j}(\mathbf{r}) . \nabla \Phi(\mathbf{r}) = 0 \tag{11}$$

This condition expresses that the mass flux has to preserve the potential energy. The simplest way to account for this is to add to Fick's law another gradient term representing the tendency for the stars to drift to a deeper potential, a bit like in Einstein–Smolukowski law of diffusion in an external potential. This involves adding to Fick's law a contribution proportional to the gradient of the energy density. This changes Eq. (9) into

$$\mathbf{j} = -D\nabla\rho - K\rho\nabla\Phi \tag{12}$$

This introduces a new unknown (and free at this stage) coefficient K. In the case of a system of molecules in a thermal bath (definitely *not* the one we consider), there is a relation between D and K such that, at thermal equilibrium, the equilibrium density in the potential Φ is given by Boltzmann's law. This requires that $K = mD/k_B T$, where k_B is Boltzmann's constant and T the absolute temperature. In Eq. (12) the factor ρ in front of $\nabla\Phi$ in $K\rho\nabla\Phi$ is to ensure that, if ρ is initially positive, it remains so later on: If ρ becomes close to zero somewhere, the contribution $-K\rho\nabla\Phi$ becomes very small, and the first term $-D\nabla\rho$ brings back ρ to positive values.

At this step, one should resist any attempt to put there a constant D and/or K. These diffusion coefficients have a physical dimension that must be related to the parameters of the gravitational plasma.

One can check first that, at least in a simple situation, positive values for K and D yield what is expected. The coefficient D has to be positive, to make the diffusion equation well-behaved for the evolution toward positive times. Suppose that the density of the cluster depends on the radius only and that it decreases from the center to infinity and that the energy density grows from the center to infinity (it grows because it goes from negative values to zero, although its absolute value is expected to decreases when r goes from zero to ∞). Then the contribution $-D\nabla\rho$ to the mass flux is directed outward, although $-K\rho\nabla\Phi$ is directed inward, if K is positive, because $\nabla\Phi$ points outward.

It remains to relate the value of D to the local parameters of the gravitational plasma, ρ and Φ. This is done as follows. To estimate the coefficients D and K, we assume that they represent the effect of binary collisions in the ordinary sense, although we have seen that these close collisions are likely only one contribution to the relaxation process, the other one being the resonant interaction. At the moment there is no good understanding of the way these resonant interactions work, so we shall not take them into account. A diffusion coefficient like D is the ratio of the square of the space shift induced by a collision to the mean-free flight time. Here collision is meant as an encounter between two stars such that the trajectories are significantly disturbed. The frequency of binary collisions is, as usual, the ratio of the velocity to the mean-free path. The order of magnitude of the velocity is given by the square root of the ratio of the (kinetic) energy to the mass $\sqrt{|\Phi|} = \sqrt{-\Phi}$, this taking into account that the kinetic energy is of the same order of magnitude as the potential energy because of the virial theorem. The mean-free path (that is not computed here along a straight line) is the distance run by a star between two close collisions. It is given by the standard expression $1/\rho a^2$, where a is the impact parameter, a length. This length is found by imposing that the distance a is such that the two-body interaction at this distance is of the same order as the energy $m\Phi$. This gives $a \sim Gm/|\Phi|$. Therefore the mean-free flight time is

$\tau \sim |\Phi|^{3/2}/\rho G^2 m^2$. Each close collision switches the orbit by a distance of the order of magnitude of the size of this orbit. Let l be this length. It is such that the energy of the star $m\Phi$ is $G\rho l^3 m^2/l$, where $m\rho l^3$ is the mass of stars contributing to the mean field defining this orbit. Therefore $l^2 \sim |\Phi|/Gm\rho$. This relates the potential energy of this orbit to ρ and Φ. If the number density of the cluster is more or less uniform inside this cluster, this would give $\rho l^3 \sim N$, the total number of stars in the cluster. Putting the local density inside the expression of the energy leaves the possibility of describing a core with a higher density than the rest of the cluster. Therefore the diffusion coefficient D is now

$$D \sim \frac{l^2}{\tau} = C_D \frac{Gm}{|\Phi|^{1/2}} \tag{13}$$

In this equation, C_D is a dimensionless positive number. This expression for D has the right physical dimension, length square divided by time: Φ scales like a velocity square, although Gm scales like a velocity square times a length. The coefficient K, as introduced into Eq. (12), is found by dimensional reasoning too. Because it appears in front of $(-\rho \nabla \Phi)$ that contributes to the same diffusion current as $-D \nabla \rho$, the two contributions to the flux $\mathbf{j}$ should be of the same order of magnitude for each collision event. This is realized by taking

$$K \sim \frac{D}{|\Phi|} = C_K \frac{Gm}{|\Phi|^{3/2}} \tag{14}$$

In Eq. (14), C_K is another dimensionless number. Up to an overall multiplicative constant the coefficients C_D and C_K are derived now by imposing that the energy is conserved. The relevant condition is written in Eq. (11). Inserting into this equation the expression of $\mathbf{j}$ given in Eq. (12) and using the expressions for D and K given in (13, 14), one obtains the condition equivalent to the conservation of energy:

$$C_D < \frac{\nabla \Phi \cdot \nabla \rho}{|\Phi|^{1/2}} > + C_K < \frac{\rho (\nabla \Phi)^2}{|\Phi|^{3/2}} > = 0 \tag{15}$$

In this equation, we introduced the notation $< F(\mathbf{r}) > = \int d^3 \mathbf{r} \, F(\mathbf{r})$. Equation (15) is a single equation for two unknowns, C_D and C_K. Therefore they depend on a single arbitrary (positive) constant C such that

$$C_D = C \tag{16}$$

and

$$C_K = C\gamma \tag{17}$$

where the dimensionless number γ depends on the density distribution in the cluster and is defined by

$$\gamma = -\frac{< \frac{\nabla\Phi \cdot \nabla\rho}{|\Phi|^{1/2}} >}{< \frac{\rho(\nabla\Phi)^2}{|\Phi|^{3/2}} >} \tag{18}$$

Let us look first at the convergence of the integrals that enter into this last expression of C_K or in Eq. (15) and then at the sign of D and K.

Concerning the convergence, consider the expression $< \rho(\nabla\Phi)^2/|\Phi|^{3/2} >$. Recall that the notation $< \ldots >$ is for the integral over the whole space of the quantity inside the brackets. At large distances, Φ tends to zero like $1/r$. Therefore the integrand decays at least like $r^{-3/2}\rho(r)$. Since the integral of ρ itself should converge at r large, the integral of ρ times a function decaying like $r^{-3/2}$ converges too at infinity. Consider now the other quantity in brackets occuring in the equations for C_D and C_K. It reads $< \nabla\Phi \cdot \nabla\rho/|\Phi|^{1/2} >$.

Since Φ decays like $1/r$ at large r, the integrand $\nabla\Phi \cdot \nabla\rho/|\Phi|^{1/2}$ decays like $r^{-3/2}\nabla\rho$ for r large. By integration by part over the radius, one can see easily that, if the integral of ρ converges, so does the integral of $r^{-3/2}\nabla\rho$ for r large.

Let us discuss now the sign of D and K. From all the previous considerations, we obtain for D

$$D = C\frac{Gm}{|\Phi|^{3/2}} \tag{19}$$

This is obviously positive for C positive. It has also the interesting property that, on the edge of the cluster where $|\Phi|$ tends to zero, the diffusion becomes more effective, because D becomes very large. This is a reasonable property of this model, since one expects elliptical orbits close to parabolic and thus large diffusion.

The case of the coefficient K is less obvious. From all the calculations above, K is

$$K = C\gamma\frac{Gm}{|\Phi|^{1/2}} \tag{20}$$

If ρ decreases continuously from $r = 0$ to smaller values and if Φ, a negative quantity, increases from a negative minimum at $r = 0$ to zero at infinity, then the scalar product $\nabla\Phi \cdot \nabla\rho$ and $(-\gamma)$ are both negative and K is positive. To show that γ is positive in general, one notices that γ has the same sign as

$$\gamma_{\text{num}} = - < \frac{\nabla\Phi \cdot \nabla\rho}{|\Phi|^{1/2}} >$$

By integration by part, one finds

$$\gamma_{\text{num}} = \left\langle \frac{\rho \nabla^2 \Phi}{|\Phi|^{1/2}} \right\rangle + \left\langle \rho \nabla \Phi \cdot \nabla \left(\frac{1}{(-\Phi)^{1/2}} \right) \right\rangle$$

The second term, $< \rho \nabla \Phi \cdot \nabla \frac{1}{(-\Phi)^{1/2}} >$ is easily shown to be equal to

$$< \frac{\rho}{2} \frac{1}{(-\Phi)^{3/2}} (\nabla \Phi)^2 >$$

and so is positive. The first one is $< \frac{\rho \nabla^2 \Phi}{|\Phi|^{1/2}} >$. From Poisson's equation $\nabla^2 \Phi = 4\pi G m \rho$, a positive quantity. Therefore, γ is positive in general, as expected.

We have found now an equation for the evolution of the density inside the cluster without any uncontrolled parameter, except for the dimensionless number C. Below we shall do two things. First, in Section V, we shall find the steady solutions for the density, that turns out to transform into a quite simple problem, mathematically equivalent to the equilibrium of self-gravitating atmosphere. Then, in Section VI we shall look at the possible existence of finite time singularities in the dynamical problem.

V.   STEADY SOLUTIONS OF THE DYNAMICAL EQUATION

An obvious question arises now: are they stationary solutions with finite mass and energy of the diffusion Eq. (8)?

In a steady state the flux $\mathbf{j}$, as given by Eq. (12) must be zero. With our expressions for D and K, this amounts to impose

$$C_D \frac{Gm}{|\Phi|^{1/2}} \nabla \rho + C_K \frac{Gm}{|\Phi|^{3/2}} \nabla \Phi = 0 \tag{21}$$

In Eq. (21), C_D and C_K are the coefficients defined in Eq. (16) and (17). Therefore one can simplify the condition of steady state into

$$\left\langle \frac{\nabla \Phi \nabla \rho}{|\Phi|^{1/2}} \right\rangle \frac{\nabla \Phi}{|\Phi|} = \left\langle \frac{|\nabla \Phi|^2 \rho}{|\Phi|^{3/2}} \right\rangle \frac{\nabla \rho}{\rho} \tag{22}$$

This equation has a simple solution (recall that the averages $< \ldots >$ yield constants, independent on r):

$$\frac{\rho(r)}{\rho_0} = \left(\frac{|\Phi(r)|}{\Phi_0} \right)^\gamma \tag{23}$$

To derive the solution (23), one has to pay attention to the fact that Φ is negative so that $|\Phi| = -\Phi$, which explains why one does find the exponent $+\gamma$, not $-\gamma$, in the final result (I am glad to thank Francis Corson for this important remark). In the solution we introduced two positive constants of integration, ρ_0 and Φ_0, although there is only one such constant. This was for the purpose of making the result dimensionless. One expects that $|\Phi|$ behaves at r large as the potential of a single mass—namely, like $\frac{1}{r}$. Therefore, to make the density distribution [as given by Eq. (23)] of finite total mass, it is necessary to have $\gamma > 3$.

A remarkable result is that γ is defined self-consistently and is a free parameter for the steady solutions. To show this point, let us introduce dimensionless quantities with overlines:

$$\overline{\rho} = \frac{\rho}{\rho_0}$$

$$\overline{\Phi} = \frac{\Phi}{\Phi_0}$$

Therefore Poisson's equation for Φ becomes

$$\nabla^2 \overline{\Phi} = 4\pi \frac{Gm\rho_0}{\Phi_0} |\overline{\Phi}|^\gamma \tag{24}$$

This can be transformed into a (numerical) parameterless equation by taking as unit length $\sqrt{\left(\frac{\Phi_0}{Gm\rho_0}\right)}$. This yields

$$\nabla^2 \overline{\Phi} = 4\pi |\overline{\Phi}|^\gamma \tag{25}$$

For spherically symmetric equilibria, such as the one we are looking at, this becomes the following numerical ODE (ordinary differential equation):

$$\frac{d}{dr}\left(r^2 \frac{d\overline{\Phi}}{dr}\right) = 4\pi \, r^2 |\overline{\Phi}|^\gamma \tag{26}$$

This equation has a dilation symmetry: If $\Phi(r)$ is a solution, $\lambda^{-\frac{2}{\gamma-1}}\Phi(r\lambda)$ is also a solution, for λ arbitrary real positive constant. Therefore one can fix arbitrarily a parameter in the solution, which may be (for instance) the value of $\Phi(r)$ for $r = 0$ that can be (-1). This defines uniquely a solution of the ODE (26). It could seem that there is too little freedom in the solution obtained in this way: If one had only one free parameter, one could only impose the total number of stars, N, without being able to change the energy. Actually this is not so, because there is a continuum of solutions, parameterized by the exponent γ. This exponent appears in the numerical equation (26). Once this equation is solved, the result should be consistent with the definition of γ in Eq. (18). If one replaces ρ by $(-\Phi)^\gamma$, one finds that the numerator is

$$\left\langle \frac{\nabla\Phi \cdot \nabla\rho}{|\Phi|^{1/2}} \right\rangle = -\gamma\langle \nabla\Phi \cdot \nabla\Phi|\Phi|^{\gamma-3/2}\rangle$$

exactly the same expression as the denominator, up to a factor γ. Therefore the condition that defines γ is always satisfied by a solution such that ρ is proportional to $(-\Phi)^\gamma$ and γ is a free parameter in the interval $[3, \infty)$. This is the second free parameter necessary to set a prescribed value for the total (negative) energy.

Equation (26) has been studied [7] as a model of equilibrium of gaseous stars under their own gravity field, where the density and pressure are related by a given power law depending on the properties of the gas. There is no physically admissible solution for γ between 3 and 5. For $\gamma = 5$ the problem is integrable. Numerical solutions have been computed for γ larger than 5. In the coming subsection, one explains how to derive a solution in the limit γ large positive.

A.   Solution of Eq. (26) in the Limit $\gamma \to \infty$

The equation we shall solve in the limit γ large positive is

$$\frac{d}{dr}\left(r^2 \frac{d\overline{\Phi}}{dr}\right) = 4\pi\, r^2 |\overline{\Phi}|^\gamma \qquad (27)$$

The boundary conditions are $\overline{\Phi}(0) = -1$ and $d\overline{\Phi}/dr = 0$ at $r = 0$, although $\overline{\Phi}(r)$ should tend to zero at infinity. As soon as $\overline{\Phi}$ differs significantly from (-1), it has to have a value somewhere between -1 and 0. Therefore the right-hand side of Eq. (27) becomes negligible there compared to the left-hand side for γ large. This gives the idea to split the range of values of r into two pieces, one where $\overline{\Phi}$ remains close enough to -1 to make both sides of (27) of the same order of magnitude and another domain farther from $r = 0$ where the left-hand side is dominant because $|\overline{\Phi}|$ is significantly less than 1. In the first domain, let us write $\overline{\Phi}$ as $\overline{\Phi} = -1 + \epsilon(r)$ where $\epsilon(r)$ is unknown but positive and close to zero. Plugging this into the equation, one obtains

$$-\frac{d}{dr}\left(r^2 \frac{d\epsilon}{dr}\right) = 4\pi r^2 e^{\gamma \ln(1-\epsilon)} \qquad (28)$$

One can approximate $e^{\gamma \ln(1-\epsilon)}$ in the limit ϵ small by $e^{-\gamma\epsilon}$. One may now rescale ϵ by introducing $\tilde{\epsilon} = \gamma\epsilon$, which gets rid of the large γ in the exponent. The resulting

γ factor in the equation is eliminated itself by rescaling the radius as $\tilde{r} = r\sqrt{\gamma}$. This yields the parameterless equation:

$$-\frac{d}{d\tilde{r}}\left(\tilde{r}^2 \frac{d\tilde{\epsilon}}{d\tilde{r}}\right) = 4\pi\,\tilde{r}^2 e^{-\tilde{\epsilon}} \tag{29}$$

This has a unique solution because there are two conditions at $\tilde{r} = 0$. For the function $\tilde{\epsilon}(\tilde{r})$, the conditions are $\tilde{\epsilon}(\tilde{r}) = d\tilde{\epsilon}(\tilde{r})/d\tilde{r} = 0$ for $\tilde{r} = 0$. This yields that near $\tilde{r} = 0$, the beginning of the Taylor expansion of $\tilde{\epsilon}$ is $\tilde{\epsilon} \approx -\frac{2\pi}{3}(\tilde{r})^2 + \dots$. The solution for $\tilde{\epsilon}$ becomes singular for $\tilde{r} = \tilde{r}_0$ like $2\ln(|\tilde{r} - \tilde{r}_0|)$, and cannot be extended beyond this value. This is where the transition occurs between values of $\overline{\Phi}$ close to (-1) and values close to zero. This is expected to be a very narrow range of values of r where one can neglect the variation of r in Eq. (27) and take it as constant. In this range the equation becomes

$$\frac{d^2\overline{\Phi}}{dr^2} = 4\pi\,(-\overline{\Phi})^{\gamma} \tag{30}$$

This is an integrable problem, with the solution

$$\int_{-1}^{\overline{\Phi}} \frac{d\overline{\Phi}}{[1 - (-\overline{\Phi})^{(\gamma+1)}]} = (r - r_0)\sqrt{\frac{8\pi}{\gamma + 1}} \tag{31}$$

This completes the solution of the present problem as we have the solution near $r = 0$ and at any finite value of r. To summarize, in a neighborhood of $r = 0$ of width of order $\gamma^{-1/2}$ the solution $\overline{\Phi}(r)$ is close to (-1) and in an exponentially narrow interval near $r = r_0$ it jumps from (-1) to practically zero.

VI.   DYNAMICAL QUESTIONS

The next (and much more difficult) question is the stability of this solution. This is a complex issue because the coefficients of the diffusion equation depend on the solution itself. To summarize the full dynamical problem, we look at the stability of steady solutions of the dynamical problem:

$$\frac{\partial\rho}{\partial t} = C_D \nabla \frac{1}{|\Phi|^{3/2}}\,(\gamma\rho\nabla\Phi - \Phi\nabla\rho) \tag{32}$$

where Φ is negative and related to ρ by the Eq. (10):

$$\Phi(\mathbf{r}) = -Gm \int d^3\mathbf{r}' \frac{\rho(\mathbf{r}')}{|\mathbf{r} - \mathbf{r}'|} \tag{33}$$

Another writing of Eq. (32) is

$$\frac{\partial \rho}{\partial t} = C_D \nabla ((-\Phi)^{\gamma - 1/2} \nabla (\rho (-\Phi)^{-\gamma})) \tag{34}$$

Recall too that the constant γ depends on ρ and Φ as given in Eq. (18). Even the writing of the equations for linear stability is already cumbersome, so that the stability analysis is possible only numerically, something we postpone to future work.

Another interesting issue is the possibility of self-similar collapse. According to Zel'dovich [8] there are two kinds of collapse, related to what he calls self-similarity of the first and second kind. In the present problem, self-similarity of the first kind implies that in the singular region there is a constant amount of either of the conserved quantities. This can be mass or energy. Such a collapse cannot imply both the density and the energy density Φ, since their physical dimensions are different. Collapse of a finite amount of mass to a single point is impossible because this would bring locally an infinite negative energy. In the present model this is obviously impossible because, if the mass distribution outside of the collapse remains smooth, it has a finite negative energy so that the total energy cannot be conserved. However, a self-similar collapse of the first kind with accumulation of a finite negative energy at a single point, with no mass, seems possible in this model, as was imagined in Ref. 4. This collapse is analyzed as follows. One assumes first that the total energy is conserved in the collapse region. Let this energy be $\mathcal{E}$ and let $R(t)$ be the typical radius of the collapse region, expected to scale like a positive power of $(-t)$, where $t = 0$ is the collapse time. Let us write as $[\rho]$ the typical value of the density inside the collapse domain, a power function of $(-t)$ to be found too. The energy per unit mass in the same region scales like $[\Phi] = \mathcal{E}/[\rho]R^3$, although $\mathcal{E}$ scales like $[\rho]^2 R^5$. Therefore $[\rho] = \mathcal{E}^{1/2} R^{-5/2}$ and $[\Phi] = [\rho]R^2 = \mathcal{E}^{1/2} R^{-1/2}$. Consider now the equation of diffusion. The left-hand side scales like $[\rho]/[t]$, where $[t]$ is for the (short) time scale before collapse. The first diffusion term, $\nabla(D\nabla\rho)$, scales like $[D]([\rho]/R^2)$. From Eq. (13), $[D] = [\Phi]^{-1/2}$. Putting together all the scaling relations, one obtains

$$R = [t]^{4/7} \mathcal{E}^{-1/7}$$
$$[\rho] = [t]^{-10/7} \mathcal{E}^{1/2}$$

and

$$[\Phi] = \mathcal{E}^{4/7} [t]^{-2/7}$$

Because of the way it was derived, the last term in the diffusion equation due to the gradient of Φ has the same scaling laws as the first one. Therefore it does not change quantitatively the scaling behavior we have just found. Indeed, as expected, the total mass inside the collapsing domain does tend to zero near the collapse time. Specifically, the total mass scales like $[\rho]R^3 = [t]^{2/7}\mathcal{E}^{1/14}$. The mass has to be lost to infinity by an outward flux. Another significant remark is that the coefficient γ, as defined by Eq. (18), is invariant under multiplication of the various quantities like ρ, Φ, and the unit length. Therefore one expects it to reach a constant value in the self-similar regime. It does not seem to be possible to find any simple expression for this coefficient, which has no reason to be free, as γ for steady solutions. The equations for the self-similar collapse are obtained as usual by inserting into the original equation a guessed form of the solution like $(-t)^{\alpha}F(r(-t)^{\beta})$, where α is an exponent that depends of the quantity under consideration, although, according to our estimate, $\beta = -\frac{4}{7}$. This makes a new set of ODE, now equivalent to the dynamical problem, with a single variable $\tilde{r} = r(-t)^{\beta}$ instead of the two original variables, r and t, for the spherically symmetric problem. The coefficient γ has to be found self-consistently, by solving the similarity equation first for an arbitrary γ and then computing γ from this solution and by imposing at the end that the value of γ is the one we started from.

Let us now look at the self-similar collapse of the second kind. Although this seems never stated explicitly in the literature on the topic, it seems to be what is found in the numerical simulation based on the solutions of the kinetic equations. Generally speaking, the numerical discovery of collapsing solutions of nonlinear partial differential equations or of integrodifferential equations as in the kinetic theory of the gravitational plasma must be considered with caution. To take an example, despite years of research and tremedous efforts, there is no general agreement yet on the occurrence or not of a singularity in the time evolution of the equations of 3D inviscid incompressible fluids. The explanation of this fact is well known: All concrete numerical schema have reasons to be unstable and to yield overflows because they lead to iterate many multiplications with a result growing very quickly as soon as it deals with numbers bigger than one! Another issue related to this possibility of self-similar solutions of the second kind in Zel'dovich classification is that they may have a quite small effect on the evolution of the system at later times. A singular solution of the second kind would have no mass and no energy in its core. This makes it more a mathematical singularity than a physical one. What actually matters is what happens afterwards. One may conjecture that this singularity, after its inception, becomes a sink of energy. This does not seem to be possible for simple reasons of scaling. The flux of energy $\mathbf{j}_{\Phi}$ scales like $\Phi\mathbf{j}$, where $\mathbf{j}$ is the mass flux as defined before. A constant flux toward a point should scale as the inverse square of the distance to this point. Therefore a constant flux of (negative) energy toward a point should scale like $\mathbf{j}_{\Phi} \sim c/d^2$, where c is a constant (independent on d, distance to the point). Suppose that ρ depends also on d with a (negative)power law. From the definition of Φ, it scales near $d = 0$ like $\Phi \sim \rho d^2$. The flux of matter scales like $D\frac{\rho}{d}$ and j_Φ scales like $D\rho^2 d^2/d$. Therefore the condition of constant flux toward the single point yields $\rho \sim d^{-3/2}$. This is not compatible with a finite total energy: Near such a singularity the energy scales like $\rho^2 d^2$, a quantity diverging like d^{-1} as d tends to zero if ρ is of order $d^{-3/2}$. Therefore, even if a singularity of the second kind exists, it cannot become a constant sink for the energy because the scalings yield a diverging potential energy at the singularity. Therefore a finite time singularity, if it exists, shall be physically irrelevant. The generality of the scaling argument against the singular region acting as a constant sink of energy could make it valid too for the full kinetic theory.

To conclude this contribution to the special issue of *Advances in Chemical Physics*, I would only recall that this question of the statistical mechanics of systems with gravitational interaction remained of interest to Professor Prigogine until the very end of his life, as it was the topic of an interesting and lively discussion between the two of us in June 2001 at a scientific meeting at Les Treilles in Southern France.

References

1. J. F. Luciani and R. Pellat, Kinetic equation of finite Hamiltonian systems with integrable mean field, *J. Phys.* **48**, 591 (1987).

2. J. Binney and S. Tremaine, *Galactic Dynamics*, Princeton University Press, Princeton, NJ, 1987.

3. H. Kohn, Numerical integration of the Fokker–Planck equation and the evolution of stars clusters, *Astrophys. J* **234**, 1036 (1979).

4. M. L. Chabanol, F. Corson, and Y. Pomeau Statistical mechanics of point particles with a gravitational interaction, *Europhys. Lett.* **50**, 148 (2000).

5. I. Prigogine and G. Severne, On the statistical mechanics of a gravitational plasma, *Physica* **32**, 1376 (1966).

6. M. Smolukowski, Drei Fortrage uber Diffusion, *Phys. Zeitschrift* **17**, 557 (1916).

7. D. Heggie and P. Hut *The Gravitational Million-Body Problem*, Cambridge University Press, Cambridge, 2003.

8. G. I. Barenblatt and Ya. B. Zel'dovich, Self-similar solutions as intermediate asymptotics, *Annu. Rev. Fluid Mech.* **4**, 285 (1972).

9. L. Landau and E. Lifchitz, paragraph 10 in *Mécanique*, Ed. Mir, Moscow, 1960.

INVERSE PROBLEMS FOR REACTION-DIFFUSION SYSTEMS WITH APPLICATION TO GEOGRAPHICAL POPULATION GENETICS

MARCEL O. VLAD

Department of Chemistry, Stanford University, Stanford California 94305, USA; and Institute of Mathematical Statistics and Applied Mathematics, Casa Academiei Romane, 76100 Bucharest, Romania

JOHN ROSS

Department of Chemistry, Stanford University, Stanford, California 94305, USA

CONTENTS

I. INTRODUCTION

Reaction-diffusion systems have been studied for about 100 years, mostly in solutions of reactants, intermediates, and products of chemical reactions [1–3]. Such systems, if initially spatially homogeneous, may develop spatial structures, called Turing structures [4–7]. Chemical waves of various types, which are traveling concentrations profiles, may also exist in such systems [2, 3, 8]. There are biological examples of chemical waves, such as in parts of glycolysis, heart

Special Volume in Memory of Ilya Prigogine: Advances in Chemical Physics, Volume 135,
edited by Stuart A. Rice

tissue, and slime mold cells. A substantial literature exists on these subjects: theoretical, experimental, and computational [9].

A parallel development, but almost independent, is that of the analogue of reaction-diffusion systems in population genetics: births, deaths, and migrations in populations, and mutations in their genetics. Fisher's fundamental theorem of natural selection [10] showed that for single locus genetic systems with pure selection, the rate of variation of the average population fitness equals the genetic variance of fitness. This result has been generalized to time and space-dependent systems [11].

This chapter, which is part new and review, presents a general type of inverse problem for reaction-diffusion systems and other reaction-transport systems which consists in the determination of the time and position of occurrence of an initial event, such as the initiation of an explosive chemical reaction or the occurrence of a genetic mutation in humans from the current space distribution of a concentration or a population density field which is accessible through observation or experiment. The method developed here is related to our previous research on response theory. The responses in time in spatially homogeneous as well as inhomogeneous (reaction-transport) systems has been investigated in some detail. First came the study of linear responses to small perturbations [12, 13] and later nonlinear responses to perturbations of arbitrary magnitude [14, 15]. We designed simple types of response experiments, which make it possible to extract mechanistic and kinetic information from complex nonlinear reaction systems. The main idea is to use "neutral" labeled compounds (tracers) [16], which have the same kinetic and transport properties as the unlabeled compounds. In our previous work [17, 18] we have shown that by using neutral tracers a class of response experiments can be described by linear response laws, even though the underlying kinetic equations are highly nonlinear. The linear response is not the result of a linearization procedure, but it is due to the use of neutral tracers. As a result, the response is linear even for large perturbations, making it possible to investigate global nonlinear kinetics by making use of linear mathematical techniques. Moreover, the susceptibility functions from the response law are related to the probability densities of the lifetimes and transit times of the various chemical species, making it easy to establish a connection between the response data and the mechanism and kinetics of the process [17].

In population genetics there is experimental evidence that many mutations are neutral, which is consistent with Kimura's theory of neutral evolution [19]. Kimura's theory is based on a "neutrality condition," that is, on the assumption that the natality and mortality functions as well as the transport (migration) coefficients are the same for the main population as well as for the mutants. For neutral mutations the nonlinear reaction-diffusion equations for the spreading of a mutation within a growing population which is expanding in space have a special structure, which makes it possible to transform them into linear evolution equations for the fractions of mutants (gene frequencies) at a given position and time, even though the global evolution equation for the total population density is nonlinear. The linearity of the evolution equations for the gene frequencies is due to Kimura's "neutrality condition" [19].

A comparative analysis of these two apparently unrelated problems has shown that, by introducing a general "neutrality condition," it is possible to derive a linear response theorem for a general class of reaction-diffusion systems, which include the case of homogeneous, space-independent chemical systems discussed in [20] as a particular case. With minor adaptations, this response theorem can be used for the description of the space and time propagation of neutral mutations. Such a general response theorem is of interest in the study of various problems in physics, chemistry, biology, and genetic anthropology.

The structure of this chapter is the following. In Section II we derive a modified form of a space-dependent generalized Fisher theorem derived in Ref. 11 and make a connection between our generalized Fisher theorem and the Fisher information metric, and we use it for formulating a general type of inverse problems. In Section III we illustrate the method by solving the problem of estimating the initial position and time of a single-locus mutation on the Y chromosome or on mitochondrial DNA in humans.

II.   SPACE-DEPENDENT GENERALIZED FISHER THEOREM

We have recently introduced a strong generalization of Fisher's fundamental theorem of natural selection [11]. In this chapter we derive a further generalization of this result, which is useful for formulating inverse problems. In the following we use the genetic terminology; however, the results derived in the following apply to physical and chemical systems as well. In this chapter we limit ourselves to space-dependent problems. We consider a pool of genetic objects (pool of alleles corresponding to a given locus, pool of sets of alleles corresponding to a multi-locus system, pool of sets of a given DNA sequence, etc.). A genetic object is described by a discrete state vector $\mathbf{u}$. For physical or chemical processes the genetic objects are usually replaced by molecular or atomic species of different types and the vector $\mathbf{u}$ identifies the species. We denote by $n_{\mathbf{u}}(\mathbf{x}; t)$ the density of genetic objects of type $\mathbf{u}$ at position $\mathbf{x}$ and time t and by $n(\mathbf{x}; t) = \sum_{\mathbf{u}} n_{\mathbf{u}}(\mathbf{x}; t)$ the total density of genetic objects at position $\mathbf{x}$ and time t. The local frequency of the genetic object with a state vector $\mathbf{u}$ is given by $\gamma_{\mathbf{u}}(\mathbf{x}; t) = n_{\mathbf{u}}(\mathbf{x}; t)/n(\mathbf{x}, t)$. According to their definition, the local frequencies $\gamma_{\mathbf{u}}(\mathbf{x}, t)$ satisfy the normalization condition $\sum_{\mathbf{u}} \gamma_{\mathbf{u}}(\mathbf{x}, t) = 1$ We introduce the vectors of relative space-specific rates of growth:

$$\chi_{\mathbf{u}}(\mathbf{x}, t) = \nabla_{\mathbf{x}} \ln \gamma_{\mathbf{u}}(\mathbf{x}, t) \tag{1}$$

and the tensor of space-specific relative rates of evolution:

$$\varphi_{\mathbf{u}}(\mathbf{x}, t) = \nabla_{\mathbf{x}} \otimes \chi_{\mathbf{u}}(\mathbf{x}, t) = \nabla_{\mathbf{x}} \otimes \nabla_{\mathbf{x}} \ln \gamma_{\mathbf{u}}(\mathbf{x}, t) \qquad (2)$$

As $\gamma_{\mathbf{u}}(\mathbf{x}; t)$ are relative frequencies normalized to unity, we can define the statistical averages:

$$\overline{F_{\mathbf{u}}(\mathbf{x}; t)} = \sum_{\mathbf{u}} \gamma_{\mathbf{u}}(\mathbf{x}; t) F_{\mathbf{u}}(\mathbf{x}; t) \qquad (3)$$

where $F_{\mathbf{u}}(\mathbf{x}; t)$ is a property of the genetic object $\mathbf{u}$. In particular, the average value of the vectors of relative space-specific rates of growth is zero:

$$\overline{\chi_{\mathbf{u}}(\mathbf{x}, t)} = \sum_{\mathbf{u}} \gamma_{\mathbf{u}}(\mathbf{x}, t) \chi_{\mathbf{u}}(\mathbf{x}, t) = \sum_{\mathbf{u}} \nabla_{\mathbf{x}} \gamma_{\mathbf{u}}(\mathbf{x}, t) = \mathbf{0} \qquad (4)$$

Starting from the definitions (2–3), we can also evaluate the average of the tensor of space-specific rates of evolution. We have

$$\overline{\varphi_{\mathbf{u}}(\mathbf{x},t)} = \sum_{\mathbf{u}} \gamma_{\mathbf{u}}(\mathbf{x},t) \varphi_{\mathbf{u}}(\mathbf{x},t) = \sum_{\mathbf{u}} \gamma_{\mathbf{u}}(\mathbf{x},t) \nabla_{\mathbf{x}} \otimes \{ [\gamma_{\mathbf{u}}(\mathbf{x},t)]^{-1} \nabla_{\mathbf{x}} \gamma_{\mathbf{u}}(\mathbf{x},t) \}$$

$$= \sum_{\mathbf{u}} \gamma_{\mathbf{u}}(\mathbf{x},t) [(\gamma_{\mathbf{u}}(\mathbf{x},t))^{-1} \nabla_{\mathbf{x}} \otimes \nabla_{\mathbf{x}} \gamma_{\mathbf{u}}(\mathbf{x},t) - (\nabla_{\mathbf{x}} \ln \gamma_{\mathbf{u}}(\mathbf{x},t)) \otimes (\nabla_{\mathbf{x}} \ln \gamma_{\mathbf{u}}(\mathbf{x},t))]$$

$$= -\sum_{\mathbf{u}} \gamma_{\mathbf{u}}(\mathbf{x},t) \chi_{\mathbf{u}}(\mathbf{x},t) \otimes \chi_{\mathbf{u}}(\mathbf{x},t) \qquad (5)$$

and thus

$$\overline{\varphi_{\mathbf{u}}(\mathbf{x}, t)} = -\overline{\chi_{\mathbf{u}}(\mathbf{x}, t) \otimes \chi_{\mathbf{u}}(\mathbf{x}, t)} \qquad (6)$$

For deriving Eqs. (4)–(6) we have used the rules of vector calculus and the normalization condition $\sum_{\mathbf{u}} \gamma_{\mathbf{u}}(\mathbf{x}, t) = 1$. From Eqs. (4) and (5) we notice that the average of the relative rate of evolution is zero and the average value of the tensor of the space-specific relative rate of evolution is equal to the average value of the tensorial product of the vector of the space-specific relative rate of growth by itself, with changed sign. Equation (5) is a more general form of the space-specific generalization of Fisher's fundamental theorem of natural selection presented in Ref. 11. The scalar transcription of the generalized Fisher theorem (6) is the following:

$$\overline{[\varphi_{\mathbf{u}}(\mathbf{x}, t)]_{\alpha\beta}} = -\overline{[\chi_{\mathbf{u}}(\mathbf{x}, t)]_{\alpha} [\chi_{\mathbf{u}}(\mathbf{x}, t)]_{\beta}} = -\text{cov}\{ [\chi_{\mathbf{u}}(\mathbf{x}, t)]_{\alpha}, [\chi_{\mathbf{u}}(\mathbf{x}, t)]_{\beta} \} \qquad (7)$$

where

$$\text{cov}\{[\boldsymbol{\chi_u}(\mathbf{x},t)]_\alpha, [\boldsymbol{\chi_u}(\mathbf{x},t)]_\beta\} = \sum_u \gamma_\mathbf{u}(\mathbf{x},t)[\boldsymbol{\chi_u}(\mathbf{x},t)]_\alpha[\boldsymbol{\chi_u}(\mathbf{x},t)]_\beta \qquad (8)$$

is the covariance matrix of the components of the vector $\boldsymbol{\chi_u}(\mathbf{x},t)$. If we take the trace of the tensors in Eq. (7), we come to the inequality

$$Tr\overline{[\varphi_\mathbf{u}(\mathbf{x},t)]} = \sum_\alpha \overline{\partial_{x_\alpha}[\boldsymbol{\chi_u}(\mathbf{x},t)]_\alpha} = -\sum_\alpha \overline{\{[\boldsymbol{\chi_u}(\mathbf{x},t)]_\alpha\}^2} \le 0 \qquad (9)$$

which is an evolution criterion of the type derived in Ref. 11 for other types of generalized Fisher theorems.

We introduce the vectors of absolute space-specific rates of growth,

$$\boldsymbol{\sigma_u}(\mathbf{x},t) = \nabla_\mathbf{x} \ln n_\mathbf{u}(\mathbf{x},t) = \nabla_\mathbf{x} \ln[n(\mathbf{x},t)\gamma_\mathbf{u}(\mathbf{x},t)] = \nabla_\mathbf{x} \ln n(\mathbf{x},t) + \boldsymbol{\chi_u}(\mathbf{x},t) \qquad (10)$$

and the tensor of absolute space-specific relative rates of evolution,

$$\begin{aligned}
\boldsymbol{\psi_u}(\mathbf{x},t) &= \nabla_\mathbf{x} \otimes \boldsymbol{\sigma_u}(\mathbf{x},t) = \nabla_\mathbf{x} \otimes \nabla_\mathbf{x} \ln[n(\mathbf{x},t)\gamma_\mathbf{u}(\mathbf{x},t)] \\
&= \nabla_\mathbf{x} \otimes \nabla_\mathbf{x} \ln n(\mathbf{x},t) + (\gamma_\mathbf{u}(\mathbf{x},t))^{-1}\nabla_\mathbf{x} \otimes \nabla_\mathbf{x}\gamma_\mathbf{u}(\mathbf{x},t) \\
&\quad - (\nabla_\mathbf{x} \ln \gamma_\mathbf{u}(\mathbf{x},t)) \otimes (\nabla_\mathbf{x} \ln \gamma_\mathbf{u}(\mathbf{x},t))
\end{aligned} \qquad (11)$$

The average of the tensor of space-specific, absolute rate of evolution $\overline{\boldsymbol{\psi_u}(\mathbf{x},t)}$ can be evaluated in a similar way. From Eqs. (3) and (11) we obtain

$$\overline{\boldsymbol{\psi_u}(\mathbf{x},t)} = \sum_u \boldsymbol{\psi_u}(\mathbf{x},t)\gamma_\mathbf{u}(\mathbf{x},t) = \Psi(\mathbf{x},t) + \overline{\varphi_\mathbf{u}(\mathbf{x},t)} = \Psi(\mathbf{x},t) - \overline{\boldsymbol{\chi_u}(\mathbf{x},t) \otimes \boldsymbol{\chi_u}(\mathbf{x},t)} \qquad (12)$$

where

$$\Psi(\mathbf{x},t) = \nabla_\mathbf{x} \otimes \nabla_\mathbf{x} \ln n(\mathbf{x},t) \qquad (13)$$

is the tensor of the absolute rate of evolution for the total population of genetic objects. We introduce the covariance matrix of the vectors of relative space-specific rates of growth

$$\begin{aligned}
&\overline{\Delta\boldsymbol{\sigma_u}(\mathbf{x},t) \otimes \Delta\boldsymbol{\sigma_u}(\mathbf{x},t)} \\
&= \sum_u \gamma_\mathbf{u}(t)\overline{\left[\boldsymbol{\sigma_u}(\mathbf{x},t) - \overline{\boldsymbol{\sigma_u}(\mathbf{x},t)}\right]} \otimes \overline{\left[\boldsymbol{\sigma_u}(\mathbf{x},t) - \overline{\boldsymbol{\sigma_u}(\mathbf{x},t)}\right]} = \overline{\boldsymbol{\chi_u}(\mathbf{x},t) \otimes \boldsymbol{\chi_u}(\mathbf{x},t)}
\end{aligned} \qquad (14)$$

For deriving Eq. (14) we used the equality

$$\boldsymbol{\sigma_u}(\mathbf{x},t) - \overline{\boldsymbol{\sigma_u}(\mathbf{x},t)} = \nabla_\mathbf{x} \ln n(\mathbf{x},t) + \boldsymbol{\chi_u}(\mathbf{x},t) - \overline{[\nabla_\mathbf{x} \ln n(\mathbf{x},t) + \boldsymbol{\chi_u}(\mathbf{x},t)]}$$
$$= \boldsymbol{\chi_u}(\mathbf{x},t) - \overline{\boldsymbol{\chi_u}(\mathbf{x},t)} = \boldsymbol{\chi_u}(\mathbf{x},t) \tag{15}$$

which follows from Eqs. (4) and (10). From Eqs. (13) and (14) we have

$$\boldsymbol{\Psi}(\mathbf{x},t) = \overline{\boldsymbol{\psi_u}(\mathbf{x},t)} + \overline{\Delta\boldsymbol{\sigma_u}(\mathbf{x},t) \otimes \Delta\boldsymbol{\sigma_u}(\mathbf{x},t)} \tag{16}$$
$$[\boldsymbol{\Psi}(\mathbf{x},t)]_{\alpha\beta} = \overline{[\boldsymbol{\psi_u}(\mathbf{x},t)]}_{\alpha\beta} + \text{cov}\{[\boldsymbol{\sigma_u}(\mathbf{x},t)]_\alpha, [\boldsymbol{\sigma_u}(\mathbf{x},t)]_\beta\} \tag{17}$$

and

$$\text{Tr}[\boldsymbol{\Psi}(\mathbf{x},t)]_{\alpha\beta} = \overline{\sum_\alpha \partial_{x_\alpha}[\boldsymbol{\sigma_u}(\mathbf{x},t)]_\alpha} + \overline{\sum_\alpha \{[\Delta\boldsymbol{\sigma_u}(\mathbf{x},t)]_\alpha\}^2} \geq \overline{\sum_\alpha \partial_{x_\alpha}[\boldsymbol{\sigma_u}(\mathbf{x},t)]_\alpha}$$
$$\tag{18}$$

where

$$\text{cov}\{[\boldsymbol{\sigma_u}(\mathbf{x},t)]_\alpha, [\boldsymbol{\sigma_u}(\mathbf{x},t)]_\beta\} = \sum_u \gamma_u(\mathbf{x},t) \left[\boldsymbol{\sigma_u}(\mathbf{x},t) - \overline{\boldsymbol{\sigma_u}(\mathbf{x},t)}\right]_\alpha \left[\boldsymbol{\sigma_u}(\mathbf{x},t) - \overline{\boldsymbol{\sigma_u}(\mathbf{x},t)}\right]_\beta$$
$$\tag{19}$$

Equations (16), (17), and (18) are equivalent to Eqs. (6),(7), and (9), respectively; they can be derived from one another by passing from relative to absolute evolutionary variables and vice versa. They are both equivalent space-dependent formulations of a generalized Fisher theorem.

For a unified representation, we introduce the space-dependent commutator operator

$$\mathbb{C}^{(\mathbf{x})} \ldots = \nabla_\mathbf{x}\overline{(\ldots)} - \overline{(\nabla_\mathbf{x} \ldots)} = \nabla_\mathbf{x}\left(\sum_u \gamma_u(\mathbf{x},t) \ldots\right) - \sum_u \gamma_u(\mathbf{x},t)\nabla_\mathbf{x}(\ldots)$$
$$\tag{20}$$

between the time differentiation and the averaging with respect to the relative frequencies. In terms of $\mathbb{C}^{(\mathbf{x})}$, Eqs. (6) and (16) can be expressed in the same form:

$$\mathbb{C}^{(\mathbf{x})}\boldsymbol{\chi_u}(\mathbf{x},t) = \overline{\Delta\boldsymbol{\chi_u}(\mathbf{x},t) \otimes \Delta\boldsymbol{\chi_u}(\mathbf{x},t)} \tag{21}$$
$$\mathbb{C}^{(\mathbf{x})}\boldsymbol{\sigma_u}(\mathbf{x},t) = \overline{\Delta\boldsymbol{\sigma_u}(\mathbf{x},t) \otimes \Delta\boldsymbol{\sigma_u}(\mathbf{x},t)} \tag{22}$$

Now we show that there is a surprising relation between Fisher's fundamental theorem of natural selection and other theory developed by Fisher, the likelihood theory in statistics and Fisher information [21]. As far as we know, the present chapter is the first publication in the literature pointing out the connections between these two problems formulated and studied by Fisher.

By interpreting the state vector $\mathbf{u}$ in $\gamma_{\mathbf{u}}(\mathbf{x}; t)$ and the position vector $\mathbf{x}$ as a parameter, $\ln \gamma_{\mathbf{u}}(\mathbf{x}, t)$ can be interpreted as a likelihood function and $\chi_{\mathbf{u}}(\mathbf{x}, t)$ as a vector of statistical scores; as expected from likelihood theory, the average score is zero, $\overline{\chi_{\mathbf{u}}(\mathbf{x}, t)} = \mathbf{0}$ [Eq. (4)]. Moreover, the covariance matrix of the relative rates of evolution plays the role of a Fisher information metric:

$$g_{\alpha\beta} = \sum_{\mathbf{u}} \gamma_{\mathbf{u}}(\mathbf{x}, t) [\chi_{\mathbf{u}}(\mathbf{x}, t)]_{\alpha} [\chi_{\mathbf{u}}(\mathbf{x}, t)]_{\beta} = \sum_{\mathbf{u}} \gamma_{\mathbf{u}}(\mathbf{x}, t) \frac{\partial}{\partial x_\alpha} [\ln \gamma_{\mathbf{u}}(\mathbf{x}, t)]_j \frac{\partial}{\partial x_\beta} [\ln \gamma_{\mathbf{u}}(\mathbf{x}, t)]$$

$$(23)$$

that is, $g_{\alpha\beta}$ is a metric tensor for a statistical differential manifold and can be used to calculate the informational difference between measurements. The expression

$$ds^2 = \sum g_{\alpha\beta} dx_\alpha dx_\beta \qquad (24)$$

can be interpreted intuitively as the distance between two points on a statistical differential manifold and is the amount of information between them—that is, the informational difference between them [21].

In order to test the interpretation of Eq. (24), we consider the relative frequencies for two close positions in space, $\mathbf{x}$ and $\mathbf{x} + \Delta\mathbf{x}$, $\gamma_{\mathbf{u}}(\mathbf{x}, t)$ and $\gamma_{\mathbf{u}}(\mathbf{x} + \Delta\mathbf{x}, t)$, respectively. The relative information can be evaluated by using the Kullback–Leibler entropy:

$$K[\gamma_{\mathbf{u}}(\mathbf{x} + \Delta\mathbf{x}, t); \gamma_{\mathbf{u}}(\mathbf{x}, t)] = \sum_{\mathbf{u}} \gamma_{\mathbf{u}}(\mathbf{x} + \Delta\mathbf{x}, t) \ln \left[\frac{\gamma_{\mathbf{u}}(\mathbf{x} + \Delta\mathbf{x}, t)}{\gamma_{\mathbf{u}}(\mathbf{x}, t)} \right] \qquad (25)$$

For small variations $\Delta\mathbf{x}$ the dominant term of $K[\gamma_{\mathbf{u}}(\mathbf{x} + \Delta\mathbf{x}, t); \gamma_{\mathbf{u}}(\mathbf{x}, t)]$ is given by a Fisher information metric of the type (23):

$$K[\gamma_{\mathbf{u}}(\mathbf{x} + \Delta\mathbf{x}, t); \gamma_{\mathbf{u}}(\mathbf{x}, t)] = \sum_{\alpha\beta} g_{\alpha\beta} \Delta x_\alpha \Delta x_\beta + \mathcal{O}((\Delta x_\alpha)^3) \qquad (26)$$

The Fisher information metric (23) or (25) is the ideal tool for solving inverse problems for reaction transport systems with incomplete knowledge of the parameters. Since the reaction-transport systems are described by local, partial differential equations, considering small space variations, the differential equations can be embedded as constraints in an optimization algorithm for the Fisher information metric. Different approaches can be derived, depending on the structure of the reaction-transport equations and on the unknown parameters corresponding to a given problem. Sometimes all essential parameters are known and there is no need for optimization. Nevertheless, even in this case the information approach clarifies the physical meaning of the procedure, as we shall see in the next section.

The generalized Fisher theorems derived in this section are statements about the space variation of the vectors of the relative and absolute space-specific rates of growth. These vectors have a simple natural (biological, chemical, physical) interpretation: They express the capacity of a species of type **u** to fill out space; in genetic language, they are space-specific fitness functions. In addition, the covariance matrix of the vector of the relative space-specific rates of growth, $g_{\alpha\beta}$, [Eq. (25)] is a Riemannian metric tensor that enters the expression of a Fisher information metric [Eqs. (24) and (26)]. These results may serve as a basis for solving inverse problems for reaction transport systems.

III. APPLICATION TO GEOGRAPHICAL POPULATION GENETICS

For illustration we consider the problem of evaluating the time and the position of an initial occurrence of an event in a reaction-diffusion system and its application to human population genetics. The spreading of a mutation in a migrating population may display enhanced (hydrodynamic) transport induced by population growth, a phenomenon that can occur not only in population genetics but also in physics and chemistry [11, 20, 22, 23]. We consider a system made up of different individuals $X_u, u = 1, 2, \ldots$ (molecules, quasiparticles, biological organisms, etc.). The species $X_u, u = 1, 2, \ldots$, replicate, transform into each other, die, and undergo slow, diffusive motion, characterized by the diffusion coefficients $D_u, u = 1, 2, \ldots$, which are constant. The replication and disappearance rates $R_u^{\pm}$ of the different species are proportional to the species densities $n_u, u = 1, 2, \ldots$; we have $R_u^{\pm} = n_u \rho_u^{\pm}(\mathbf{n})$, where the rate coefficients $\rho_u^{\pm}(\mathbf{n})$ are generally dependent on the composition vector $\mathbf{n} = (n_u)$; similarly, the rate $R_{u \to v}$ of transformation of species X_u into the species X_v is given by $R_{u \to v} = n_u k_{uv}(\mathbf{n})$, where $k_{uv}(\mathbf{n})$ are composition-dependent rate coefficients. The process can be described by the following reaction-diffusion equations:

$$\frac{\partial}{\partial t} n_u = n_u \rho_u^+(\mathbf{n}) - n_u \rho_u^-(\mathbf{n}) + \sum_{v \neq u} [n_v k_{vu}(\mathbf{n}) - n_u k_{uv}(\mathbf{n})] + D_u \nabla^2 n_u \qquad (27)$$

We study the time and space evolution of the fractions of the different species present in the system: $\gamma_u = n_u/n$, with $1 = \sum_u \gamma_u$, where $n = \sum_u n_u$ is the total population density. For example, in chemistry γ_u are molar fractions whereas in population genetics they are gene frequencies. Equation (27) leads to the following evolution equations for the total population density n and for the fractions γ_u:

$$\frac{\partial}{\partial t} n = n[\tilde{\rho}^+(n, \gamma) - \tilde{\rho}^-(n, \gamma)] + \nabla^2(n\tilde{D}(\gamma)) \tag{28}$$

$$\frac{\partial}{\partial t}\gamma_u + \nabla(\mathbf{v_u}\gamma_u) = D_u\nabla^2\gamma_u + \varepsilon_u\gamma_u + \sum_{v \neq u}[\gamma_v k_{vu}(n\gamma) - \gamma_u k_{uv}(n\gamma)] + \delta\mathcal{R}_u \tag{29}$$

where

$$\tilde{\rho}^\pm(n, \gamma) = \sum_u \gamma_u \rho_u^\pm(n\gamma), \quad \tilde{D}(\gamma) = \sum_u \gamma_u D_u \tag{30}$$

are average rate and transport coefficients,

$$\delta\rho_u^\pm(n, \gamma) = \rho_u^\pm(n\gamma) - \tilde{\rho}^\pm(n, \gamma), \qquad \delta D_u(\gamma) = D_u - \tilde{D}(\gamma) \tag{31}$$

are deviations of the individual rate and transport coefficients from the corresponding average values,

$$\mathbf{v_u} = -2D_u\nabla \ln n, \qquad \varepsilon_u = \mathrm{div}(\mathbf{v_u}) \tag{32}$$

are transport (hydrodynamic) speeds and expansion coefficients attached to different population fractions,

$$\delta\mathcal{R}_u = \gamma_u[\delta\rho_u^+(n, \gamma) - \delta\rho_u^-(n, \gamma)]$$
$$- \gamma_u\left\{\sum_v \delta D_v(\gamma)[\nabla^2\gamma_v + 2(\nabla \ln n) \bullet \nabla\gamma_v]\right\} + \delta D_u(\gamma)\gamma_u n^{-1}\nabla^2 n \tag{33}$$

are the components of the rates of change of the population fractions due to the individual variations of the rate and transport coefficients, and $\gamma = (\gamma_1, \gamma_2, \ldots)$ is the vector of population fractions.

Although the different species are undergoing slow, diffusive motions, the corresponding population fractions move faster: In the evolution equations (29), there are both diffusive terms and convective transport (hydrodynamic) terms depending on the transport speeds $\mathbf{v_u}$ given by Eqs. (32). According to Eqs. (32), these transport speeds are generated by the space variations of the total population density and have the opposite sign of the gradient of the total population densities. For a growing population the population cloud usually expands from an original area and tries to occupy all space available. The population density decreases toward the edge of the population cloud; thus the population gradient is negative and the transport velocities are positive, oriented toward the directions of propagation of the population cloud. The cause of enhanced transport of the species fractions is the net population growth. Since the gradient tends to increase towards the edge of the population wave, an initial perturbation of the species fractions generated in the propagation front has good chances of undergoing enhanced transport and spreading all over the system. An initial perturbation, produced close to the initial area where the population originates, has poor chances of undergoing sustained enhanced transport. The mathematical and physical significance of the hydrodynamic transport terms $\nabla(\mathbf{v_u}\gamma_u)$ in Eq. (29) can be easily understood. From the mathematical point of view, the terms $\nabla(\mathbf{v_u}\gamma_u)$ emerge as a result of a nonlinear transformation of the state variables, from species densities to species fractions. The physical interpretation of the transport terms $\nabla(\mathbf{v_u}\gamma_u)$ depends on the direction and orientation of the speed vectors: For expanding populations, $\mathbf{v_u}$ are generally oriented toward to direction of expansion of the population cloud, resulting in enhanced transport. For shrinking population clouds the terms $\nabla(\mathbf{v_u}\gamma_u)$ lead to the opposite effect: The transport process slows down.

The above-mentioned results can be easily extended to anisotropic diffusion characterized by a diffusion tensor $\mathbb{D}_u$. In this case the hydrodynamic speed is given by

$$\mathbf{v}_u = 2\mathbb{D}_u \bullet \nabla_\mathbf{x} \ln n(\mathbf{x}, t), \qquad (\mathbf{v}_u)_\alpha = 2\sum_\beta (\mathbb{D}_u)_{\alpha\beta}\partial_\beta \ln n(\mathbf{x}, t) \qquad (34)$$

which is a generalization of Eq. (32). The generalized Fisher theorem developed in the previous section leads to a simple physical (or biological) interpretation of Eqs. (32) and (34). From Eq. (10) it follows that the vector of the total rate of growth for the whole population

$$\boldsymbol{\sigma}_\Sigma(\mathbf{x}, t) = \nabla_\mathbf{x} \ln n(\mathbf{x}, t) \qquad (35)$$

is given by

$$\boldsymbol{\sigma}_\Sigma(\mathbf{x}, t) = \sum_\mathbf{u} \boldsymbol{\sigma}_\mathbf{u}(\mathbf{x}, t) \qquad (36)$$

Equations (35) and (36) lead to

$$\mathbf{v}_u = 2\mathbb{D}_u \bullet \boldsymbol{\sigma}_\Sigma^{(\mathbf{x})}(\mathbf{x}, t) = \sum_v 2\mathbb{D}_u \bullet \boldsymbol{\sigma}_v(\mathbf{x}, t) \qquad (37)$$

that is, the hydrodynamic speed of the subpopulation u is twice the contraction of the diffusion tensor $\mathbb{D}_u$ and the vector of the space-specific rate of growth of the total population $\boldsymbol{\sigma}_\Sigma(\mathbf{x}, t)$. Thus, the hydrodynamic speed of the subpopulation u is proportional to the diffusion tensor $\mathbb{D}_u$, which expresses the intrinsic mobility of the subpopulation u, and to the vector of space-specific rate of growth of the total population $\boldsymbol{\sigma}_\Sigma(\mathbf{x}, t)$; this, in turn, expresses the capability of the total population to fill out the available space. We notice that we can also derive a generalized Fisher relation, which connects the average of the rate of space variation of the hydrodynamic speed to the covariance matrix of the space-specific rates of growth. We differentiate Eq. (34) with respect to x_α and take an average with respect to the different subpopulations. We have

$$\overline{\partial_{x_\alpha}(v_u)_\beta} = 2 \sum_\gamma (\overline{\mathbb{D}_u})_{\beta\gamma} \left\{ \overline{[\boldsymbol{\psi}_u(\mathbf{x}, t)]}_{\gamma\alpha} + \mathrm{cov}\left\{ [\boldsymbol{\sigma}_u(\mathbf{x}, t)]_\gamma, \ [\boldsymbol{\sigma}_u(\mathbf{x}, t)]_\alpha \right\} \right\} \quad (38)$$

The Fisher relation (38) has a structure similar to a fluctuation dissipation relation in statistical mechanics: It relates a macroscopic transport coefficient, the hydrodynamic speed, to the diffusion tensor and to the statistical properties of the genetic process.

We consider a particular case, for which the replication and disappearance rate coefficients and the diffusion coefficients are the same for all species and depend only on the total population density $\rho_u^\pm(\mathbf{n}) = \rho^\pm(n)$, $D_u = D$. We also assume that the transformation rates are constant $k_{uv}(\mathbf{n}) = k_{uv}$. These conditions are fulfilled in chemistry by tracer experiments, for which the variation of the rate and transport coefficients due to the kinetic isotope effect can be neglected. Similar restrictions also hold in population genetics for neutral mutations, for which the demographic and transport parameters are the same for mutants and nonmutants, respectively. The evolution equations become

$$\frac{\partial}{\partial t} n = n\mu(n) + D\nabla^2 n \quad (39)$$

$$\frac{\partial}{\partial t} \gamma_u + \nabla(\mathbf{v}\gamma_u) = D\nabla^2\gamma_u + \varepsilon\gamma_u + \sum_{v \neq u}(\gamma_v k_{vu} - \gamma_u k_{uv}) \quad (40)$$

where $\mu(n) = \rho^+(n) - \rho^-(n)$ is the net production rate of the total population. The total population density obeys a separate equation, which is independent of the species fractions, and the evolution equations for the fractions become linear.

In the particular case of geographical spreading of single locus mutations on mitochondrial DNA or on Y chromosome [22, 23], we consider a growing population that diffuses slowly in time and assume that the net rate of growth is a linear function of population density, $\mu(n) = \varkappa_L(1 - n/n_\infty)$, where $\varkappa_L$ is

Lotka's intrinsic rate of growth of the population. At an initial position and time, a neutral mutation occurs and afterwards no further identical mutations occur (infinite allele model). We are interested in the time and space dependence of the local fractions of the individuals, which are the offspring of the individual that carried the initial mutation. The goal of this analysis is the evaluation of the position and time where the mutation originated from measured data representing the current geographical distribution of the mutation. We limit our analysis to one-dimensional systems, for which a detailed theoretical analysis is possible. Eqs. (39) and (40) turn into a simpler form:

$$\frac{\partial}{\partial t} n = \varkappa_L n \left(1 - \frac{n}{n_\infty}\right) + D\nabla^2 n \tag{41}$$

$$\frac{\partial}{\partial t}\gamma + \nabla(\mathbf{v}\gamma) = D\nabla^2\gamma + \varepsilon\gamma \tag{42}$$

where γ is the local fraction of mutants. Equation (41) for the total population growth is the well-known Fisher equation, which has solitary solutions. We use an asymptotic solitary wave solutions for Eq. (41) [23 and 24], which gives an excellent representation of the population wave from the top saturation level to the front edge where the total population tends to zero. In our notation the asymptotic solution is

$$n(x,t) = n_\infty \left\{1 + \exp\left[\frac{c}{2D}(x - ct)\right]\right\}^{-1}, \qquad \text{with } c = 2\sqrt{\varkappa_L D} \tag{43}$$

the transport speed v and the expansion factor ε are given by

$$\frac{v}{c} = 1 - \frac{n}{n_\infty} = \frac{2\exp\left[\frac{c}{4D}(x - ct)\right]}{\cosh\left[\frac{c}{4D}(x - ct)\right]} \tag{44}$$

$$\varepsilon = \frac{c^2}{4D}\left\{\cosh\left[\frac{c}{4D}(x - ct)\right]\right\}^{-2} \geq 0 \tag{45}$$

For a mutation that occurs at the very edge of the total population wave, $n \sim 0$, the transport speed is equal to the speed of propagation front, $v = c$; the two waves, of the total population and of the mutation, are synchronized.

The probability density $P(\mathbf{r}, t)$ of the position of the center of gravity of the mutant population can be roughly estimated by normalizing the mutant gene frequency:

$$P(x,t) \sim \gamma(x,t) / \int \gamma(x,t)\, dx \tag{46}$$

Eqs. (42) and (46) lead to a Gaussian distribution for the position x of the center of gravity of the mutant population both for the diffusive regime and for the enhanced transport:

$$P(x,t) \sim [4\pi D(t - t_0)]^{-1/2} \exp\{-(x - x_0 - v_{\mathrm{eff}}(t - t_0))^2 / [4D(t - t_0)]\} \quad (47)$$

where x_0 and t_0 are the initial position and time where the mutation occurred and v_{eff} is an effective transport rate. In Eq. (47) we have $v_{\mathrm{eff}} \sim 0$ for diffusive transport ($n \sim n_\infty$) and $v_{\mathrm{eff}} \sim c$ for the enhanced transport ($n \sim 0$). Although for intermediate cases between these two extremes the probability density $P(x,t)$ is generally not Gaussian, a Gaussian can be used as a reasonable approximation, where the effective propagation speed v_{eff} has an intermediate value between zero (diffusive transport) and the maximum values for enhanced transport corresponding to the two solutions of the Fisher equation. The average position of the center of gravity of the mutant population increases linearly in time:

$$\langle x(t) \rangle = \int xP(x,t)\ dx = x_0 + v_{\mathrm{eff}}(t - t_0) \quad (48)$$

If we examine the current geographical distribution of a mutation, it is hard to estimate the value of the population density n at the position and time where the mutation originates. It makes sense to treat n as a random variable selected from a certain probability density $p(n)$. The constraints imposed on $p(n)$ are the conservation of the normalization condition $\int p(n)\ dn = 1$ and the range of variation, $n_\infty \geq n \geq 0$. The maximum information entropy approach leads to a uniform distribution

$$p(n) = (n_\infty)^{-1}[h(n) - h(n - n_\infty)] \quad (49)$$

where $h(n)$ is the Heaviside step function. The effective speed v_{eff} can be evaluated as an average value:

$$v_{\mathrm{eff}} = \int v(n)p(n)\ dn \quad (50)$$

which corresponds to a large sample of initial distributions and where $v(n)$ is given by Eq. (44). We have

$$v_{\mathrm{eff}} = c \int_0^{n_\infty} \left(1 - \frac{n}{n_\infty}\right) \frac{dn}{n_\infty} = \frac{1}{2}c \quad (51)$$

For the estimation of the initial position of a mutation, it is useful to consider the ratio

$$\zeta = \frac{x(t_L) - x(t_0)}{\langle x(t_L) \rangle - x(t_0)} \tag{52}$$

where $x(t_L)$ is the position of the limit of expansion, t_L is the time necessary for reaching the limit of expansion, $\langle x(t_L) \rangle$ is the position of the center of gravity of the mutant population for $t = t_L$, and $x(t_0) = x_0$ is the position where the mutation originates. In Eq. (52) both $x(t_L)$, the position of the limit of expansion, and $\langle x(t_L) \rangle$, the current average position of the center of gravity of the mutant population, are accessible experimentally. It follows that if the ratio ζ can be evaluated from theory, then $x(t_0) = x_0$, the point of origin of the mutation, can be evaluated from Eq. (52). By taking into account that the total population wave moves with the speed c and that the average center of gravity moves with the speed v_{eff}, we have

$$\zeta = \frac{ct_L - ct_0}{v_{\text{eff}}(t_L - t_0)} = \frac{c}{v_{\text{eff}}} \tag{53}$$

From Eqs. (51) and (53) it follows that $\zeta = 2$, a value in good agreement with the numerical simulations of Edmonds, Lillie, and Cavalli-Sforza [22], which lead to $\zeta = 2.2$. The difference of 0.2 between theory and simulations is due to the random drift, which was taken into account in the simulations but is neglected in our theory. By including the random drift, our theory provides information about the details of the motion of the propagation front [23].

In this section we studied the phenomenon of enhanced (hydrodynamic) transport, induced by population growth in reaction-diffusion systems. Based on our Fisher theorem approach, we have shown that the expressions for the emerging hydrodynamic speeds have a simple physical interpretation: They are proportional to space specific fitness functions, which express the ability of a population to fill out space. Based on our approach, we came up with simple rules for solving inverse problems in geographical population genetics.

IV.   CONCLUSIONS

In this chapter we have shown that space-dependent generalized Fisher theorems are effective tools for formulating and solving inverse problems for reaction-diffusion systems. For the simple example presented here, there was no need for the extremization of the Fisher information; nevertheless, our Fisher approach was of great help in formulating the problem and clarifying the physical (biological) significance of the results. Further research will focus on more complicated problems of population genetics, biology, and chemistry, such as the determination of the original, ancestral of a protein or DNA sequence, the determination of the initial position, and time of infection for "zero" patients of an epidemic disease or the determination of the initial time and position of occurrence of a fire or explosion.

Acknowledgments

We dedicate the article to the memory of I. Prigogine. This project has been supported in part by the CEEX GRANT-2006 "BIOMAT" of the Romanian Government and by the National Science Foundation. The authors thank G. Zbaganu for useful suggestions concerning the simplified derivation of the space-dependent generalized Fisher theorem.

References

1. R. Luther, *Z. Elektrochem.* **12**, 596–600 (1906).

2. R. Field and M. Burger, eds., *Oscillations and Waves in Chemical Systems*, John Wiley & Sons, New York, 1985.

3. J. Ross, S. C. Mueller, and C. Vidal, *Science* **240**, 460–465 (1988).

4. A. Turing, *Philos. Trans. R. Soc. London Ser. B* **237**, 37–72 (1952).

5. M. Flicker and J. Ross, *J. Chem. Phys.* **60**, 3458–3456 (1974).

6. V. Castets, E. Dulos, J. Boissonade, and P. De Kepper, *Phys. Rev. Lett.* **64**, 2953–2956 (1990).

7. S. Mueller and J. Ross *J. Phys. Chem. A.* **107**, 7997 (2003).

8. A. T. Winfree, *Science* **175**, 634 (1972).

9. S. Sawai, D. A. Thomason, and Z. C. Cox, *Nature* **433**, 323–326 (2005); M. Kaem, M. Menzinger, R Satnoianu, and A. Hunding, *Faraday Disc.* **120**, 295–312 (2001).

10. R. A. Fisher, (1930) *The Genetical Theory of Natural Selection*, Clarendon, Oxford, 1930; reprinted by Oxford University Press, Oxford, 1999, pp. 22–47.

11. M. O. Vlad, S. E. Szedlacsek, N. Pourmand, L. L. Cavalli-Sforza, P. Oefner, and J. Ross, *PNAS* **102**, 9848–9853 (2005).

12. M. Eigen, *Disc. Faraday Soc.* **24**, 25 (1957).

13. T. Chevalier, I. Schreiber, and J. Ross, *J. Phys. Chem.* **97**, 6776–6787 (1993).

14. A. Arkin and J. Ross, *J. Phys. Chem.* **99**, 970–979 (1995).

15. W. Vance, A. Arkin, and J. Ross, *PNAS* **99**, 5816–5821 (2002).

16. M. B. Neiman and D. Gál, *The Kinetic Isotope Method and Its Application*, Akademic Kiado, Budapest, 1971.

17. F. Moran, M. O. Vlad, and J. Ross, *J. Phys. Chem.* **101**, 9410 (1997); M. O. Vlad, F. Moran, and J. Ross, *J. Phys. Chem. ibid.* **102**, 4598(1998); *J. Phys. Chem. B* **103**, 3965 (1999); *Physica A* **278**, 504 (2000); *J. Phys. Chem. B* **105**, 11 710 (2001).

18. J. Ross, I. Schreiber, and M. O. Vlad, *Determination of Complex Reaction Mechanisms*, Oxford University Press, Oxford, 2006.

19. M. Kimura, *The Neutral Theory of Molecular Evolution*, Cambridge University Press, Cambridge, 1983.

20. M. O. Vlad, F. Moran, M. Tsuchiya, L. L. Cavalli-Sforza, P. J. Oefner, and J. Ross, *Physical Rev. E*, **65**, 061110-17 (2002).

21. B. Roy Frieden, *Science from Fisher Information*, 2nd ed., Cambridge University Press, Cambridge, 2004.

22. C. A. Edmonds, A. S. Lillie, and L. L. Cavalli-Sforza, *PNAS* **101**, 975–979 (2004).

23. M. O. Vlad, L. L. Cavalli-Sforza, and J. Ross, *PNAS* **101**, 10249–10253 (2004).

24. J. D. Murray, *Mathematical Biology*, Vol. 1, 3rd ed., Springer, New York, Berlin, 2002, pp. 451–452.

CARNOT EFFICIENCY REVISITED

C. VAN ᴅᴇɴ BROECK

Hasselt University, Dept. WN1, B-3590 Diepenbeek, Belgium

CONTENTS

I. INTRODUCTION

Carnot efficiency is one of the cornerstones of thermodynamics. This concept was derived by Carnot from the impossibility of a *perpetuum mobile* of the second kind [1]. It was used by Clausius to define the most basic state function of thermodynamics, namely the entropy [2]. The Carnot cycle deals with the extraction, during one full cycle, of an amount of work W from an amount of heat Q, flowing from a hot reservoir (temperature T_1) into a cold reservoir (temperature $T_2 \leq T_1$). The efficiency η for doing so obeys the following inequality:

$$\eta = \frac{W}{Q} \leq 1 - \frac{T_2}{T_1} \tag{1}$$

The equality sign is reached for a reversible process, entailing overall zero entropy production. Concomitantly, efficiency will be below Carnot in the presence of dissipative entropy producing fluxes. While this statement is strictly speaking correct, we will show below that it can be misleading. The delicate

Special Volume in Memory of Ilya Prigogine: Advances in Chemical Physics, Volume 135,
edited by Stuart A. Rice

point is that only the *overall* entropy production has to vanish. It turns out that this can be achieved in the presence of dissipative fluxes—like, for example, a heat flux — if the latter can be completely eliminated by an opposite heat flux of equal amplitude. The opposing flux can appear, not in response to a thermal gradient, but resulting from the application of another type of (thermodynamic) force. The main object of this chapter will be to establish the precise conditions under which this can be realized. Furthermore, the issue of efficiency is of particular relevance to Brownian motors [3–5], especially since conflicting reports have been published about the possibility of reaching Carnot efficiency.

II. BROWNIAN MOTORS

A prototypical example of a Brownian motor [6] is the so-called Feynman ratchet, reproduced in Fig. 1. The setup consists of two reservoirs: One contains blades, and the other one contains a ratchet and pawl mechanism, reminiscent of the mechanical rectifier used in clockworks of all kinds. Both units form a single entity that can rotate along the common axis by which the units are rigidly linked together. The construction generates the impression that fluctuations of the torque, acting on the blades, will always lead to a one-sided rotation of the entire setup, so that work can be extracted. This construction was discussed on the basis of a phenomenological model in the Feynman Lectures [7]. Feynman shows that the alluded rectification will not take place if the temperature is the same in both reservoirs. This observation is in agreement with the Carnot principle [cf. Eq. (1)], with $\eta = 0$ for $T_1 = T_2$. Turning to the case of different temperatures, Feynman goes on to prove that one can indeed extract work and that the Brownian motor can reach Carnot efficiency. The latter conclusion was later criticized by Parrondo and Espagnol [8] and by Sekimoto [9]. Indeed, in contrast to the Carnot engine, the auxiliary work performing system, namely the ratchet–pawl–vane

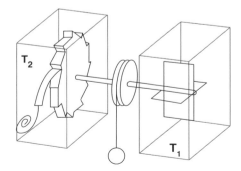

Figure 1.   Schematic representation of the Feynman ratchet.

unit, is *at all times* in contact with both thermal baths. Its corresponding rotational degree of freedom will, in its "attempt" to relax to the temperature of each bath separately, achieve an intermediate temperature [10]. As a result, it will, on average, absorb energy from the hot reservoir and lose it to the cold reservoir. Clearly, the resulting heat conduction is an irreversible entropy-producing step and Carnot efficiency cannot be attained. Furthermore, Parrondo and Espagnol show, on the basis of a linear Langevin description for the interaction with each of the heat baths, that the heat current $\dot{Q}$ from reservoir 1 to reservoir 2 is given by a Fourier law:

$$\dot{Q} = \kappa(T_1 - T_2)$$
$$\kappa = \frac{k_B}{M}\frac{\gamma_1\gamma_2}{\gamma_1 + \gamma_2} \quad (2)$$

Here M is the mass of the motor, while γ_1 and γ_2 are the friction coefficients of the motorparts in reservoir 1 and 2, respectively. The above conclusion was discussed and largely confirmed in a number of subsequent studies [11–17].

As a second example, we cite the problem of the so-called adiabatic piston (cf. Fig. 2) [18]. An insulating piston, free to move without friction in a cylinder, separates two gases initially at equilibirum at the same pressure, but at different temperatures. The question is whether the piston will move, and allow the system to relax to full equilibrium, with overall equal temperature and pressure. This entails the appearance of a heat flux through the "adiabatic" piston. This issue is also discussed in the Feynman Lectures. Feynman gives an incisive, albeit handwaving, argument to show that a piston of finite mass will indeed, through its asymmetric fluctuations, transfer energy from the hot side to the cold side. In fact, the adiabatic piston can be regarded as a variant of the ratchet/pawl/vane construction with the minor difference that the motion is translational rather than rotational and that the asymmetry of the ratchet is replaced by the

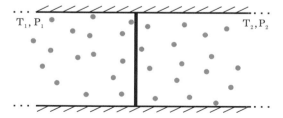

Figure 2. The adiabatic pistion separates compartments to its left and right filled with gases at temperature T_1 and T_2 and pressure P_1 and P_2, respectively. Does the piston move when $P_1 = P_2$ but $T_1 \neq T_2$?

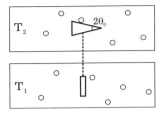

Figure 3. Two-dimensional simplification of the Feynman ratchet, consisting of one vane (flat sheet in lower reservoir) and one ratchet (triangular shape in upper reservoir), that is free to move as a rigid whole along the horizontal direction x. The boundary conditions are periodic both left and right and up and down in each container.

asymmetry in the contact with the heat reservoirs. In this respect, it is surprising that Feynman failed to make the connection between these two problems. The adiabatic piston has been the object of a large number of theoretical and numerical studies, all confirming that the adiabatic piston moves and conducts heat. The conductivity has been measured and even calculated analytically in a specific case study [19].

As a third and last example, we cite a recently introduced model [20] (cf. Fig. 3) that can be considered as the epitome of both previous Brownian motors. The motor consists of two rigidly linked parts, each residing in a different reservoir and moving as a single unit, under the result of the perfectly elastic collisions with the gas particles in each compartment. Instead of a rotational motion as in the Feynman ratchet, the degree of freedom is translational along a horizontal axis x, akin to the adiabatic piston. One or both of the units can be asymmetric. The choice of units in Fig. 3 is inspired by the Feynman ratchet, with the flat unit representing the blades and the triangular unit representing the asymmetric ratchet. This model can, like the adiabatic piston, be investigated in great detail by both analytic theory and hard-sphere molecular dynamics. In fact, in the limit of infinitely large reservoirs filled with ideal gases, the heat conductivity, stationary speed, and other properties of the motor can be calculated exactly by a perturbation expansion in the ratio of the mass M of the motor and the mass m of the gas particles [20]. We cite the following results (for the motor represented in Fig. 3, ρ_1 and ρ_2 are the densities of the gas particles in reservoir 1 and 2, respectively, and $2\theta_0$ is the opening angle of the triangle). The conductivity is given by Eq. (2), while the stationary speed reads

$$
\langle V \rangle = \rho_1 \rho_2 (1 - \sin^2 \theta_0) \sqrt{\frac{m}{M}} \sqrt{\frac{\pi k_B}{2M}}
$$
$$
\times \frac{(T_1 - T_2)\sqrt{T_1}}{[2\rho_1 \sqrt{T_1} + \rho_2 \sqrt{T_2}(1 + \sin \theta_0)]^2}.
$$

(3)

Note that the latter vanishes at equilibrium $T_1 = T_2$, which is also a point of flux reversal ($\langle V \rangle > 0$ for $T_1 > T_2$, $\langle V \rangle < 0$ for $T_1 < T_2$). Both the heat conductivity κ and the average speed $\langle V \rangle$ also vanish in the macroscopic limit $M \to \infty$, in agreement with the fact that the fluctuations, being the driving force behind both features, average out in this limit. The above predictions are in very good agreement with molecular dynamics simulations for a dilute hard-sphere (Enskog) gas, despite the correction of nonideality and, more surprisingly, despite the strong finite size effects [20]. Note furthermore that in a finite system, the heat conductivity of the motor will in time restore full equilibrium with overall equal temperature. This return to full equilibrium will, as for the case of the adiabatic piston, be accompanied by the disappearance of the systematic motion.

III. BROWNIAN REFRIGERATOR

Carnot was concerned with steam engines, and his efficiency refers to the motor function: heat flow as input and work as output. However, Carnot efficiency is reached for a reversible process—that is, a process for which the direction of operation can be changed at no cost. Hence the same construction can be used to run the process backward with work as input and, as output, a resulting heat flow, serving either as a refrigerator on the cold side or as a heat pump on the hot side. The application of a similar idea to Brownian motors appears at first sight impossible, since the motor function has an intrinsic heat leak. In fact, the Brownian motor can only operate thanks to this irreversible heat flux. The situation appears therefore to be far removed from the ideal reversible Carnot cycle. However, it is crucial to include in the analysis the effect of loading. We briefly review the discussion presented in Ref. 21. For concreteness, we focus again on the Brownian motor from Fig. 3, but with the same temperature in both reservoirs; that is, the system is initially in thermodynamic equilibrium. Upon the application on the motor unit of an external force F, the system will react according to the Chatelier–Braun principle [22]: It will create processes that resist the force. One such process is quite familiar—namely, friction causing Joule heating in both reservoirs. To lowest order in F, the amplitude of this effect is independent of the direction of the force; that is, it is an F^2 effect. However, the system possesses another mechanism to counteract the force, namely by creating a Brownian motor acting against it. One thus expects that the application of the force will generate a heat flow, which tends to warm one reservoir and cool the other one, such that it activates a Brownian motor working against the applied force. Clearly this effect will depend on the direction of the force. Hence, to lowest order, it will be proportional to F and will be dominant over the Joule heating for F small!

This handwaving discussion is confirmed by invoking the famous concept of Onsager reciprocity [23]. To lowest order in the temperature difference $\Delta T = T_2 - T_1$, the speed given in Eq. (3) can be rewritten as follows:

$$\langle V \rangle = J_1 = L_{12} X_2 \tag{4}$$

with

$$L_{12} = \rho_1 \rho_2 (1 - \sin^2 \theta_0) \sqrt{\frac{m}{M}} \sqrt{\frac{\pi k_B}{2M}}$$
$$\times \frac{T^{3/2}}{[2\rho_1 + \rho_2(1 + \sin \theta_0)]^2} \tag{5}$$

where we introduced the conventional notations $J_1 = \langle V \rangle = \langle \dot{x} \rangle$ for a flux, $X_2 = \Delta T / T^2$ is a thermodynamic force, and L_{12} is the Onsager coefficient corresponding to the cross effect of particle motion induced by a temperature gradient. According to Onsager symmetry, microscopic time reversibility implies the existence of a mirror process, with the force associated to particle motion, namely $X_1 = F/T$, inducing a flux associated to a temperature gradient, namely a heat flux $J_2 = Q$, with amplitude $J_2 = L_{21} X_1$ and $L_{21} = L_{12}$. We arrive at the surprising conclusion that the thermal Brownian motors can be used as heat pumps or refrigerators. This conclusion is again confirmed by an exact analytic calculation and corroborated in molecular dynamics simulations [21].

IV. CARNOT EFFICIENCY

Thermal Brownian motors can function as a motor, a heat pump, and a refrigerator. But can they reach Carnot efficiency? As we explained in some detail above, arguments have been raised against this possibility [11–17]. On the other hand, a number of alternative constructions have been proposed for which it is claimed that Carnot efficiency can be reached [24–29].

To clarify the issue, we will show below that the question can be answered affirmatively using the general framework of linear irreversible thermodynamics [30]. Before we do so, we anticipate the final conclusion by an intuitive argument that builds on the previous discussion and highlights the differences and similarities with the traditional Carnot engine. Suppose that an amount of heat Q leaves a hot reservoir (temperature T_1) and is transformed by an auxiliary system in an amount of work W, while the remainder energy $Q - W$ flows to a cold reservoir (temperature T_2). In order to avoid an entropy-producing step, the reservoirs and auxiliary system are kept at the same temperature during the heat exchange, and the auxiliary system changes its temperature during adiabatic phases while being disconnected from the heat baths. The second law of thermodynamics stipulates that the above process will take place spontaneously, provided that the total entropy does not increase. Since the auxiliary system is returned to its original state after each entire cycle, with its entropy (and energy) being unchanged, the total entropy production per cycle reads $\Delta S_{\text{tot}} = -Q/T_1 + (Q - W)/T_2 \geq 0$. Clearly, the maximum of work will be reached for a reversible process, $\Delta S_{\text{tot}} = 0$, in which case Carnot efficiency (1) is attained. The thermal Brownian motors discussed above are very different from the Carnot machine since we deal with a steady-state situation in which there exists, at all times, a thermal contact between the two heat reservoirs. To avoid this problem, one could pursue the idea of decoupling the Brownian motor during a part of the cycle from its heat bath following the original Carnot construction. However, an isolated nonextensive system cannot be cooled down from an initial canonical distribution to a canonical distribution at lower temperature by merely performing work. As a result, there will be an inherent irreversible, entropy-producing step upon putting the system in contact with the cold reservoir (cf. the detailed discussion in Ref. 31). There is, however, an alternative. We introduced examples of Brownian motors, activated by a temperature difference and driven by the resulting heat flux. While discussing the refrigerator function in Section III, we showed that these motors produce an opposing heat flux upon loading by external force. There is no principal reason why both fluxes could not cancel each other. The resulting process would have zero overall entropy production, and Carnot efficiency would be attained. The thermodynamic analysis given below confirms this general observation but also reveals that such a state of affairs is not at all automatic: It implies a stringent condition on the structural parameters, a condition that is not met in a typical construction.

   Our starting point is a generic construction for the extraction of work from a flow of heat (cf. Fig. 4). The auxiliary system—for example, a Brownian motor—performs work, $W = -Fx$, against an external force F, where x is the corresponding variation of the thermodynamically conjugated variable. The system is at a temperature T and we introduce the corresponding thermodynamic force, $X_1 = F/T$, and flux $J_1 = \dot{x}$ (the dot referring to the time derivative).

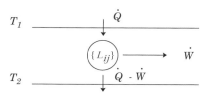

Figure 4.   Generic construction for the extraction of power $\dot{W}$ from a heat flux $\dot{Q}$.

The power (work by the system per unit time) is thus $\dot{W} = -F\dot{x} = -J_1X_1T$. The work is performed under the influence of a heat flux $\dot{Q}$ leaving the hot reservoir at temperature T_1. The cold reservoir is at temperature T_2 (where $T_1 \geq T_2$). The corresponding thermodynamic force is $X_2 = 1/T_2 - 1/T_1$, and the flux is $J_2 = \dot{Q}$. The temperature difference $T_1 - T_2 = \Delta T$ is assumed to be small compared to $T_2 \approx T_1 \approx T$, so one can also write $X_2 = \Delta T/T^2$. Linear irreversible thermodynamics is based on the assumption of local equilibrium with the following linear relationship between the fluxes and forces:

$$J_1 = L_{11}X_1 + L_{12}X_2$$
$$J_2 = L_{21}X_1 + L_{22}X_2 \tag{6}$$

The positivity of the entropy production, $dS/dt = J_1X_1 + J_2X_2 \geq 0$, which is a quadratic form in the thermodynamic forces, implies for the Onsager coefficients L_{ij} that

$$L_{11} \geq 0, \ L_{22} \geq 0$$
$$L_{11}L_{22} - L_{12}L_{21} \geq 0 \tag{7}$$

Furthermore, the Onsager symmetry resulting from the time reversibility of the microscopic dynamics stipulates

$$L_{12} = L_{21} \tag{8}$$

The basic thermodynamic inequality contained in Eq. (7) can conveniently be expressed in terms of the dimensionless coupling strength,

$$q = L_{12}/\sqrt{L_{11}L_{22}} \tag{9}$$

namely,

$$-1 \leq q \leq +1. \tag{10}$$

Note that $|q| = 1$ is equivalent to saying that the Onsager matrix L_{ij} has a determinant equal to zero, implying that $J_1 = J_2 = 0$ can be attained for nonzero values of the thermodynamic forces X_1 and X_2 [cf. Eq. (6)].

After these preliminaries, the question of efficiency to lowest order in $\Delta T/T$ can be addressed following the analysis given in Ref. 32. First we rewrite the efficiency, output work over input heat, as the output power over input heat flux. In Eq. (1) this corresponds to dividing nominator and denominator by the time required for the process under consideration. When operating under steady-state conditions, as we are doing here, the latter quantities are also the time derivatives of work and heat, respectively. Using the general expressions for power and heat flux obtained from Eq. (6), the efficiency is found to be the following function of the thermodynamic force ratio $\kappa = X_1/X_2$:

$$\eta = \frac{\dot{W}}{\dot{Q}} = -\frac{\Delta T}{T} \frac{J_1 X_1}{J_2 X_2} = -\frac{\Delta T}{T} \kappa \frac{L_{11}}{L_{21}} \frac{\kappa + L_{12}/L_{11}}{\kappa + L_{22}/L_{21}} \qquad (11)$$

We next note that the efficiency is further simplified under the condition $|q| = 1$: In this case the last fraction in Eq. (11) reduces to one. To fix the ideas, we consider $q = -1$, with $L_{12} < 0$. While applying a given thermal gradient, with $X_2 > 0$, we start increasing the load X_1. As the ratio κ increases, so does the efficiency, until eventually we reach the stopping force ratio, $\kappa = \kappa_{\text{stop}}$ $= -L_{12}/L_{11}(= -L_{22}/L_{21})$. At this point, both fluxes vanish. The simplified expression for the efficiency immediately reduces to $\eta = \Delta T/T$. This result, valid to lowest order in $\Delta T/T$, is identical to Carnot efficiency [cf. Eq. (1)]. For $\kappa < \kappa_{\text{stop}}$, the efficiency is clearly below Carnot. For values $\kappa > \kappa_{\text{stop}}$, both fluxes reverse sign (i.e., become negative) and we are in the regime of the heat pump or refrigerator: Work is now injected with a resulting heat flow from the cold to the hot reservoir. The relevant definitions for the efficiency in these cases are different, namely $\dot{Q}/\dot{W}$ and $(\dot{Q} - \dot{W})/\dot{W}$ for the heat pump and refrigerator function, respectively. For completeness, note finally that Eq. (11) predicts an efficiency always below the Carnot value when the coupling is not complete, $|q| < 1$.

To go beyond the linear approximation in $\Delta T/T$, while staying within the framework of linear irreversible thermodynamics, we follow the analysis presented in Ref. 33, with the introduction of a cascade construction as in Fig. 5. We introduce, between the hot reservoir at T_1 and the cold one at T_2, a continuous set of auxiliary heat reservoirs, labeled by the coordinate y, at decreasing temperatures $T(y)$ $(T(1) = T_1, T(2) = T_2)$. These reservoirs will play a role akin to a catalyst, serving merely as temporary repositories of energy. We furthermore assume that we have at our disposal an infinite set of identical copies of the auxiliary system, operating between the successive pairs of reservoirs. For simplicity, we assume from the onset that these machines operate at Carnot efficieny, implying that the coupling strength $|q|$ equals 1. The heat flux traversing the reservoir located at y (at temperature $T(y)$), will be denoted by $\dot{Q}(y)$ $(\dot{Q}(y = 1) = \dot{Q}$ is the heat flux leaving the hot reservoir). The incremental power delivered by the system located between y and $y + dy$ is denoted by $d\dot{W}(y)$. Since the power is derived solely from the transfer of the heat, and not from the internal energy of the system, conservation of energy implies that $\dot{Q}(y + dy) = \dot{Q}(y) - d\dot{W}(y)$, whence (a) $\dot{Q}(y) = \dot{Q} - \int_1^y d\dot{W}(y')$. Assuming that the system operates under the above-mentioned conditions

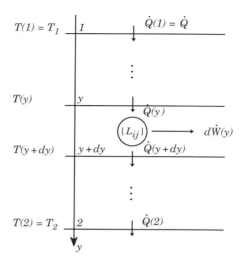

Figure 5. Cascade construction with a continuum of copies of the system with Onsager matrix L_{ij} located in $[y, y + dy]$, along with corresponding heat baths for the conversion of a heat flow $\dot{Q}(y)$ into a power contribution $d\dot{W}(y)$.

(Carnot efficiency, perfect coupling $|q| = 1$), we have furthermore: (b) $d\dot{W}(y)/\dot{Q}(y) = [T(y) - T(y + dy)]/(T(y))$. By combining (a) and (b), we obtain the following closed equation for $d\dot{W}(y)/dy$, as a function of the prescribed temperature profile:

$$\frac{d\dot{W}(y)}{dy} = -\frac{d\ln T(y)}{dy}\left(\dot{Q} - \int_1^y dy' \, \frac{d\dot{W}(y')}{dy'}\right) \tag{12}$$

Differentiating this equation with respect to y, one obtains a first-order differential equation for $d\dot{W}(y)/dy$. Straightforward integration [with the appropriate boundary condition that follows from the application of Eq. (12) at $y = 1$) leads to

$$\frac{d\dot{W}(y)}{dy} = -\dot{Q}\frac{T(y)}{T(1)}\frac{d\ln T(y)}{dy} = -\frac{\dot{Q}}{T(1)}\frac{dT(y)}{dy} \tag{13}$$

from which the result $\dot{W}(y) = \dot{Q}[1 - T(y)/T(1)]$ immediately follows. We conclude that the efficiency is given by

$$\eta = \frac{\int_1^2 d\dot{W}(y)}{\dot{Q}} = \frac{\dot{W}(2)}{\dot{Q}} = 1 - \frac{T(2)}{T(1)} \tag{14}$$

which is independent of the prescribed temperature profile and identical to Carnot effciency [cf. Eq. (1)]! The above analysis reveals the conditions for reaching Carnot efficiency. The entropy production of the heat flow needs to be exactly compensated by the (negative) entropy production of the loading $(J_1 X_1 = -J_2 X_2)$. This can only be achieved in the limit of perfectly coupled systems $(|q| = 1)$. In addition, the appropriate operational condition is, for a given value of the thermal force X_2, the application of the stopping force $X_1 \to X_1^{stop}$, defined as the force for which the particle motion stops $J_1 = 0$. In this limit the (total) heat flux J_2, entailing both the direct heat flux resulting from the thermal conductivity L_{22} and the refrigerator flux due to the loading (Onsager coefficient L_{22}), also vanishes, rendering the entire process infinitely slow and reversible. The value of $|q|$ for the mechanical thermal Brownian motors discussed above has not been calculated explicitly, but it is presumably not even close to one, which would explain the reported low efficiencies. There is, however, no principal reason why such a Brownian motor could not be designed. Such a construction would also be of more than academic importance for the reason explained below. We first point out that maximal efficiency per se is not the right criterion from an operational point of view. Typically, both engineering design and natural selection of motors aim at minimizing a weighted cost comprising various factors including efficiency, but also power and material constraints [33]. Interestingly, maximizing efficiency subject to maximum power leads to the same requirement $|q| = 1$ [34]; hence a construction with this property could also be of technological interest. In this respect, it is also worth mentioning that values of $|q|$ close to one have been reported in several bio- and electrochemical processes [35].

V.   DISCUSSION AND PERSPECTIVES

The above discussion about reaching Carnot efficiency by coupling an existing heat flow to another process is completely general, even though we have used it to settle the issue about the efficiency of thermal Brownian motors. It is reassuring that Brownian motors are, from the basic thermodynamic point of view, not really different from macroscopic (thermodynamic) machines. In fact, one of the powerfull messages of thermodynamics is precisely that it can treat thermodynamic forces, deriving from thermal motion, in the same way as external forces. At the same time, we have to stress that thermodynamic concepts have only a meaning in a statistical sense. They do not apply to a single realization of a process in stochastic entities such as Brownian motors. For example, a harmonically bound Brownian particle (or a Brownian motor) at equilibrium with its surrounding fluid will continuously lose energy by friction but will equally gain energy by fluctuation. In a single short realization, the particle can occasionally contribute potential energy to the spring, but will equally often also dissipate it. On average, no work is performed. The fluctuation dissipation theorem is an expression of this perfect balance. Over the past years, various new results have shown that this relation can be extended beyond the regime of linear irreversible thermodynamics. More spectacularly, it seems that a probabilistic approach, taking into account precisely those fluctuations in which a system receives more work than traditional thermodynamics prescribes, allows us to replace the inequalities (typical for thermodynamics) by equalities [13, 36–40]. Such fluctuations become extremely rare as the system size goes macroscopic. Hence, while these new ideas are practically irrelevant for macroscopic systems, they may be of great interest in small-scale systems—in particular, in nanotechnology and molecular biology [41].

Acknowledgments

We thank R. Kawai, P. Gaspard, J. M. R. Parrondo, and C. Jarzynski for stimulating discussions.

References

1. S. Carnot, *Réflexions sur la Puissance Motrice du Feu, et sur les Machines Propres à Développer cette Puissance*, 1824.

2. R. Clausius, *Mechanical Theory of Heat*, John van Voorst, London, 1867.

3. J. M. R. Parrondo and B. J. de Cisneros, *Appl. Phys.* **A75**, 179 (2002).

4. H. Linke, M. Downton, and M. Zuckermann, *Chaos* **15**, 026111 (2005).

5. R. D. Astumian, *J. Phys.: Condens. Matter* **17**, S3753 (2005).

6. P. Reimann, *Phys. Rep.* **361**, 57 (2002).

7. R. P. Feynman, R. B. Leighton, and M. Sands, *The Feynman Lectures on Physics* I, Addison-Wesley, Reading, MA, 1963.

8. J. M. R. Parrondo and P. Espagnol, *Am. J. Phys.* **64**, 1125 (1996).

9. K. Sekimoto, *J. Phys. Soc. Japan* **66**, 1234 (1997).

10. To be more precise, the distribution of the rotational speed will be Maxwellian, to leading order in the ratio of the mass m of the gas particles over the mass M of the motor, at a temperature intermediate between that of both baths.

11. M. Magnasco and G. Stolovitzky, *J. Stat. Phys.* **93**, 615 (1998).

12. T. Hondou and F. Takagi, *J. Phys. Soc. Japan* **67**, 2974 (1998).

13. C. Jarzynski and O. Mazonka, *Phys. Rev. E* **59**, 6448 (1999).

14. T. Hondou and K. Sekimoto, *Phys. Rev. E* **62**, 6021 (2000).

15. S. Velasco1, J. M. M. Roco, A. Medina, and A. Calvo Hernandez, *J. Phys. D Appl. Phys.* **34**, 10001006 (2001).

16. K. Pesz, B. J. Gabrys, and S. J. Bartkiewicz, *Phys. Rev. E* **66**, 061103 (2002).

17. D. Suzuki and T. Munakata, Stationary probability flow and vortices for the Feynman ratchet, unpublished.

18. H. S. Leff, *Am. J. Phys.* **62**, 120 (1994); Ch. Gruber, *Eur. J. Phys.* **20**, 259 (1999); Ya. G. Sinai, *Theor. Math. Phys.* **125**, 1351 (1999); E. Kestemont, C. Van den Broeck, and M. Malek

Mansour, *Europhys. Lett.* **49**, 143 (2000); B. Crosignani and P. Di Porto, *Europhys. Lett.* **53**, 290 (2001); T. Munakata and H. Ogawa, *Phys. Rev. E* **64**, 036119 (2001); M. Malek Mansour, C. Van den Broeck, and E. Kestemont, *Europhys. Lett.* **69**, 510 (2005).

19. C. Van den Broeck, M. Malek Mansour, and E. Kestemont, *Europhys. Lett.* **56**, 771 (2001).

20. C. Van den Broeck, R. Kawai, and P. Meurs, *Phys. Rev. Lett.* **93**, 090601 (2004); P. Meurs, C. Van den Broeck, and A. Garcia, *Phys. Rev. E* **70**, 051109 (2004); C. Van den Broeck, P. Meurs, and R. Kawai, *New J. Phys.* **7**, 10 (2005); P. Meurs and C. Van den Broeck, *J. Phys. Condens. Matter* **17** S3673 (2005).

21. C. Van den Broeck and R. Kawai, *Phys. Rev. Lett.* **96**, 210601 (2006).

22. H. Callen, *Thermodynamics and an Introduction to Thermostatistics*, 2nd ed., John Wiley & Sons, New York, 1985.

23. L. Onsager, *Phys. Rev.* **37**, 405 (1931); **38**, 2265 (1932).

24. I. Derenyi and R. D. Astumian, Phys. Rev. **E59**, R6219 (1999).

25. T. E. Humfrey, R. Newbury, R. P. Taylor, and H. Linke, *Phys. Rev. Lett.* **89**, 116801 (2002); T. E. Humfrey and H. Linke, *Phys. Rev. Lett.* **94**, 096601 (2005).

26. H. Sakaguchi, *J. Phys. Soc. Japan* **67**, 709 (1998).

27. I. M. Sokolov, *Europhys. Lett.* **44**, 278 (1998).

28. M. Matsuo and S. Sasa, *Physica* **A276**, 188 (2000).

29. J. V. Hernandez, E. R. Kay, and D.A. Leigh, *Science* **306**, 1532 (2004).

30. I. Prigogine, *Etude Thermodynamique des Phénomènes Irréversibles*, Desoer, Liège, 1947; S. R. de Groot and P. Mazur, *Non-Equilibrium Thermodynamics*, Dover, New York, 1984.

31. K. Sato, K. Sekimoto, T. Hondou, and F. Takagi, *Phys. Rev.* **E66**, 016119 (2002).

32. O. Kedem and S. R. Caplan, *Trans. Faraday Soc.* **61**, 1897 (1965).

33. A. De Vos, *Endoreversible Thermodynamics of Solar Energy Conversion*, Oxford University Press, Oxford, 1992; R. S. Berry, V. A. Kazakov, S. Sieniutycz, Z. Szwast, and A. M. Tsvilin, *Thermodynamic Optimization of Finite-Time Processes*, John Wiley & Sons, Chichester, 2000); P. Salamon, J. D. Nulton, G. Siragusa, T. R. Andersen, and A. Limon, *Energy* **26**, 307 (2001).

34. C. Van den Broeck, *Phys. Rev. Lett.* **95**, 190602 (2005).

35. J. C. Aledo and A. E. del Valle, *J. Biol. Chem.* **279**, 55372 (2004); V. M. Rozenbaum, D. Y. Yang, S. H. Lin, et al., *J. Phys. Chem.* **B108**, 15880 (2004); J. W. Stucki, *Proc. Roy. Soc. London*, Biol. Sci. **244**, 197 (1991); S. Koter and C. H. Hamann, *J. Non-equilibrium Thermodynamics* **15**, 315 (1990).

36. G. Gallavotti and E. G. D. Cohen, *Phys. Rev. Lett.* **74**, 2694 (1995); D. Evans, E. G. D. Cohen, and G. P. Morris, *Phys. Rev. Lett.* **71**, 2401 (1993).

37. C. Jarzynski, *Phys. Rev. Lett.* **78**, 2690 (1997).

38. G. E. Crooks, *Phys. Rev.* **E60**, 2721 (1999).

39. F. Ritort, Séminaire Poincaré **2**, 195 (2003).

40. B. Cleuren, C. Van den Broeck, and R. Kawai, *Phys. Rev. Lett.* **96**, 050601 (2006).

41. D. Collin, F. Ritort, C. Jarzynski, S. B. Smith, I. Tinoco, Jr., and C. Bustamante, *Nature* **437**, 231 (2005).

DNA IN CHROMATIN: FROM GENOME-WIDE SEQUENCE ANALYSIS TO THE MODELING OF REPLICATION IN MAMMALS

A. ARNEODO, B. AUDIT, E. B. BRODIE OF BRODIE, S. NICOLAY, AND P. St. JEAN

Laboratoire Joliot-Curie et Laboratoire de Physique, CNRS, Ecole Normale Supérieure de Lyon, 69364 Lyon Cedex 07, France

Y. d'AUBENTON-CARAFA, C. THERMES, AND M. TOUCHON

Centre de Génétique Moléculaire, CNRS, 91198 Gif-sur-Yvette, France

C. VAILLANT

Laboratoire Statistique et Génome, 91000 Evry, France

CONTENTS

Special Volume in Memory of Ilya Prigogine: Advances in Chemical Physics, Volume 135,
edited by Stuart A. Rice
Copyright © 2007 John Wiley & Sons, Inc.

INTRODUCTION

The dynamics of folding and unfolding of DNA within living cells plays a major role in regulating many biological processes, such as gene expression, DNA replication, recombination, and DNA damage repair [1–4]. As sketched in Fig. 1, the genomic DNA of eukaryotic cells is tightly packaged into nucleosomes that constitute the basic units of chromatin [5]. As experimentally detailed by high-resolution X-ray analysis [6–8], each nucleosome consists of almost two turns of DNA wrapped around an octamer of core histone proteins. An additional fragment of DNA associated with a linear histone separates successive nucleosomes that are disposed as beads-on-a-string along the DNA. This nucleosomal array is further organized into successive higher-order structures [4] including the condensation into the 30-nm chromatin fiber and the formation of chromatin loops, up to a full extent of condensation in metaphase chromosomes. Actually, the structure and dynamics of chromatin are under the control of a number of mechanisms involving DNA–protein interactions, which may depend upon the nucleotide sequence since DNA is an heteropolymer with locally sequence-dependent physical (mechanical, geometrical, etc.) properties. The precise influence of the so-called primary structure (i.e., the sequence) on the organization of chromatin at all scales remains controversial. On a local scale, specific sequence elements have been identified to interact with protein components of chromatin. For instance, some sequence motifs that favor the formation and positioning of nucleosomes were found to be regularly spaced— for example, the 10-bp periodicity [9, 10] exhibited by the AA dinucleotide. Alternatively, similar motifs were shown to present long-range correlations along the genome that are a signature of nucleosomes [11–13]. Other DNA regions, the scaffold or matrix attachment regions that constitute the anchor points of chromatin loop domains, are constituted by ∼1-kbp AT-rich sequence patterns [14, 15]. On larger scales, the folding of the nucleosomal strings into higher-order structures has been the issue of various models involving, for example, random packing, coiling into hierarchical helical structures (solenoids) [16–18], or loop-models [14, 15, 19–21], but the DNA sequence itself was not taken into

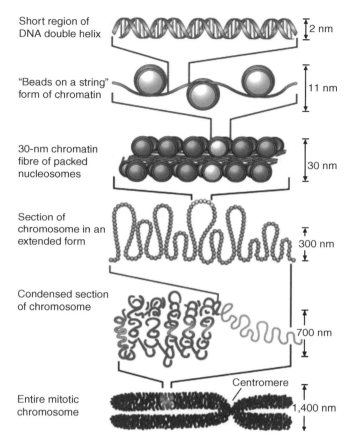

Short region of
DNA double helix — 2 nm

"Beads on a string"
form of chromatin — 11 nm

30-nm chromatin
fibre of packed
nucleosomes — 30 nm

Section of
chromosome in an
extended form — 300 nm

Condensed section
of chromosome — 700 nm

Centromere

Entire mitotic
chromosome — 1,400 nm

Figure 1. Hierarchical structure of eukaryotic DNA. Each DNA molecule is packed into a mitotic chromosome that is 1/50,000 shorter than its extended length. (Adapted with permission from Macmillan Publishers Ltd. [25], copyright 2003).

account. Some recent experimental results suggest that loops might be organized by the active transcription complexes [22–24]. Accordingly, gene positions and transcriptional activity would constitute major determinants of the microscopic structure of chromatin that would self-organize in a rather predictable way: The 3D structure would then result to some extent from the DNA primary sequence.

Actually, there is much more to be learned about the different stages of compaction of DNA inside the cell nucleus (Fig. 1) from the DNA sequence than commonly thought. The originality of the approach described in this chapter relies on the fact that we are going to extract structural, dynamical, and functional informations from the primary DNA sequence using concepts coming from statistical and nonlinear physics and methodologies issued from physics and signal processing [13, 26, 27]. More precisely, we will mainly use a mathematical microscope, namely the continuous wavelet transform [28, 29], to explore the structural complexity of signals generated from adequate codings of the DNA sequences. In a preliminary work [11–13], we have used the space-scale decomposition provided by the wavelet transform to reveal and analyze the scale invariance properties of eukaryotic, eubacterial, and archaeal genomic sequences. This study suggests that the existence of long-range correlations, up to distances $\sim$ 20 kbp, is the signature of the nucleosomal structure and dynamics of the 30-nm chromatin fiber. Actually, these long-range correlations are mainly observed in the DNA bending profiles obtained when using some structural coding of the DNA sequences that accounts for the fluctuations of the local double-helix curvature within the nucleosome complex. Because of the approximate planarity of nucleosomal DNA loops, we have developed some modeling of the thermodynamics of 2D DNA loops in the presence of long-range correlated structural disorder induced by the sequence [30, 31]. These long-range correlations clearly favor the autonomous formation of small (i.e., few hundred base paris) DNA loops, larger correlations, smaller size of the loop, and in turn the propensity of eukaryotic DNA to interact with histones to form nucleosomes. In addition, we have shown that this long-range correlated structural disorder is likely to induce local hyperdiffusion of these loops, which provides a very attractive interpretation to the nucleosome repositioning dynamics. Let us emphasize that a recent statistical analysis [32] of nucleosome positioning data obtained recently by Yuan et al. [33] for chromosome III of *S. cerevisiae* has provided a convincing experimental confirmation that long-range correlations in the genomic sequence strongly influence the overall formation and positioning of nucleosomes. In this chapter, we will keep decreasing the magnification of our mathematical wavelet transform microscope to investigate the complexity of DNA sequences at scales larger than 30 kbp. Our goal is to show that at these large scales, the primary sequence still contains structural and mechanical information, no longer on DNA but rather on the 30-nm chromatin fiber and its propensity to form loop and multi-loop structural patterns (Fig. 1) that are likely to stabilize chromatin domains of autonomous DNA replication and gene expression. This study leads us to propose some universal physical mechanism accounting for the self-organization of multi-looped rosettes that would favor the so-called tertiary chromatin structure, prior to the replication and transcription proteic complexes coming into play.

The chapter is organized as follows. Section II is devoted to materials and methods. In Section III, we show [34, 35] that the GC content displays rather regular nonlinear oscillations with two main periods of 110 ± 20 kbp and 400 ± 50 kbp, which are well-recognized characteristic scales of chromatin loops and loop domains involved in the hierarchical folding of chromatin fibers.

These frequencies are also remarkably similar to the size of mammalian replicons and replicon clusters. When further investigating deviations from intrastrand equimolarities between T and A and between G and C, the so-called TA and GC skews, we corroborate the existence of these rhythms as the footprints of replication and/or transcription mutation bias and we show that the observed relaxational oscillations enlighten a remarkable cooperative gene organization [34, 35]. In Section IV, with the specific goal to disentangle the replication and transcription contributions to the TA and GC skews, we analyze 14,854 intron-containing genes annotated in the human genome and we show [36, 37] that these skews are correlated to each other and display a characteristic step-like profile exhibiting sharp transitions between transcribed (finite bias) and nontranscribed (zero bias) regions. In most sequences, we observe an excess of T over A and of G over C, reflecting transcriptionally coupled mutational processes in germ line cells. In Section V, we reveal [38, 39] the actual existence of replication-associated strand asymmetries by further studying the behavior of the TA and GC skews around the origins of replication experimentally identified. We find that the (TA + GC) skew displays rather sharp upward jumps from negative to positive values at the origin locations. When using the wavelet transform to perform a multiscale analysis of the 22 human autosomal chromosomes skew profiles, we reveal numerous sharp upward jumps that allow us to identify a set of 1012 putative replication initiation zones. Between two neighboring sharp upward jumps, the skew displays a linear decreasing profile leading to a characteristic jagged pattern also observed in mouse and dog genomes [38]. In Section VI, we propose [38, 39] a model of replication in mammalian cells with well-positioned replication origins and random terminations. A systematic analysis of the gene content around the putative replication origins enlightens a remarkable gene organization with clusters of genes mostly co-oriented with the progression of the replication fork. This observation suggests that these replication initiation zones are likely to correspond to regions where the chromatin fiber is more open so that DNA be more easily accessible. In Section VII, we show that, in the crowded environment of the cell nucleus, the presence of such intrinsic (sequence-dependent) decondensed structural defects actually predisposes the chromatin fiber to spontaneously form rosette-like structures. Prior to any external factors coming into play, these multi-looped rosettes self-organize from the entropy-driven assembling of neighboring defects into clusters by depletive forces. These rosettes provide an attractive description of the compartmentalization of the genome into replication foci that are observed in interphase mammalian nuclei as stable chromatin domains of autonomous DNA replication and gene expression. We conclude, in Section VIII, by discussing some new experimental perspectives including *in vivo* visualization of the rosette-like organization of the tertiary chromatin structure via the clustering of replication origins.

II.   MATERIALS AND METHODS

A.   Data Sets

Sequences.   Sequence and gene annotation data were retrieved from the Genome Browser of the University of California, Santa Cruz for the human (July 2003 in Sections III and IV, May 2004 in Sections V and VI), mouse (May 2004), and dog (July 2004) genomes. To delineate the most reliable intergenic regions, transcribed regions were retrieved from "`all_mrna`", one of the largest sets of annotated transcripts. Among transcribed sequences, *sense* (respectively *antisense*) genes have the same orientation as the Watson (respectively Crick) strand. To obtain intronic sequences, we used the KnownGene annotation (containing only protein-coding transcripts); when several transcripts presented common exonic regions, only common intronic sequences were retained. For the dog genome, only preliminary gene annotation were available, precluding the analysis of intergenic and intronic sequences.

Sequence Repeats.   In Section IV, V, and VI, to exclude repetitive elements that might have been inserted recently and would not reflect long-term evolutionary patterns, we used REPEATMASKER [40], leading to a reduction of $\sim$ 40–50% of the human sequence length.

Human Intron Sequences.   In Section IV, human intron sequences were downloaded from RefGene (April 2003) at UCSC. When several genes presented identical exonic sequences, only the longest one was retained; repeated elements were removed with REPEATMASKER. The introns of each gene were taken as a single sequence; introns without repeats were also taken as a single sequence; to avoid the skew associated with splicing signals, 560 bp were removed at both intron extremities. When the resulting intron sequences were shorter than 1120 bp, they were not considered for the analysis, leading to 14,854 intron-containing genes.

Human Replication Origins.   Nine replication origins were examined—namely, those situated near the genes MCM4 [41], HSPA4 [42], TOP1 [43], MYC [44], SCA-7 [45], AR [45], DNMT1 [46], Lamin B2 [47], and β-globin [48].

Sequence Alignments.   Mouse and dog regions homologous to the six human regions shown in Fig. 9 were retrieved from University of California, Santa Cruz (Human Synteny). Mouse intergenic sequences were individually aligned by using PIPMAKER [49], leading to a total of 150 conserved segments longer than

100 bp ($>70\%$ identity) and corresponding to a total of 26 kbp (5.3% of intergenic sequences).

B. Coding Rules

GC Content.   GC content fluctuations in the human genome were computed in adjacent (nonoverlapping) 1-kbp windows.

Strand Asymmetries.   The TA and GC skews were calculated in nonoverlapping windows of size 1-kbp as

$$S_{\text{TA}} = \frac{n_{\text{T}} - n_{\text{A}}}{n_{\text{T}} + n_{\text{A}}}, \qquad S_{\text{GC}} = \frac{n_{\text{G}} - n_{\text{C}}}{n_{\text{G}} + n_{\text{C}}} \tag{1}$$

where n_{A}, n_{C}, n_{G}, and n_{T} are, respectively, the numbers of A, C, G, and T in the windows. Because of the observed correlation between the TA and GC skews (Section IV), we also considered the total skew

$$S = S_{\text{TA}} + S_{\text{GC}} \tag{2}$$

From the skews $S_{\text{TA}}(n)$, $S_{\text{GC}}(n)$, and $S(n)$, obtained along the sequences, where n is the position (in kbp units) from the origin, we also computed the cumulative skew profiles:

$$\Sigma_{\text{TA}}(n) = \sum_{j=1}^{n} S_{\text{TA}}(j), \qquad \Sigma_{\text{GC}}(n) = \sum_{j=1}^{n} S_{\text{GC}}(j) \tag{3}$$

and

$$\Sigma(n) = \sum_{j=1}^{n} S(j) \tag{4}$$

C.   Space-Scale Analysis Based on the Continuous Wavelet Transform

The continuous wavelet transform (WT) is a space-scale analysis that consists in expanding signals in terms of wavelets that are constructed from a single function, the analyzing wavelet ψ, by means of dilations and translations [13, 27–29]. When using the successive derivatives of the Gaussian function as analyzing wavelets, namely

$$g^{(N)}(x) = (-1)^{N} d^{N} g^{(0)}(x) / dx^{N} \tag{5}$$

where

$$g^{(0)}(x) = \frac{1}{\sqrt{2\pi}} e^{-x^2/2} \tag{6}$$

then the WT of a signal s takes the following simple expression:

$$\begin{aligned} W_{g(N)}[s](x, a) &= \frac{1}{a} \int_{-\infty}^{+\infty} s(y) g^{(N)}\left(\frac{y - x}{a}\right) dy \\ &= \frac{d^N}{dx^N} W_{g(0)}[s](x, a) \end{aligned} \tag{7}$$

where x and $a(> 0)$ are the space and scale parameters, respectively. Equation (7) shows that the WT computed with $g^{(N)}$ at scale a is nothing but the Nth derivative of the signal $s(x)$ smoothed by a dilated version $g^{(0)}(x/a)$ of the Gaussian function. This property is at the heart of various applications of the WT microscope as a very efficient multi-scale singularity tracking technique [13, 50]. Actually, the skeleton of the WT provides a space-scale partitioning that is likely to contain all the information on the singularities of the signal considered. The WT skeleton is defined, at each scale a, by the set of all the points x_i that correspond to a local maximum of $|W_\psi[s](x, a)|$ and then by connecting these points across scales into the so-called maxima lines [13, 27, 29]. In Section VI, the ability of identifying sharp jumps in noisy skew profiles from the WT skeleton will be at the heart of the methodology we will propose to detect the origins of replication in mammalian genomes [38, 39].

One of the main advantages of the WT is its adaptative ability to perform time–frequency analysis [28, 29] when using complex analyzing wavelets like the Morlet's wavelet:

$$\psi_M(x) = \frac{1}{\sqrt{2\pi}} e^{i\omega x}(e^{-x^2/2} - \sqrt{2} e^{-\omega^2/4} e^{-x^2}) \tag{8}$$

where the second term in the r.h.s. is negligible for large ω values ($\omega \gtrsim 5$). The *scale-spectrum* of a signal s of total length L is defined as

$$\Lambda(a) = \frac{1}{L} \int_0^L |W_{\psi_M}[s](x, a)| dx \tag{9}$$

III.   LOW-FREQUENCY RHYTHMS IN HUMAN DNA SEQUENCES

A.   GC Content

The recent sequencing of the human genome [51] has opened the door to the statistical analysis of genomic sequence complexity on a chromosomal scale.

One of the most striking features of eukaryotic chromosomes is their large-scale variations in base composition. In particular, an extraordinary large heterogeneity of the GC content is observed in mammalian genomes; this has led Bernardi [52, 53] to propose a description of these genomes in terms of a mosaic organization of domains of relatively constant GC levels, originally called *isochores*. The isochore model is a topic of controversial discussions [51, 54–60]. Nevertheless, there is definite evidence that the compositional heterogeneity in a DNA sequence correlates with its GC content [57], which is unanimously recognized as a fundamental property of the chromosomal DNA and is likely to be one of the possible key to the understanding of the organization of eukaryotic genomes [52, 53, 56, 57]. Indeed the GC content has a taxonomy value [61]; it determines the amino acid composition of the encoded proteins and is also related to codon usage in genes [62]. Moreover, there is conspicuous evidence [56, 57, 63] that GC-rich and GC-poor regions, respectively, match the cytogenic R and G bands and correlate well with early and late replicating domains in the cell cycle. GC-rich regions correspond to regions of very high density of genes including the housekeeping genes and associated CpG islands and also of short interdispersed repetitive DNA elements (SINES, Alu) [51]. In contrast, GC-poor regions are definitively poor in genes, predominantly tissue-specific genes containing rather long introns, but are relatively rich in long interdisperse repetitive DNA elements (LINES) [51] that are significantly more abundant in these regions. The GC content has also some impact on the structure of chromatin. For example, it has been suggested [15] that the proteic chromosomal scaffold that serves as a structural skeleton for the organization of chromatin loops is much less tightly folded (to the benefit of replication and transcription processes) in GC-rich than in GC-poor regions.

Figure 2 reports the results [34, 35] of a space-scale decomposition of the GC content fluctuations of a 10-Mbp-long fragment of human chromosome 22, when using the Gaussian $g^{(0)}(x)$ [Eq. (6)] as smoothing filter. This decomposition reveals that for distances larger than $\sim$20–30 kbp, the GC content can no longer be considered as fluctuating homogeneously (at smaller scales the fluctuations cannot be distinguished from a monofractal long-range correlated noise [35]); it instead displays rather regular nonlinear oscillatory behavior. The corresponding scale spectrum $\Lambda(a)$ [Eq. (9)], shown vertically in Fig. 2b, reveals the existence of two main broad peaks corresponding to the scales $l_1 = 100 \pm 20$ kbp and $l_2 = 400 \pm 50$ kbp, respectively, that emerge from a continuous background. The former is the characteristic length of the basic oscillations obtained with the low-pass filtering scale $a_1^* = 40$ kbp, although one may observe from time to time oscillations that have a larger length ($\sim 2l_1 = 200$ kbp). If one uses a larger filtering scale $a_2^* = 160$ kbp, in order to smooth both the "small scales" (high frequencies) colored noise component and the basic oscillations of scale l_1, one gets some oscillatory profile with a

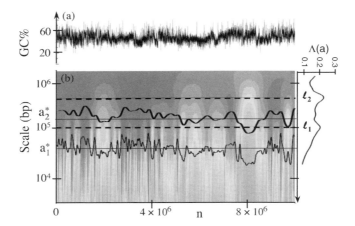

Figure 2. Space-scale representation of the GC content of a 10-Mbp-long fragment of human chromosome 22 when using a Gaussian smoothing filter $g^{(0)}(x)$ [Eq. (6)]. (a) GC content fluctuations computed in adjacent 1 kbp intervals. (b) Color coding of the convolution product $W_{g(0)}[GC](n,a) = (GC * g^{(0)}(\cdot/a))(n)$ using 256 colors from black (0) to red (max); superimposed are shown the smoothed GC profiles obtained at scales $a_1^* = 40$ kbp and $a_2^* = 160$ kbp. On the right-hand side we see vertically the scale (frequency^{-1}) spectrum $\Lambda(a)$ [Eq. (9)] computed with the complex Morlet wavelet [Eq. (8)] over the entire chromosome 22. The horizontal dashed lines in the color picture correspond to the two main characteristic oscillations length $l_1 = 100$ kbp and $l_2 = 400$ kbp. See color insert.

fundamental length $l_2 = 400$ kbp as illustrated in Fig. 3a. Let us point out that the investigation of large GC-rich fragments in various human chromosomes reveals similar periodicities [34], namely $l_1 = 120 \pm 30$ kbp, $l_2 = 410 \pm 60$ kbp for chromosome 11 (24 Mbp, NT_033899.3), $l_1 = 130 \pm 30$ kbp, $l_2 = 420 \pm 60$ kbp for chromosome 14 (68 Mbp, NT_02637.9), and $l_1 = 110 \pm 20$ kbp, $l_2 = 390 \pm 50$ kbp for chromosome 21 (29 Mbp, NT_011512.7).

A possible interpretation of these low-frequency rhythms observed in the GC content is of structural nature and is related to recent experimental and numerical observations of the high-order hierarchical folding of chromatin into fibers and loops of different sizes. 100 kbp is typically the size of DNA loops that are observed by electron microscopy [14, 64, 65] to be held together by a longitudinal network of scaffolding proteins in histone-depleted chromosomes, favoring a radial loop/scaffold model [15, 66, 67] of metaphase chromosome structure. 100 kbp is also very close to the chromosome loop size ($\sim$80 kbp) estimated from physical measurements of the dynamics of force relaxation in single mitotic chromosomes [68]. Furthermore, 400 kbp is likely to be the size of larger chromatin loops made of a few basic loops that have some coherent dynamical behavior during the cell cycle possibly governed by the mechanisms

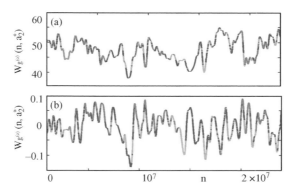

Figure 3. Compositional oscillations observed in the human chromosome 22 fragment (23 Mbp, NT_011520.8) after low-pass filtering at scale $a_2^* = 160\,\mathrm{kbp}$ (see Fig. 2). (a) GC content. (b) Total skew $S = S_{TA} + S_{GC}$ [Eq. (2)]. The red (blue) portions of the profiles correspond to the location of sense (antisense) genes that have the same (opposite) orientation than the sequence. The location of the immunoglobulin locus is shown in pink. See color insert.

that underlie replication and transcription processes [69–73]. As an alternative to the loop/scaffold [15, 66, 67] and multi-loop subcompartment [74] models, 100 kbp and 400 kbp might well be characteristic scales involved into the successive levels of helical coiling of chromatin into fibers or tubes of diameters ranging from 30 to 700 nm observed during interphase using either electron or light microscopy [16, 75–80].

B.   Strand Asymmetries

An alternative interpretation of the rhythms observed in the GC content is of functional nature and is a direct consequence of the observation that 100 kbp and 400 kbp correlate well with the replicon sizes observed in warm-blooded vertebrate organisms [81–83]. Early experimental investigations [84, 85] of replicon size by fiber auto-radiography or fluorography have led to the classical view that mammalian replicons are heterogeneous in size but that most fall into the range of 30–450 kbp with the most frequent sizes in the range 75–150 kbp. Furthermore, there is experimental evidence [81, 83, 84] that replicons are likely to be in groups, the so-called replicon foci, with all the replicons in each group firing at similar time in the S-phase. Newer results obtained with modern replicon-mapping methods clearly show the existence of much larger replicons (the largest ones being as large as a few Mbp) than previously thought, requiring most or all the S-phase to be completed [83, 86, 87]. In particular, the average size of a mammalian replicon has been reconsidered to be more likely ~500 kbp [86, 87].

According to the second parity rule [88, 89], under no strand-bias conditions, each genomic DNA strand should present equimolarities [90, 91] of A and T and of G and C. Deviations from intrastrand equimolarities have been extensively studied in prokaryote, organelle, and viral genomes for which they have been used to detect the origins of replication [92–95]. During replication, mutational events can affect the leading and lagging strands (see Section IV.A) differently, and one strand can be more efficiently repaired than the other one, leading to strand compositional asymmetry. In eukaryotes, the existence of compositional biases has been debated, and most attempts to detect the replication origins from strand compositional asymmetry have been inconclusive. When using our WT microscope to perform a space-scale analysis of both the S_{TA} and S_{GC} skews of human chromosome 22, one gets oscillatory profiles similar to those obtained for the GC content, with still two main characteristic lengths $l_1 = 140 \pm 20$ kbp and $l_2 = 400 \pm 40$ kbp as shown in Fig. 4, where the corresponding scale spectrum of the total skew S [Eq.(2)] displays two main bumps, the latter at 400 kbp being the most pronounced. Figure 3b shows the oscillatory skew profile obtained for the smoothing scale $a_2^* = 160$ kbp. This profile displays rather regular oscillation trends of basic length ~400 kbp. Quite remarkably, this oscillatory skew profile provides a guide for the organization of the spatial location and orientation of the (largest) genes [34, 35]: *sense* genes with the same orientation as the sequence are located around the positive maxima of the oscillations (among transcribed sequences, this corresponds to $79.6 \pm 1.9\%$ (ch.22), $84.0 \pm 2.6\%$ (ch.11), $89.2 \pm 1.2\%$ (ch.14), and $88.1 \pm 2.4\%$ (ch.21) of 1-kbp fragments that have the same orientation as the Watson strand), while *antisense* genes are quite symmetrically located around the minima (mainly

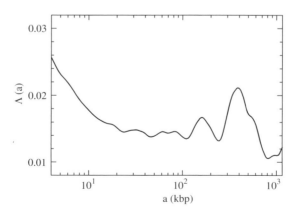

Figure 4.   Scale spectrum $\Lambda(a)$ versus (a) of the total skew S [Eq. (2)] computed over the entire chromosome 22 of the human genome.

negative). Let us point out that since these skew oscillations are also observed in large intergenic regions (but with smaller amplitude), they may arise from replication mutation biases. Indeed, as we will elaborate about in the next sections, these oscillations of rather marked relaxational character are likely to reflect some correlation between gene organization into clusters with preferential gene orientation and replication.

For more details on the existence of low-frequency rhythms in DNA sequences, we refer the reader to the Ph.D. manuscripts of E. B. Brodie of Brodie [96] and S. Nicolay [97].

IV.   TRANSCRIPTION-COUPLED STRAND ASYMMETRIES IN THE HUMAN GENOME

During genome evolution, mutations do not occur at random as illustrated by the diversity of the nucleotide substitution rate values [98–101]. This nonrandomness is considered as a byproduct of the various DNA mutation and repair processes that can affect each of the two DNA strands differently. Deviations from intrastrand equimolarities, the so-called Chargaff's second parity rule [88, 89], have been extensively studied during the past decade, and the observed skews have been attributed to asymmetries intrinsic to the replication or to the transcription processes. Asymmetries of substitution rates coupled to transcription have been mainly observed in prokaryotes [102–104], with only preliminary results in eukaryotes. In the human genome, excess of T was observed in a set of gene introns [105] and some large-scale asymmetry was observed in human sequences, but they were attributed to replication [106]. Only recently, a comparative analysis of mammalian sequences demonstrated a transcription-coupled excess of G+T over A+C in the coding strand [107]. In contrast to the substitution biases observed in bacteria presenting an excess of $C \rightarrow T$ transitions, these asymmetries are characterized by an excess of purine ($A \rightarrow G$) transitions relatively to pyrimidine ($T \rightarrow C$) transitions. These might be a by-product of the transcription-coupled repair mechanism acting on uncorrected substitution errors during replication [108]. In this section, we report the results of a genome-scale analysis of human genes that definitely establish the existence of transcription-coupled nucleotide biases [36, 37].

A.   Strand Asymmetries in Human Gene Sequences

We have started examining nucleotide compositional strand asymmetries in transcribed regions of human sequences [36]. We have computed the S_{TA} and S_{GC} skews [Eq. (1)] for intron sequences since, in contrast to exonic sequences, they can be considered as weakly selected sequences. For each gene, we have concatenated all the introns in a unique sequence (see Section II.A). The distributions of the TA and GC skews, computed on the 14,854 intro-containing

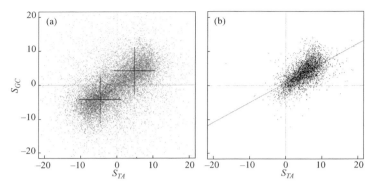

Figure 5. (a) S_{TA} and S_{GC} skews in human introns [36]: Each point corresponds to one of the 14,854 intron-containing genes; repeated elements are removed from the analysis (Section 2.1); red points correspond to sense genes (7508) with the same orientation as the Watson strand; blue points correspond to antisense (7346) with opposite orientation; black crosses represent the standard deviations of the distributions. (b) Correlation between S_{TA} and S_{GC} skews determined on the coding strand from intronic regions without repeats: Each point corresponds to a gene for which the total length of intronic regions is $l > 25$ kbp (7797 genes); Pearson's correlation coefficient r equals 0.61 (the slope of the regression line is 0.58). See color insert.

genes, present positive mean values for sense genes (7508), namely $\bar{S}_{TA} = 4.72 \pm 0.07\%$ and $\bar{S}_{GC} = 2.97 \pm 0.07\%$, and nearly opposed values for antisense genes (7346), namely $\bar{S}_{TA} = -4.56 \pm 0.07\%$ and $\bar{S}_{GC} = -3.05 \pm 0.07\%$. When removing the repeated sequences from the analysis (Section II.A), the TA and GC biases are not strongly altered (Fig. 5a). When examined on the coding strand, the mean values for all intron sequences without repeats present significant excess of T over A, namely $\bar{S}_{TA} = 4.49 \pm 0.01\%$, and excess of G over C, namely $\bar{S}_{GC} = 3.29 \pm 0.01\%$ (after appropriately removing intron extremities, see Section II.A). The corresponding probability density functions (pdf) of S_{TA} and S_{GC} skews when computed from the intronic sequences of the whole set of 14,854 intron-containing genes (after removing the noncoding regions closer than 560 bp to an exon) are shown in Fig. 6 in a semilogarithmic representation. For both sense (Fig. 6a) and antisense (Fig. 6b) genes, one observes the presence, for both skews, of large tails that clearly indicates some departure from a parabolic profile, the signature of Gaussian statistics.

A question of interest is the possible existence of correlations between S_{TA} and S_{GC} skews in intronic sequences without repeats. When all genes are considered, only small correlation is observed (Pearson's correlation coefficient r equals 0.09). However, the values of the skews from small genes turn out to be highly noisy. When one excludes these small genes, S_{TA} and S_{GC} present larger correlation (e.g., $r = 0.45$) for genes with total intron length $l > 10$ kbp and

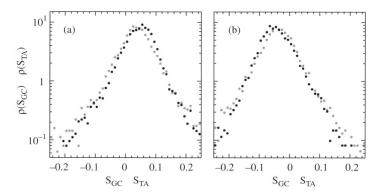

Figure 6. Probability density functions of the skews S_{TA} (●) and S_{GC} (●) values computed from the intronic sequences of the 14,854 intron-containing genes after removing repeated sequences. (a) Sense genes; (b) antisense genes.

$r = 0.61$ for genes with $l > 25$ kbp as illustrated in Fig. 5b. Let us point out that S_{TA} and S_{GC} present weak correlation with the intronic GC content as well as with the sequence length, and this even if only the large genes are considered [36].

B. Transcription-Induced Step-like Skew Profiles in the Human Genome

In order to compare the TA and GC asymmetry values in transcribed regions to those in the neighboring intergenic sequences, we have computed S_{TA} and S_{GC} in adjacent 1-kbp windows along the genome sequence [36, 37]. Figure 7 reports the mean values of these skews for all genes as a function of the distance to the 5′

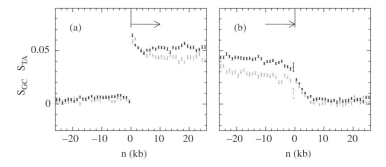

Figure 7. TA (●) and GC (●) skew profiles in the regions surrounding 5′ and 3′ gene extremities [36]. S_{TA} and S_{GC} were calculated in 1-kbp windows starting from each gene extremities in both directions. The abscissa reports the distance (n) of each 1-kbp window to the indicated gene extremity; zero values of abscissa correspond to 5′ (a) or 3′ (b) gene extremities. In ordinate is reported the mean value of the skews over our set of 14,854 intron-containing genes for all 1-kbp windows at the corresponding abscissa. Error bars represent the standard error of the means.

or 3′ end. At the 5′ gene extremities (Fig. 7a), a sharp transition of both skews is observed from about zero values in the intergenic regions to finite positive values in transcribed regions ranging between 4 and 6% for $\bar{S}_{TA}$ and between 3 and 5% for $\bar{S}_{GC}$. At the gene 3′ extremities (Fig. 7b), the TA and GC skews also exhibit transitions from significantly large values in transcribed regions to very small values in untranscribed regions. However, in comparison to the steep transitions observed at 5′-ends, the 3′-end profiles present a slightly smoother transition pattern extending over ∼5 kbp and including regions downstream of the 3′ end likely reflecting the fact that transcription continues to some extent downstream of the polyadenylation site. In pluricellular organisms, mutations responsible for the observed biases are expected to occur in germ-line cells. It could happen that gene 3′ ends annotated in the databank differ from the poly-A sites effectively used in the germ-line cells. Such differences would then lead to some broadening of the skew profiles.

C.   A Model for Transcription-Coupled TA and GC Skews

As shown in Fig. 7, TA and GC biases are specifically observed in transcribed sequences indicating that each of them clearly results from transcription-coupled processes acting in germ-line cells. This observation is reinforced by the observed correlation between S_{TA} and S_{GC} (Fig. 5b). Indeed, according to this hypothesis, S_{TA} and S_{GC} are likely to increase simultaneously with transcription. How many genes have biased sequences? When comparing [36] the observed biases to those expected for random sequences with same length and same (T+A) composition, 64% of genes are found to present significant TA bias (p-values $< 10^{-2}$). When considering only larger genes, this proportion increases to 82% (total intron length $l > 10\,\mathrm{kbp}$) and 86% ($l > 25\,\mathrm{kbp}$), respectively. These results indicate that in germ-line cells, a large majority of genes are expressed.

A recent study [107] showed a transcription-coupled excess of purine transitions and a deficit of pyrimidine transitions in a small set of human genes. To examine if these transition rates might explain the strand asymmetries measured in Figs. 5 to 7, for the whole set of genes, we have performed numerical calculations of the composition at equilibrium of a DNA sequence (given the substitution rates) [36]. When supposing that transcription alters transition rates only, we obtained a value of $S_{TA} = 4.7\%$ similar to our observations, while the value of $S_{GC} = 7.8\%$ significantly exceeds the value found in our study. This led us to suppose that GC transversions might also produce strand asymmetry in eukaryotes [36]. During evolution, both processes would have been active in germ-line cells.

More recently, we have extended this study of strand asymmetries in intron sequences to evolutionarily distant eukaryotes [37]. When appropriately examined, all genomes present transcription-coupled excess of T over A ($S_{TA} > 0$) in the coding strand. In contrast, GC skew is found positive in mammals and plants but negative in invertebrates, suggesting different mutation repair mechanisms associated to transcription in vertebrates and invertebrates. For more details on the existence of transcription-coupled strand asymmetries in eukaryotic genomes, we refer the reader to the Ph.D. manuscript of M. Touchon [109].

The results reported in Fig. 7 suggest that S_{TA} and S_{GC} are constant along introns. Since introns account for about 80% of gene sequences, this means that skew profiles induced by transcription processes have a characteristic step-like shape [96, 97, 109]. However, the absence of asymmetries in intergenic regions does not exclude the possibility of additional replication associated biases. Such biases would present opposite signs on leading and lagging strands and would cancel each other in our statistical analysis as a result of the spatial distribution of multiple unknown replication origins [83]. The following sections will be devoted to the study of replication-associated strand asymmetries in mammalian genomes.

V. REPLICATION-ASSOCIATED STRAND ASYMMETRIES IN THE HUMAN GENOME

DNA replication is an essential genomic function responsible for the accurate transmission of genetic information through successive cell generations. According to the so-called "replicon" paradigm derived from prokaryotes [110], this process starts with the binding of some "initiator" protein to a specific "replicator" DNA sequence called *origin of replication*. The recruitment of additional factors initiate the bi-directional progression of two divergent replication forks along the chromosome. As illustrated in Fig. 8a, one strand is replicated continuously (leading strand), while the other strand is replicated in discrete steps toward the origin (lagging strand). In eukaryotic cells, this event is initiated at a number of replication origins and propagates until two converging forks collide at a terminus of replication [111]. The initiation of different replication origins is coupled to the cell cycle, but there is a definite flexibility in the usage of the replication origins at different developmental stages [112–116]. Also, it can be strongly influenced by the distance and timing of activation of neighboring replication origins, by the transcriptional activity, and by the local chromatin structure [113–116]. Actually, sequence requirements for a replication origin vary significantly between different eukaryotic organisms. In the unicellular eukaryote *Saccharomyces cerevisiae*, the replication origins spread over 100–150 bp and present some highly conserved motifs [111]. However, among eukaryotes, *S. cerevisiae* seems to be the exception that remains faithful to the replicon model. In the fission yeast *Schizosaccharomyces pombe*, there is no clear consensus sequence and the replication origins spread over at least 800 to 1000 bp [111]. In multicellular organisms, the nature of initiation sites of DNA

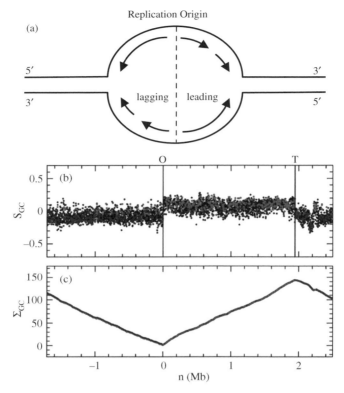

Figure 8. (a) Schematic representation of the divergent bi-directional progression of the two replication forks from the replication origin. (b) S_{GC} calculated in 1-kbp windows along the genomic sequence of *Bacillus subtilis*. (c) Cumulated skew Σ_{GC}. The vertical lines correspond respectively to the replication origin (O) and termination (T) positions. In (b) and (c), red (blue) points correspond to sense (antisense) genes that have the same (opposite) orientation than the sequence. See color insert.

replication is even more complex. Metazoan replication origins are rather poorly defined, and initiation may occur at multiple sites distributed over a thousand of base pairs [117]. The initiation of replication at random and closely spaced sites was repeatedly observed in *Drosophila* and *Xenopus* early embryo cells, presumably to allow for extremely rapid S phase, suggesting that any DNA sequence can function as a replicator [112, 118, 119]. A developmental change occurs around midblastula transition that coincides with some remodeling of the chromatin structure, transcription ability, and selection of preferential initiation sites [112, 119]. Thus, although it is clear that some sites consistently act as replication origins in most eukaryotic cells, the mechanisms that select these sites and the sequences that determine their location remain elusive in many cell types [120, 121]. As recently proposed by many authors [122–124], the need to fulfill specific requirements that result from cell diversification may have led high eukaryotes to develop various epigenetic controls over the replication origin selection rather than to conserve specific replication sequence. This might explain that only very few replication origins have been identified so far in multicellular eukaryotes, namely around 20 in metazoa and only about 10 in human [41–48]. Along the line of this epigenetic interpretation, one might wonder what can be learned about eukaryotic DNA replication from DNA sequence analysis.

A.   Replication-Associated Strand Asymmetries in Prokaryotic Genomes: The Replicon Model

As mentioned in Section III.B, the existence of replication associated strand asymmetries has been mainly established in bacterial genomes [91–95]. As illustrated in Fig. 8, the GC and TA skews abruptly switch sign (over few kbp) from negative to positive values at the replication origin and in the opposite direction from positive to negative values at the replication terminus. This step-like profile is characteristic of the replicon model [110]. In *Bacillus subtilis*, as in most bacteria, the leading (respectively, lagging) strand (Fig. 8a) is generally richer (respectively, poorer) in G than in C (Fig. 8b), and to a lesser extent in T than in A (data not shown). This typical pattern is particularly clear when plotting the cumulated skews Σ_{GC} (Fig. 8c) and Σ_{TA} [Eq. (3)]; both present decreasing (or increasing) profiles in regions situated 5′ (or 3′) to the origin, displaying a characteristic ∨-shape pointing to the replication origin position (similarly a characteristic ∧-shape is observed at the terminus position). The research of ∨ patterns in the cumulated skews has been extensively used as a strategy to detect the position of the (unique) replication origin in (generally circular) bacterial genomes [92–95].

As shown in Fig. 8b and 8c, when looking at the gene organization around the replication origin of *Bacillus subtilis*, one observes that most of the sense (respectively, antisense) genes are preferentially on the right (respectively, left) of the replication origin. This suggests that the replication forks progression is co-oriented with transcription, as to minimize the risk of frontal collision between DNA and RNA polymerases [125–128].

B.   Analysis of Strand Asymmetries Around Experimentally Determined Replication Origins in the Human Genome

In eukaryotes, the existence of compositional biases is unclear and most attempts to detect the replication origins from strand compositional asymmetry have been inconclusive. Several studies have failed to show compositional biases related to replication, and analysis of nucleotide substitutions in the region of the β-*globin* replication origin in primates do not support the existence of mutational bias between the leading and the lagging strands [92, 129, 130]. Other studies have led to rather opposite results. For instance, strand asymmetries associated with replication have been observed in the subtelomeric regions of *Saccharomyces cerevisiae* chromosomes, supporting the existence of replication-coupled asymmetric mutational pressure in this organism [131]. With the same methodology as the one developed in Section IV for gene extremities, we present in this section analyses of strand asymmetries flanking experimentally determined human replication origins [38, 39].

As shown in Fig. 9, most of the known replication origins in the human genome correspond to rather sharp (over several kbp) transitions from negative to positive S_{TA} and S_{GC} skew values that clearly emerge from the noisy background. This is reminiscent of the behavior observed in Fig. 8 for *Bacillus subtilis*, except that the leading strand is relatively enriched in T over A and in G over C. This observation is even more patent when looking at the cumulated skew Σ_{TA} and Σ_{GC} profiles that both display characteristic $\vee$-shapes pointing to the experimentally identified initiation zones. According to the gene environment, the amplitude of the jump observed in the skew profiles can be more or less important and its position more or less localized (from a few kbp to a few tens of kbp). Indeed, we have seen in Section IV that transcription generates positive TA and GC skews on the coding strand [36, 37, 132], which explains that larger jumps are observed when the sense and/or the antisense genes are on the leading strand so that replication and transcription biases add to each other. To measure compositional asymmetries that would result from replication only, we have calculated the skews in intergenic regions on both sides of the origins [38]. The total skew S definitely shifts from negative ($S = -6.2 \pm 0.4\%$) to positive ($S = 11.1 \pm 1\%$) values when crossing the replication origin. This result strongly suggests the existence of mutational pressure associated with replication, leading to the mean compositional biases $S_{TA} = 4.0 \pm 0.4\%$ and $S_{GC} = 3.0 \pm 0.5\%$ (Table I). Let us note that the value of the skew could vary from one origin to another, possibly reflecting different initiation efficiencies. From the calculation of the intron skew values on the leading and lagging strands reported in Table I, one can estimate the mean skew associated with transcription by subtracting intergenic skews from S_{lead} values giving $S_{TA} = 3.6 \pm 0.7\%$ and $S_{GC} = 3.8 \pm 0.9\%$. These estimations are remarkably consistent with those obtained with our large set of human introns in Section IV, further supporting the existence of replication-coupled strand asymmetries. Overall, these results indicate that the mean replication bias on the leading strand and the mean transcriptional bias on the coding strand are of the same order of magnitude, namely $S = S_{TA} + S_{GC} \sim 7\%$ (Table I).

In that context, one can wonder to which extent the biases observed in intergenic regions may result from the possible presence of still undetected genes. Two pieces of evidence argue against this eventuality. First, we have been

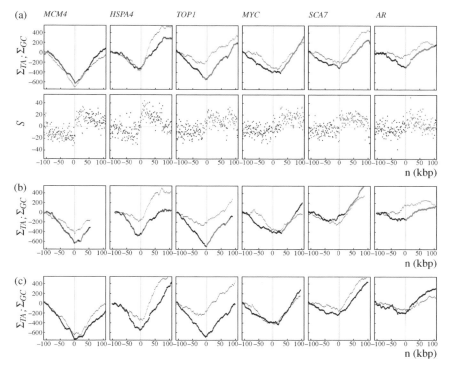

Figure 9. TA and GC skew profiles around experimentally determined human replication origins [38]. (a) The skew profiles were determined in 1-kbp windows in regions surrounding ($\pm$100 kbp without repeats) experimentally determined human replication origins (see Section II.A). (Upper) TA and GC cumulated skew profiles Σ_{TA} (thick line) and Σ_{GC} (thin line). (Lower) Skew S calculated in the same regions. The ΔS amplitude associated with these origins, calculated as the difference of the skews measured in 20-kbp windows on both sides of the origins, are: MCM4 (31%), HSPA4 (29%), TOP1 (18%), MYC (14%), SCA7 (38%), and AR (14%). (b) Cumulated skew profiles calculated in the six regions of the mouse genome homologous to the human regions analyzed in (a). (c) Cumulated skew profiles in the six regions of the dog genome homologous to human regions analyzed in (a). The abscissa (n) represents the distance (in kbp) of a sequence window to the corresponding origin; the ordinate represents the values of S given in percent. The colors have the following meaning: red, sense genes (coding strand identical to the Watson strand); blue, antisense genes (coding strand opposite to the Watson strand); black, intergenic regions. In (c), genes are not represented. See color insert.

careful enough to retain as transcribed regions one of the largest sets of transcripts available, resulting in a stringent definition of intergenic regions. Second, several studies have demonstrated the existence of hitherto unknown transcripts in regions where no protein coding genes have been previously identified [133–136]. Taking advantage of the set of non-protein-coding RNAs

TABLE I

Strand Asymmetries Associated with Human Replication Origins [38][a]

	S_{TA}	S_{GC}	S	I	$G + C$, %
Intergenic (H. s.) all	3.9 ± 0.4	3.0 ± 0.4	6.9 ± 0.4	487	42
Intergenic (H. s.) ncr.	4.0 ± 0.4	3.0 ± 0.5	7.0 ± 0.5	461	42
Intergenic (M. m.) ncr.	3.6 ± 0.4	2.2 ± 0.5	5.8 ± 0.5	441	42
S_{lead} (H. s. introns)	7.5 ± 0.3	6.8 ± 0.4	14.3 ± 0.4	358	40
S_{lag} (H. s. introns)	-1.9 ± 1.0	-0.3 ± 1.4	-2.2 ± 1.3	49	44

[a]The skews were calculated in the regions flanking the six human replication origins (Fig. 9a) and in the corresponding homologous regions of the mouse genome. Intergenic sequences were always considered in the direction of replication fork progression (leading strand); they were considered in totality (all) or after elimination of conserved regions (ncr.) between human (Homo sapiens, H.s.) and mouse (Mus musculus, M.m.) (see Section II.A). To calculate the mean skew in introns, the sequences were considered on the nontranscribed strand. For S_{lead}, the orientation of transcription was the same as the replication fork progression; for S_{lag}, the situation was the opposite. The mean values of the skews S_{TA}, S_{GC}, and S are given in percent ($\pm$SEM). I, total sequence length in kbp.

identified in the "H-Inv" database [137], we have checked that none of them are present in the intergenic regions studied here. Finally, we have eliminated the possibility that intergenic skews are due to conserved sequences by checking that the removal of homologous segments found in the mouse genome ($\sim$5.3% of all intergenic sequences) does not change significantly the skews in intergenic regions [38].

C. Conservation of Replication-Associated Strand Asymmetries in Mammalian Genomes

As a next step of our study, we have analyzed [38] the S_{TA} and S_{GC} skew profiles in DNA regions of mammalian genomes homologous to the six human origin investigated in Fig. 9a. As shown in Figs. 9b and 9c, the human, mouse and dog cumulated skew profiles look strikingly similar to each other, suggesting that in mouse and dog, these regions also correspond to replication initiation zones (note that they are very similar in primate genomes). For each replication origin, one robustly observes a $\vee$-shape characteristic of a sharp upward jump from negative to positive skew values. A detailed examination of the mouse intergenic regions suggests the existence of a compositional bias associated with replication $S = S_{TA} + S_{GC} \sim 5.8 \pm 0.5\%$ (Table I). Let us point out that, at these homologous loci, human and mouse intergenic sequences present almost no ($\sim$5.3%) conserved elements. Hence, the presence of strand asymmetry in regions that have strongly diverged during evolution further supports the existence of compositional bias associated with replication in both organisms. In the absence of such a process, intergenic sequences would have lost a significant fraction of their strand asymmetry.

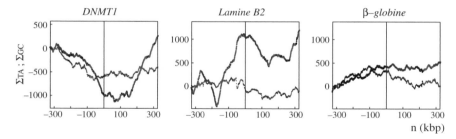

Figure 10. Cumulated skew profiles calculated around the origin of replication DNMT1, Lamin B2, and β-globin in the human genome: Σ_{TA} (thick line) and Σ_{GC} (thin line). The colors have the same meaning as in Fig. 9. See color insert.

Altogether, these results establish the existence of strand asymmetries associated with replication in mammalian germ-line cells [38]. They show that most replication origins experimentally detected in somatic cells coincide with sharp upward transitions of the skew profile. They also imply that for the majority of experimentally determined origins, the position of initiation zones are conserved in mammalian genomes as recently confirmed by the identification of a replication origin in the mouse MYC locus [138]. Let us emphasize that among nine human origins known experimentally, three do not present typical ∨-shape cumulated profiles as reported in Fig. 10. For DNMT1 (left panel in Fig. 10), the sharp central part of the ∨ profile is replaced by a large horizontal plateau (few tens of kbp), possibly reflecting the presence of several origins dispersed over the whole plateau. Note that dispersed origins have been observed, for example, in the hamster DHFR initiation zone [139]. By contrast, the cumulated skew profiles of the Lamin B2 (central panel of Fig. 10) and β-globin (right panel of Fig. 10) origins present no ∨ profile, suggesting that they might be inactive in germ-line cells or less active than neighboring origins.

D. Factory-Roof Skew Profiles in the Human Genome

As illustrated in Fig. 11a, for TOP1 replication origin, when examining the behavior of the skews at larger distances from the origin, one does not observe a step-like pattern with upward and downward jumps at the origin and termination positions respectively as expected for the bacterial replicon model (Fig. 8b). Surprisingly, on both sides of the upward jump, the noisy S profile decreases steadily in the 5' to 3' direction without clear evidence of pronounced downward jumps. As shown in Figs. 11b–d, sharp upward jumps of amplitude $\Delta S \gtrsim 15\%$, similar to the ones observed for the known replication origins (Fig. 9), seem to exist also at many other locations along the human chromosomes. But the most striking feature is the fact that in between two neighboring major upward jumps,

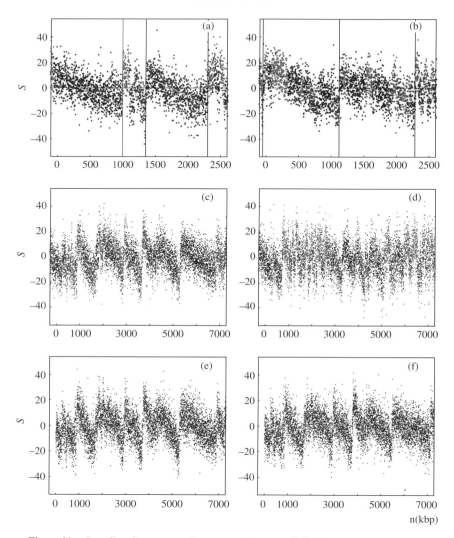

Figure 11. *S* profiles along mammalian genome fragments [38]. (a) Fragment of chromosome 20 including the TOP1 origin (red vertical line). (b and c) Chromosome 4 and chromosome 9 fragments, respectively, with low GC content (36%). (d) Chromosome 22 fragment with larger GC content (48%). In (a) and (b), vertical lines correspond to selected putative origins (see Section VI.A); yellow lines are linear fits of the *S* values between successive putative origins. Black, intergenic regions; red, sense genes; blue, antisense genes. Note the fully intergenic regions upstream of TOP1 in (a) and from positions 5290–6850 kbp in (c). (e) Fragment of mouse chromosome 4 homologous to the human fragment shown in (c). (f) Fragment of dog chromosome 5 syntenic to the human fragment shown in (c). In (e) and (f), genes are not represented. See color insert.

not only the noisy S profile does not present any comparable downward sharp transition, but it displays a remarkable decreasing linear behavior. At chromosome scale, one thus gets jagged S profiles that have the aspect of "factory roofs" [38–39]. For comparison, we show in Fig. 12, the S_{TA}, S_{GC}, and S profiles obtained for a large fragment of the human chromosome 12 after (Figs. 12a–c) and before (Figs. 12a′–c′) removing the repeated sequences (Section II.A). There is no doubt that repeated sequences increase the level of noise in the skew profiles. Indeed factory roofs are more easily seen on the masked sequences and specially on the total skews $S = S_{TA} + S_{GC}$. As reported in Fig. 13, the pdfs of S_{TA}, S_{GC}, and S are nearly Gaussian for the masked sequences; some large tails are present but for skew amplitudes larger than 40%. The fact that the skew pdfs of the native sequences, and more particularly the S_{TA} pdf, significantly depart from Gaussian distributions justifies, *a posteriori*, the need of removing repeated sequences prior to our statistical analysis. Most of these sequences have been inserted recently in the human genome and do not reflect long-term evolutionary skew patterns.

The jagged S profiles shown in Figs. 11a–d and 12a–c look somehow disordered because of the extreme variability in the distance between two successive upward jumps, from spacing ∼50–100 kbp (∼100–200 kbp for the native sequences) up to 2–3 Mbp (∼4–5 Mbp for the native sequences) in

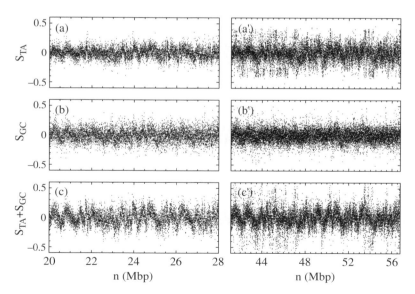

Figure 12. Skew profiles along a large fragment of the human chromosome 12. Repeat-masked sequence (8 Mb): S_{TA} (a), S_{GC} (b), and $S = S_{TA} + S_{GC}$ (c). Native sequence (15.7 Mb): S_{TA} (a'), S_{GC} (b'), and S (c').

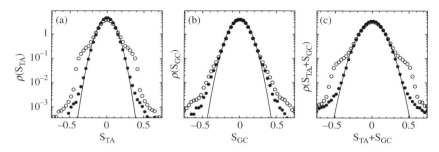

Figure 13. Probability density functions of the skews S_{TA} (a), S_{GC} (b), and $S = S_{TA} + S_{GC}$ (c) values computed in nonoverlapping 1-kbp windows from the DNA sequences of the 22 human autosomal chromosomes. Symbols have the following meaning: ($\circ$) native sequences and ($\bullet$) repeat-masqued sequences.

agreement with recent experimental studies [83] that have shown that mammalian replicons are heterogeneous in size with an average size $\sim$500 kbp, the largest ones being as large as a few Mbp. But what is important to notice is that some of these segments between two successive upward jumps of the skew are entirely intergenic (Figs. 11a and 11c), clearly illustrating the particular profile of a strand bias resulting solely from replication [38, 39]. In most other cases, one observes the superimposition of this replication profile and of the step-like profiles of sense and antisense genes, appearing as upward and downward blocks standing out from the replication pattern (Fig. 11c). Importantly, as illustrated in Figs. 11e and 11f, the factory-roof pattern is not specific to human sequences but is also observed in numerous regions of the mouse and dog genomes [38].

VI. FROM THE DETECTION OF PUTATIVE REPLICATION ORIGINS TO THE MODELING OF REPLICATION IN THE HUMAN GENOME

A. A Wavelet-Based Method to Detect Putative Replication Origins

We have shown in Section V that experimentally determined human replication origins coincide with large-amplitude upward transitions in noisy skew profiles. The corresponding ΔS ranges between 14% and 38%, owing to possible different replication initiation efficiencies and/or different contributions of transcriptional biases (Fig. 9). To predict replication origins, one thus needs a methodology to detect discontinuities in noisy signals. As introduced in Section II.C, the continuous wavelet transform is a mathematical microscope that is well adapted for singularity tracking [13, 26–29]. The basic principle of the detection of jumps in the skew profiles with the WT is illustrated in Fig. 14. From Eq. (7), when

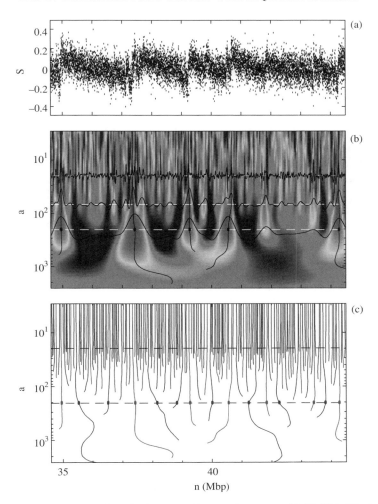

Figure 14. (a) Skew profiles of a fragment of human chromosome 12. (b) WT of S using $g^{(1)}$; $W_{g^{(1)}}[S](n, a)$ is coded from black (min) to red (max); three cuts of the WT at constant scale $a = a^* = 200$ kbp, 70 kbp and 20 kbp are superimposed together with five maxima lines identified as pointing to upward jumps in the skew profile. (c) WT skeleton defined by the maxima lines in red (respectively, blue) when corresponding to positive (respectively, negative) values of the WT. At the scale $a^* = 200$ kbp, one thus identify 7 upward (red dots) and 8 downward (blue dots) jumps. The black dots in (b) correspond to the five WTMM of largest amplitude that have been identified as putative replication origins; it is clear that the associated maxima lines point to the five major upward jumps in the skew profile in the limit $a \to 0^+$. See color insert.

using the first derivative of the Gaussian function as analyzing wavelet, it is obvious that at a fixed scale a, a large value of the modulus of the WT coefficient corresponds to a strong derivative of the smoothed skew profile. In particular, jumps manifest as local maxima of the WT modulus as illustrated for three different scales in Fig. 14b. The main issue when dealing with noisy signals like the skew profile in Fig. 14a, is to distinguish the local WT modulus maxima (WTMM) associated to the jumps from those induced by the noise. In this respect, the freedom in the choice of the smoothing scale a is fundamental since, whereas the noise amplitude is reduced when increasing the smoothing scale, an isolated jump contributes equally at all scales.

As shown in Fig. 14c, our methodology consists in computing the WT skeleton [13, 27, 38] defined by the set of maxima lines obtained by connecting the WTMM across scales. Then we select a scale $a^* = 200$ kbp, which is smaller than the typical replicon size and larger than the typical gene size. In this way, we not only reduce the effect of the noise but we also reduce the contribution of the upward (5′ extremity) and backward (3′ extremity) jumps associated to the step-like skew pattern induced by transcription only (Fig. 7), to the benefit of maintaining a good sensitivity to replication induced jumps. The maxima lines that exist at that scale a^* are likely to point to jump positions at small scale (Fig. 14c). The detected jump locations are estimated as the positions at scale 20 kbp of the so-selected maxima lines. According to Eq. (7), upward (respectively, downward) jumps are indentified by the maxima lines corresponding to positive (respectively, negative) values of the WT as illustrated in Fig. 14c by the red (respectively, blue) maxima lines. When applying this methodology to the total skew S along the repeat-masked DNA sequences of the 22 human autosomal chromosomes, 2415 upward jumps are detected and, as expected, a similar number (namely 2686) of downward jumps. Figure 15a shows the histograms of the amplitude $|\Delta S|$ of the so-identified upward ($\Delta S > 0$) and downward ($\Delta S < 0$) jumps, respectively. These histograms do not superimpose, with the former being significantly shifted to larger $|\Delta S|$ values. When plotting $N(|\Delta S| > \Delta S^*)$ versus ΔS^* in Fig. 15b, one can see that the number of large-amplitude upward jumps overexceeds the number of large-amplitude downward jumps. These results [38, 39] confirm that most of the sharp upward transitions in the S profiles in Figs. 11 and 14a have no sharp downward transition counterpart. This excess likely results from the fact that, contrasting with the prokaryote replicon model (Fig. 8) where downward jumps result from precisely positioned replication terminations, termination in mammals appears not to occur at specific positions but to be randomly distributed [38, 39] (this point will be detailed in Section VI.C). Accordingly, the small number of downward jumps with large $|\Delta S|$ is likely to result from transcription (Fig. 7) and not from replication. These jumps

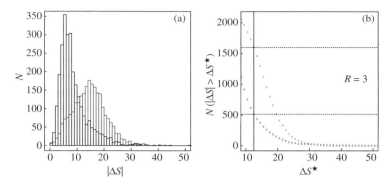

Figure 15. Statistical analysis of the sharp jumps detected in the S profiles of the 22 human autosomal chromosomes by the WT microscope at scale $a^* = 200$ kbp for repeat-masked sequences [38, 39]. $|\Delta S| = |\bar{S}(3') - \bar{S}(5')|$, where the averages were computed over the two adjacent 20-kbp windows, respectively, in the $3'$ and $5'$ direction from the detected jump location. (a) Histograms $N(|\Delta S|)$ of $|\Delta S|$ values. (b) $N(|\Delta S| > \Delta S^*)$ versus ΔS^*. In (a) and (b), the black (respectively, grey) line corresponds to downward $\Delta S < 0$ (respectively, upward $\Delta S > 0$) jumps. $R = 3$ corresponds to the ratio of upward over downward jumps presenting an amplitude $|\Delta S| \geq 12.5\%$ (see text).

are probably due to highly biased genes that also generate a small number of large-amplitude upward jumps, giving rise to false-positive candidate replication origins. In that respect, the number of large downward jumps can be taken as an estimation of the number of false positives. In a first step, we have retained as acceptable a proportion of 33% of false positives. As shown in Fig. 15b, this value results from the selection of upward and downward jumps of amplitude $|\Delta S| \geq 12.5\%$, corresponding to a ratio of upward over downward jumps, $R = 3$. Let us note that the value of this ratio is highly variable along the chromosome (Fig. 16). In G+C poor regions, namely G+C < 37%, we observe in Fig. 16a,a$'$ the largest R value, namely $R = 6.5$. In regions with $37\% \leq$ G+C $\leq 42\%$, we obtain $R = 3.9$ (Figs. 16b,b$'$) which contrasts with small R values, $R = 1.9$ (Figs. 16c,c$'$) found in regions with G+C > 42%. In these latter regions (accounting for $\sim$40% of the genome) with high gene density and small gene length [51], the skew profiles oscillate rapidly with large upward and downward amplitudes (Fig. 11d), resulting in a too large estimate of the number of false positives ($\sim$53%).

In a final step, we have decided [38] to retain as putative replication origins upward jumps with $|\Delta S| \geq 12.5\%$ detected in regions with G+C $\leq 42\%$. This selection leads to a set of 1012 candidates among which our estimate of the proportion of true replication origins is 79% ($R = 4.76$). Some of these putative replication origins are illustrated in Fig. 11.

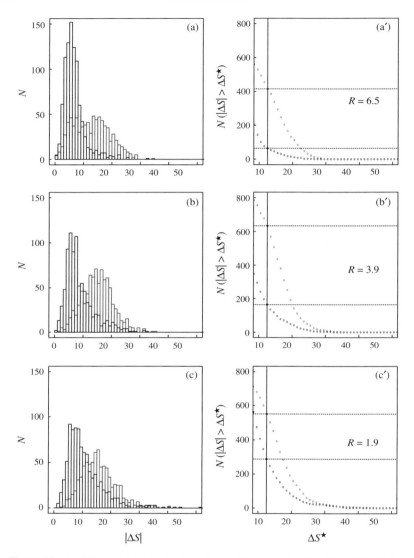

Figure 16. Statistical analysis of the sharp jumps detected in the S profiles of the 22 human autosomal chromosomes by the WT microscope at scale $a^* = 200$ kbp for repeat-masked sequences [109]. The detected jumps have been classified into three categories according to the GC content found in a 100-kbp window centered at the position of the jump. Same as in Fig. 15: (a, a')
$G + C < 37\%$; (b, b') $37\% \leq G + C \leq 42\%$; (c, c') $42\% < G + C$.

B. Gene Organization Around the 1012 Putative Replication Origins in the Human Genome

The mean amplitude of the upward jumps associated with the 1012 putative origins is 18%, consistent with the range of values observed for the six experimentally known origins in Fig. 9. Let us remark that all six origins have been identified by our detection methodology. When investigating the gene content around these putative origins [109], one finds that in a close vincinity (± 20 kbp), most DNA sequences (55% of the analyzing windows) are transcribed in the same direction as the progression of the replication fork (namely sense genes on the 3' side of the origin and antisense genes on the 5' side). By contrast, only 7% of the sequences are transcribed in the opposite direction (38% are intergenic). These results show that the $|\Delta S|$ amplitude at putative origins mostly results from superimposition of biases (i) associated with replication and (ii) with transcription of the genes proximal to the origin. Determining whether transcription is co-oriented with replication at larger distances is the subject of current study.

In Fig. 17 is shown the mean skew profile calculated in intergenic windows on both sides of the 1012 putative replication origins [38]. This mean skew profile presents a rather sharp transition from negative to positive values when crossing the origin position. To avoid any bias in the skew values that could

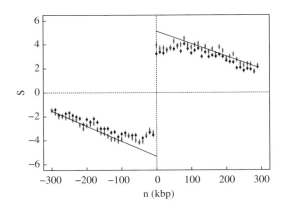

Figure 17.   Mean skew profile of intergenic regions around putative replication origins [38]. The skew S was calculated in 1 kbp windows (Watson strand) around the position (± 300 kbp without repeats) of the 1012 detected upward jumps; 5' and 3' transcript extremities were extended by 0.5 and 2 kbp, respectively (●), or by 10 kbp at both ends (∗). The abscissa represents the distance (in kbp) to the corresponding origin; the ordinate represents the skews calculated for the windows situated in intergenic regions (mean values for all discontinuities and for 10 consecutive 1-kbp window positions). The skews are given in percent (vertical bars, SEM). The lines correspond to linear fits of the values of the skew (∗) for $n < -100$ kbp and $n > 100$ kbp.

result from incompletely annotated gene extremities (e.g., 5′ and 3′ UTRs), we have removed 10-kbp sequences at both ends of all annotated transcripts. As shown in Fig. 17, the removal of these intergenic sequences does not significantly modifies the mean skew profile, indicating that the observed values do not result from transcription. On both sides of the jump, we observe a linear decrease of the bias with some flattening of the profile close to the transition point. Note that, due to (i) the potential presence of signals implicated in replication initiation and (ii) the possible existence of dispersed origins [139], one might question the meaningfulness of this flattening that leads to a significant underestimate of the jump amplitude. Furthermore, according to our detection methodology, the numerical uncertainty on the putative origin position estimate may also contribute to this flattening. As illustrated in Fig. 17, when extrapolating the linear behavior observed at distances $> 100\,$kbp from the jump, one gets a skew of 5.3%—that is, a value consistent with the skew measured in intergenic regions around the six experimentally known replication origins namely $7.0 \pm 0.5\%$ (Table I). Overall, the detection of sharp upward jumps in the skew profiles with characteristics similar to those of experimentally determined replication origins and with no downward counterpart further supports the existence, in human chromosomes, of replication-associated strand asymmetries, leading to the identification of numerous putative replication origins active in germ-line cells.

C.   A Model of Replication in Mammalian Genomes

Following the observation of jagged skew profiles similar to factory roofs in Section V.4, and the quantitative confirmation of the existence of such (piecewise linear) profiles in the neighborhood of 1012 putative origins in Fig. 17, we have proposed, in Touchon et al. [38] and Brodie of Brodie et al. [39], a rather crude model for replication in the human genome that relies on the hypothesis that the replication origins are quite well positioned while the terminations are randomly distributed. Although some replication terminations origins have been found at specific sites in *S. cerevisiae* and to some extent in *Schizosaccharomyces pombe* [140], they occur randomly between active origins in *Xenopus* egg extracts [141, 142]. Our results indicate that this property can be extended to replication in human germ-line cells. As illustrated in Fig. 18, replication termination is likely to rely on the existence of numerous termination sites distributed along the seq-uence. For each termination site (used in a small proportion of cell cycles), strand asymmetries associated with replication will generate a step-like skew profile with a downward jump at the position of termination and upward jumps at the positions of the adjacent origins (as in bacteria, Fig. 8b). Various termination positions will thus correspond to classical replicon-like skew profiles (Fig. 18, left panel). Addition of those profiles will generate the intermediate profile (Fig. 18, central panel). In a simple picture, we

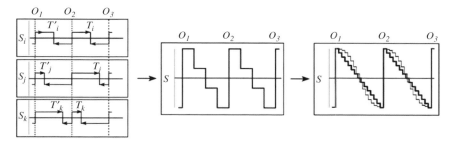

Figure 18. Model of replication termination [38, 39]. Schematic representation of the skew profiles associated with three replication origins O_1, O_2, and O_3; we suppose that these replication origins are adjacent, bi-directional origins with similar replication efficiency. The abscissa represents the sequence position; the ordinate represents the S value (arbitrary units). Upward (or downward) steps correspond to origin (or termination) positions. For convenience, the termination sites are symmetric relative to O_2. (Left) Three different termination positions T_i, T_j, and T_k, leading to elementary skew profiles S_i, S_j, and S_k. (Center) Superposition of these three profiles. (Right) Superposition of a large number of elementary profiles leading to the final factory-roof pattern. In the simple model, termination occurs with equal probability on both sides of the origins, leading to the linear profile (thick line). In the alternative model, replication termination is more likely to occur at lower rates close to the origins, leading to a flattening of the profile (gray line).

can reasonably suppose that termination occurs with constant probability at any position on the sequence. This behavior can, for example, result from the binding of some termination factor at any position between successive origins, leading to a homogeneous distribution of termination sites during successive cell cycles. The final skew profile is then a linear segment decreasing between successive origins (Fig. 18, right panel). Let us point out that firing of replication origins during time interval of the S phase [143] might result in some flattening of the skew profile at the origins as sketched in Fig. 18 (right panel, gray curve). In the present state, our results [38, 39] support the hypothesis of random replication termination in humans, and more generally in mammalian cells (Figs. 9 and 11), but further analyses will be necessary to determine what scenario is precisely at work.

In conclusion, we have revealed a factory roof skew profile as an alternative in mammalian genomes to the replicon step-like profile observed in bacteria (Fig. 8). This pattern is displayed by a set of 1012 upward transitions, each flanked on each side by DNA segments of ∼300 kbp (without repeats), which can be roughly estimated to correspond to 20–30% of the human genome. In these regions, which are characterized by low and medium G+C content (G+C≤42%), skew profiles reveal a portrait of germ-line replication consisting of putative origins separated by rather long DNA segments (∼1–3 Mbp on the native sequences). Although such segments are much larger than expected from the classical view [83–85] (∼100–500 kbp on the native sequences), they are not incompatible with estimations showing that replicon size can reach up to 1 Mbp

[83, 86] and that replicating units in meiotic chromosomes are much longer than those engaged in sommatic cells [144]. Finally, it is not unlikely that in G+C-rich (gene-rich) regions (Fig. 11d), replication origins would be closer to each other than in other regions, further explaining the greater difficulty in detecting origins in these regions. Indeed, the wavelet-based methodology described in Section VI.A remains efficient as long as there exists a clear separation between the characteristic size of a replicon and the characteristic size of a gene; while this separation is unquestionable at low and medium G+C content, this is no longer obvious in high GC regions.

For more details on the existence and modeling of replication associated strand asymmetries in mammalian genomes, we refer the reader to the Ph.D. thesis manuscripts of E. B. Brodie of Brodie [96], S. Nicolay [97], and M. Touchon [109].

VII. FROM SEQUENCE ANALYSIS TO THE MODELING OF THE CHROMATIN TERTIARY STRUCTURE

Some 50 years ago, Asakura and Oosawa [145] pointed out that two large rigid spheres immersed in a solution of smaller spheres are subject to an attractive force due to the depletion induced by increasing the space available to small spheres as the large ones come close to one another. Snir and Kamien [146] have shown recently that short molecular chains, modeled as stiff (but not rigid) impenetrable tubes, are driven to a helix configuration using the same depletion argument. However, this holds only for uniform and relatively short tubes of length of the order of a few persistence lengths l_p of the rod. On longer chains, the picture rapidly grows in complexity with a plethora of optimal configurations (e.g., hairpin, beta-sheet, superhelix, torus) leading to an overwhelmingly rich phase diagram. In this section, our goal is to show that the presence of chromatin fiber rosettes can be explained using a depletion argument for long tubes with "frozen," heterogeneously distributed elastic and/or geometric properties. By frozen we mean that these fluctuations are imprinted on the 30-nm chromatin fiber by the sequence itself. Indeed, the fiber is known to be dependent upon the properties of the nucleosomal string-of-beads [147–150], which in turn is influenced by the double-helix intrinsic structural disorder induced by the sequence. In essence, it comes down to ask the following question: Is there a topological configuration in which the fiber is most likely to self-organize reproducibly?

For a semi-flexible tube in a dilute environment, local repulsive potentials among parts of the fiber induce a self-avoiding random walk configuration (swollen coil [151]). In a crowded environment, the depletive action may dominate and the fiber will tend to collapse on itself, forming a globular phase. We know from standard statistical physics of polymers that this latter phase

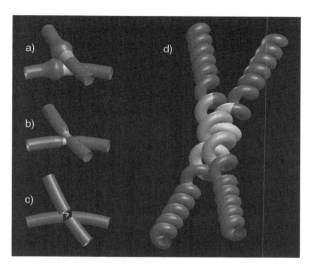

Figure 19. Examples of possible local defects along the fiber. (a) Local swelling or attachment of an external agent (e.g., RNA polymerases in the model of Cook [22, 23]); (b) local shrinking; (c) any form of fiber denaturation inducing a depletive potential well, according to the position along the fiber and the entry–exit angle; (d) as an example of (c), the fiber seen as a compact helix (condensed nucleosomal array) with local partial decondensation illustrating a situation where the excluded volume gain is quite important and the entry-exit angle is fixed. See color insert.

does not admit a universal description in terms of macroscopic parameters (such as total length, Kuhn length and virial coefficients) but rather depends on a detailed understanding of the interaction potential. However, an important feature of the depletive potential lies in its simplistic geometrical nature. We thus consider a system constituted of a dense fluid of hard spheres bathing a semi-flexible tube. The tube is assumed to be nonuniform, with localized geometrical defects (e.g., local thickening or thinning of the crosssection, see Fig. 19). The elastic nature of the tube prevents the appearance of too high curvature points; consequently, the first step in the condensation of the tube is the formation of loops. Loop formation involves a competition between the bending energy of the tube and the entropic gain of the hard-sphere fluid. The free energy cost is dominated by elastic energy for small loops and by entropy for large ones. This results in a preferential length of $3.4l_p$ in the worm-like-chain (WLC) model [152, 153].

Once a loop is formed, contact will be maintained by depletive forces; hence the loop will preferentially relax through local gliding of the two contact points (Fig. 20). This is where local defects come into play: When they meet from this gliding process, they act as local geometrical wells and "stick" together. This defect-induced stabilization is important since it prevents further depletive

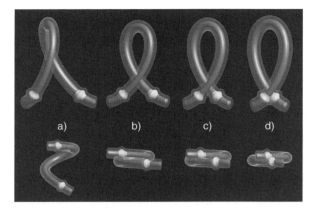

Figure 20. Steps involved in loop formation. (a) Free evolution of the tube in depletive environment; (b) formation of an unstable loop at around 3.4 l_p; (c) gliding of the loop governed by the positions of the two contact points along the fiber and the entry–exit angle; (d) trapping of the loop by local defects. The translucent green surface represents the excluded volume for the fluid of hard spheres; in (b,c,d) one sees that some of the excluded volume is reduced from the overlap resulting from formation of the loop. See color insert.

mechanisms to take place. Indeed by modifying locally the angle of tangent vectors at the contact points, the depletion force could drive them to align in opposite directions, forming the first turn of an helix or toroidal condensate; alternatively, it could align them in the same direction, favoring the formation of hairpins. The presence of defects, by favoring a specific contact geometry, *breaks* the symmetries (translational, axial) essential to the formation of these compact structures, drastically modifying the phase diagram. The condensation rather occurs via the aggregation of defects, inducing rosette-like patterns.

In that context we propose [154] to characterize the distribution of the number of leaves per rosette from minimal parametrization of the system. We consider a dense fluid composed of a large number N_s of identical spheres, bathing a tube which, for simplicity, contains N equidistant defects, separated by a distance l along the tube. We assume that rosettes are formed while respecting sequential order of defects along the tube. Let n denote the number of rosettes along the tube; solitary defects are also considered as trivial "rosettes" with zero leaves. Obviously, the case where $n = N$ represents the absence of clustering since all defects are then solitary. On the other hand, the case where $n = 1$ corresponds to a single large rosette assembling all defects.

We separate in a natural fashion the system in two parts, namely the hard-sphere fluid and the tube itself. Let F, F_t, and F_s denote the free energies of the system, the tube, and the spheres, respectively (all of which depend on n), such that we can write $F = F_t + F_s$. The most probable value for n is obtained by taking the derivative of F with respect to n and equating to zero. This derivative is simply the *chemical potential* μ_r of a rosette:

$$\mu_r \equiv \frac{\partial F}{\partial n} = \frac{\partial F_t}{\partial n} + \frac{\partial F_s}{\partial n} = 0 \tag{10}$$

The equation of state of a fluid of hard spheres has been extensively studied in the past [155]. We follow the method used by Dinsmore et al. [156] and make use of the Carnahan–Starling approximation [157]:

$$\frac{P_s(n)V_s(n)}{N_s k_B T} = \frac{1 + \varphi + \varphi^2 - \varphi^3}{(1 - \varphi)^3} \tag{11}$$

where $\varphi = N_s v_s / V_s(n)$ is the density of the spheres, $P_s(n)$ the fluid osmotic pressure, v_s the volume of each sphere, and $V_s(n)$ is the volume available to the spheres. The F_t term can be expressed as

$$F_t = E_0 + (N - n) \cdot \Delta F_l - k_B T \cdot \log \binom{n}{N} \tag{12}$$

where $\Delta F_l > 0$ is the free energy cost for the formation of a single loop, and E_0 is an energy term assumed to be independent of n. The last term on r.h.s. of Eq. (12) corresponds to the number of arrangements of n rosettes from N defects and contributes to the entropy of the tube. From Eqs. (11) and (12) and from the thermodynamical identity $\frac{\partial F_s}{\partial V_s} = -P_s$, we get

$$\mu_r = -k_B T \cdot \log\left(\frac{N - n}{n}\right) - \Delta F_l$$
$$+ k_B T \cdot \left(\frac{v_{ovl}}{v_s}\right) \cdot \left(\frac{\varphi + \varphi^2 + \varphi^3 - \varphi^4}{(1 - \varphi)^3}\right) \tag{13}$$

where $v_{ovl} = -\partial V_s / \partial n$ represents the overlapping excluded volume of two interleaved defects, *i.e.* the volume gain for the spheres due to the interaction between two defects. From Eq. (13), we see that this simple model depends on three parameters: the free energy cost of a loop ΔF_l, the normalized overlap volume per loop v_{ovl}/v_s, and the sphere density φ. The free energy cost of a loop can be approximated by

$$\Delta F_l = \frac{1}{2} k_b T \cdot l \cdot l_p \kappa^2 \lesssim 6 k_B T \tag{14}$$

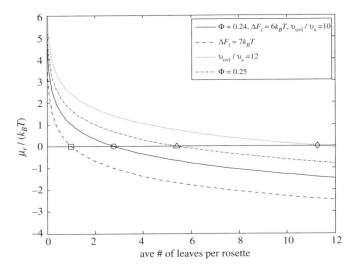

Figure 21. The chemical potential is shown as a function of the number of leaves per rosette $(N/n - 1)$. The solid curve serves as a reference and corresponds to the following parameter values: $\Phi = 0.24$, $\Delta F_l = 6k_BT$, and $v_{ovl}/v_s = 10$. For these values, an average value of 2.77 leaves ($\circ$) per rosette is expected. Increasing Φ to 0.25 increases the average to 5.39 ($\triangle$); similarly, increasing v_{ovl}/v_s to 12 results in an average value of 11.27 ($\diamond$). Increasing ΔF_l to $7k_BT$ reduces the average to 1.02 ($\square$).

where $l \simeq 3.4l_p$ is the length of a typical loop and $\kappa \simeq 2\pi/l$ its average curvature. Physiological values for the hard-sphere fluid density $\Phi = \varphi(n = N)$ vary between [158] $\simeq 0.2 \sim 0.3$. For a 30-nm fiber we can expect $v_{ovl} \sim (10\,\text{nm})^3$; the typical size of proteins is $v_s \sim (5\,\text{nm})^3$, leading to values of v_{ovl}/v_s around 10.

As illustrated in Fig. 21, increasing the sphere density Φ or the normalized overlap volume results in an increase of the average number of leaves per rosette $(N/n - 1)$, thus in a more compact structure. On the other hand, increasing the free energy cost of loop formation (stiffening of the fiber) decreases the average number of leaves. Interestingly, we find that the average number of leaves per rosette can be regulated by fine tuning the values of these parameters within physiological range (Fig. 21). For instance, for $v_{ovl}/v_s = 10$ and $\Delta F_l = 6k_BT$ and varying Φ between 0.24 and 0.25 results in the average number of leaves per rosette running from 2.77 (i.e., low clustering), to 6.39. This provides attractive scenarios for the spontaneous emergence of chromatin rosettes in the nucleus milieu prior to their possible further stabilization by external factors (e.g., specific DNA binding proteins) [23, 70, 83, 123].

Various models of interphase chromatin based on a multi-looped structure of the 30-nm fiber have been proposed in the literature [19–21], but they all involve interaction with some nucleoproteic complexes to organize the structure—for example, the scaffolding proteins that interact specifically at certain DNA regions (scaffold-associated regions) to fold the fiber [14, 15] or the transcription complexes strung along the genome that clusterize and consequently fold the chromatin fiber [22, 23]. The main message of the present work is the possibility that the chromatin fiber self-organizes into rosette-like patterns in the crowded environment of the nucleus thanks to its heterogeneous structure. Recent modeling [148, 149] has revealed an extreme sensitivity of the internal fiber conformation to the local structural and mechanical properties of the nucleosomal string—for example, the linker length, the entry–exit angle between the linkers, or the twist angle along a linker. The fiber local structure is known to be controlled by epigenetic modifications of these architectural nucleosomal parameters [147, 149] (DNA methylation, histone modifications, etc.). Yet as suggested by recent modeling of the thermodynamics of DNA loops [30, 31], the local properties of the nucleosomal string are also conditioned by the primary DNA sequence that codes for the structural disorder intrinsic to the DNA double helix. Therefore the structural defects of the fiber can be encoded in the sequence. The entropy-driven fiber folding mechanism described above [154] leads to the aggregation of neighboring defects into clusters that ensures high local concentration of distant DNA target sites. This clustering is likely to favor the recruiting of protein complexes involved in the activation of replication and transcription. In this context, the set of 1012 putative replication initiation zones identified in the human genome [38, 39] (Section VI) provides privileged locations for some intrinsic decondensated fiber defects. The spontaneous emergence of rosette patterns (likely stabilized by the origin replication complexes) provides a very attractive description of the so-called replication foci [81, 83, 84, 123] that have been observed in interphase mammalian nuclei as stable structural domains of autonomous replication that persist during all cell-cycle stages. Furthermore, the remarkable gene organization discovered around the putative replication origins [38, 39] strongly suggests that these rosettes contribute to the compartmentalization of the genome into autonomous domains of gene transcription. Via the self-organizing structural role of the replication origins, the DNA sequence might therefore code, to some extent, for the tertiary chromatin structure. Even though one expects to observe, from one cell cycle to the next, fluctuations in the number of loops contained in each rosette as illustrated in Fig. 22, the perennity of defects is likely to ensure the inheritance of the interphase chromatin rosette organization. As an illustration, we present in Fig. 23 the picture of a rosette-like pattern of the chromatin fiber in a crowded, heterogeneous environment mimicking the cell nucleus.

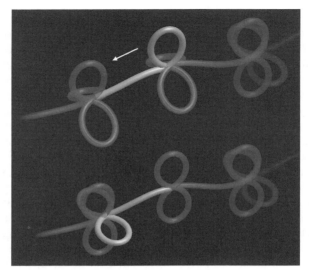

Figure 22. Illustration of the fiber defects clustering dynamics. The number of leaves per rosette fluctuates from one rosette to the next; this is due to both statistical fluctuations and variations in the local environment. From one cell cycle (up) to the next (down), leaves can be exchanged between neighboring rosettes. See color insert.

Figure 23. Illustration of the spontaneous emergence of rosette-like folding of the chromatin fiber in the crowded environment of the cell nucleus. See color insert.

VIII. PERSPECTIVES

In the recent past, the DNA double helix was simply considered as a biological macromolecule (a polymer) that contains our genetic heritage (genotype). The regulation and control of DNA replication and expression was supposed to be fully delegated to proteins. Nowadays, DNA is more and more recognized as a complex heteropolymer whose structural and mechanical properties play a relevant part in the management of the gene information it carries. The results reported in this Chapter concerning the analysis at the genome scale of mammalian DNA sequences [36–39], together with the results obtained in a previous study [11–13, 30–32] of the long-range correlations exhibited by eukaryotic DNA sequences up to distances of a few tens of kbp, demonstrate that there is a lot of information encoded in the DNA sequences concerning the different stages of compaction of DNA inside the nucleus of mammalian cells (Fig. 1). Surprisingly, if the sequence codes for the local structural and mechanical properties of the double helix and, in turn, influences the formation and dynamics of the nucleosomal string, our results show that it also conditions to some extent the next levels of compaction, via the self-organized formation of chromatin fiber rosette patterns that are likely to define structural domains of autonomous DNA replication and gene expression. Since introns and intergenic regions constitute more than 95% of the human genome, our study therefore contributes to give a role to the noncoding regions in eukaryotic genomes. These regions actually play a driving role in the condensation and decondensation processes of the chromatin architecture as well as in many related regulative functions.

The results reported in this chapter open new perspectives in DNA sequence analysis, modeling, and experiment. For a methodological point of view, there is some hope to improve the efficiency of our wavelet-based method to detect the replication origins (Section VI.A). So far our strategy was based on the use of the first-derivative of the Gaussian function as analyzing wavelet to detect sharp upward jumps in the noisy skew profiles. Along the line of the model of replication we propose in Section VI.C to account for the observed factory roof skew profiles in mammalian genomes, we consider the use of an analyzing wavelet that has exactly the jagged shape predicted by this model as illustrated in Fig. 18c. By adapting the optics of our mathematical WT microscope, we should be in better position to face the observed variability in size of the replication domains. The implementation of a replication pattern matching algorithm in the space-scale representation provided by the WT is in current progress [97]. From a bioinformatics and modeling point of view, we plan to study the lexical and structural characteristics of our set of putative origins. In particular, we will search for conserved sequence motifs in these replication initiation zones. Using a sequence-dependent model of

DNA–histone interactions, we will develop physical study of nucleosome formation and diffusion along the DNA fiber around the putative replication origins. From an experimental point of view, our study raises new opportunities for future experiments. The first one concerns the experimental validation of the predicted replication origins (e.g., by molecular combing of DNA molecules [159]), which will allow us to determine precisely the existence of replication origins in given genome regions. Large-scale study of all candidate origins is in current progress in the laboratory of O. Hyrien (ENS, Ulm). The second experimental project consists in using Atomic Force Microscopy (AFM) [160] and Surface Plasmon Resonance Microscopy (SPRM) [161] to visualize and study the structural and mechanical properties of the DNA double helix, the nucleosomal string, and the 30-nm chromatin fiber around the predicted replication origins. This work is in current progress in the experimental group of F. Argoul and C. Moskalenko at the Laboratoire Joliot-Curie (ENS, Lyon). Finally, the third experimental perspective concerns *in situ* studies of replication origins. Using fluorescence techniques (FISH chromosome painting [80]), we plan to study the distributions and dynamics of origins in the cell nucleus, as well as chromosome domains potentially associated with territories and their possible relation to nuclear matrix attachment sites. This study is likely to provide evidence of chromatin rosette patterns as suggested in Section VII. This study is under progress in the molecular biology experimental group of F. Mongelard at the Laboratoire Joliot-Curie.

Acknowledgments

We thank F. Argoul, M. Castelnovo, P. Cook, O. Hyrien, F. Mongelard, and C. Moskalenko for interesting discussions. This work was supported by the Action Concertée Incitative Informatique, Mathématiques, Physique en Biologie Moléculaire 2004 under the project "ReplicOr," the Agence Nationale de la Recherche under the project "HUGOREP," the program "Emergence" of the Conseil Régional Rhônes-Alpes and by the Natural Science and Engineering Research Council of Canada (NSERC).

References

1. K. E. van Holde, *Chromatin*, Springer-Verlag, New York, 1988.
2. A. P. Wolffe, *Chromatin Structure and Function*, 3rd ed., Academic Press, London, 1998.
3. C. R. Calladine, and H. R. Drew, *Understanding DNA*, Academic Press, San Diego, 1999.
4. B. Alberts, D. Bray, J. Lewis, M. Raff, K. Roberts, and J. D. Watson, *Molecular Biology of the Cell*, 3rd ed., Garland Publishing, New York, 1994.
5. J. Widom, Structure, dynamics and function of chromatin *in vitro*. *Annu. Rev. Biophys. Biomol. Struct.* **27**, 285–327 (1998).
6. K. Luger, A. W. Mäder, R. K. Richmond, D. F. Sargent, and T. J. Richmond, Crystal structure of the nucleosome core particle at 2.8 å resolution. *Nature* **389**, 251–260 (1997).

7. L. Chantalat, J. M. Nicholson, S. J. Lambert, A. J. Reid, M. J. Donovan, C. D. Reynolds, C. M. Wood, and J. P. Baldwin, Structure of the histone-core octamer in KCl/phosphate crystals at 2.15 Å resolution. *Acta Crystallogr. D Biol. Crystallogr.* **59**, 1395–1407 (2003).

8. T. J. Richmond, and C. A. Davey, The structure of DNA in the nucleosome core. *Nature* **423**, 145–150 (2003).

9. I. Ioshikhes, A. Bolshoy, K. Derenshteyn, M. Borodovsky, and E. N. Trifonov, Nucleosome DNA sequence pattern revealed by multiple alignment of experimentally mapped sequences. *J. Mol. Biol.* **262**, 129–139 (1996).

10. H. Herzel, O. Weiss, and E. N. Trifonov, 10–11 bp periodicities in complete genomes reflect protein structure and DNA folding. *Bioinformatics* **15**, 187–193 (1999).

11. B. Audit, C. Thermes, C. Vaillant, Y. d'Aubenton-Carafa, J.-F. Muzy, and A. Arneodo, Long-range correlations in genomic DNA: A signature of the nucleosomal structure. *Phys. Rev. Lett.* **86**, 2471–2474 (2001).

12. B. Audit, C. Vaillant, A. Arneodo, Y. d'Aubenton-Carafa, and C. Thermes, Long-range correlations between DNA bending sites: Relation to the structure and dynamics of nucleosomes. *J. Mol. Biol.* **316**, 903–918 (2002).

13. A. Arneodo, B. Audit, N. Decoster, J.-F. Muzy, and C. Vaillant, Wavelet based multifractal formalism: Application to DNA sequences, satellite images of the cloud structure and stock market data, in *The Science of Disasters: Climate Disruptions, Heart Attacks, and Market Crashes*, Springer-Verlag, Berlin, 2002, pp. 26–102.

14. U. K. Laemmli, E. Käs, L. Poljak, and Y. Adachi, Scaffold-associated regions: Cis-acting determinants of chromatin structural loops and functional domains. *Curr. Opin. Genet. Dev.* **2**, 275–285 (1992).

15. Y. Saitoh, and U. K. Laemmli, From the chromosomal loops and the scaffold to the classic bands of metaphase chromosomes. *Cold Spring Harb. Symp. Quant. Biol.* **58**, 755–765 (1993).

16. A. S. Belmont, and K. Bruce, Visualization of G1 chromosomes: A folded, twisted, supercoiled chromonema model of interphase chromatid structure. *J. Mol. Biol.* **127**, 287–302 (1994).

17. A. S. Belmont, S. Dietzel, A. C. Nye, Y. G. Strukov, and T. Tumbar, Large-scale chromatin structure and function. *Curr. Opin. Cell Biol.* **11**, 307–311 (1999).

18. P. J. Horn, and C. L. Peterson, Molecular biology. Chromatin higher order folding—wrapping up transcription. *Science* **297**, 1824–1827 (2002).

19. R. K. Sachs, G. van den Engh, B. Trask, H. Yokota, and J. E. Hearst, A random-walk/giant-loop model for interphase chromosomes. *Proc. Natl. Acad. Sci. USA* **92**, 2710–2714 (1995).

20. J. Ostashevsky, A polymer model for the structural organization of chromatin loops and minibands in interphase chromosomes. *Mol. Biol. Cell* **9**, 3031–3040 (1998).

21. C. Münkel, R. Eils, S. Dietzel, D. Zink, C. Mehring, G. Wedemann, T. Cremer, and J. Langowski, Compartmentalization of interphase chromosomes observed in simulation and experiment. *J. Mol. Biol.* **285**, 1053–1065 (1999).

22. P. R. Cook, A chromomeric model for nuclear and chromosome structure. *J. Cell. Sci.* **108**, 2927–2935 (1995).

23. P. R. Cook, Predicting three-dimensional genome structure from transcriptional activity. *Nat. Genet.* **32**, 347–352 (2002).

24. N. L. Mahy, P. E. Perry, and W. A. Bickmore, Gene density and transcription influence the localization of chromatin outside of chromosome territories detectable by FISH. *J. Cell Biol.* **159**, 753–763 (2002).

25. G. Felsenfeld, and M. Groudine, Controlling the double helix. *Nature* **421**, 448–453 (2003).

26. A. Arneodo, E. Bacry, P. V. Graves, and J.-F Muzy, Characterizing long-range correlations in DNA sequences from wavelet analysis. *Phys. Rev. Lett.* **74**, 3293–3296 (1995).

27. A. Arneodo, Y. d'Aubenton-Carafa, E. Bacry, P. V. Graves, J.-F. Muzy, and C. Thermes, Wavelet based fractal analysis of DNA sequences. *Physica D* 96, 291–320 (1996).

28. Y. Meyer, ed. *Wavelets and Their Applications*, Springer, Berlin, 1992.

29. S. Mallat, *A Wavelet Tour of Signal Processing*, Academic Press, New York, 1998.

30. C. Vaillant, B. Audit, and A. Arneodo, Thermodynamics of DNA loops with long-range correlated structural disorder. *Phys. Rev. Lett.* **95**, 068101 (2005).

31. C. Vaillant, B. Audit, C. Thermes, and A. Arneodo, Formation and positioning of nucleosomes: effect of sequence dependent long-range correlated structural disorder. *Eur. Phys. J. E,* **19**, 263–277 (2006).

32. C. Vaillant, B. Audit and A. Arneodo, First experimental evidence of nucleosome positioning by genomic long-range correlations. Preprint. (2006).

33. G.-C. Yuan, Y.-J. Liu, M. F. Dion, M. D. Slack, L. F. Wu, S. J. Altschuler, and O. J. Rando, Genome-scale identification of nucleosome positions in S. cerevisiae. *Science* **309**, 626–630 (2005).

34. S. Nicolay, F. Argoul, M. Touchon, Y. d'Aubenton-Carafa, C. Thermes, and A. Arneodo, Low frequency rhythms in human DNA sequences: A key to the organization of gene location and orientation? *Phys. Rev. Lett.* **93**, 108101 (2004).

35. S. Nicolay, E.-B. Brodie de Brodie, M. Touchon, Y. d'Aubenton-Carafa, C. Thermes, and A. Arneodo, From scale invariance to deterministic chaos in DNA sequences: towards a deterministic description of gene organization in the human genome. *Physica A* **342**, 270–280 (2004).

36. M. Touchon, S. Nicolay, A. Arneodo, Y. d'Aubenton-Carafa, and C. Thermes, Transcription-coupled TA and GC strand asymmetries in the human genome. *FEBS Lett.* **555**, 579–582 (2003).

37. M. Touchon, A. Arneodo, Y. d'Aubenton-Carafa, and C. Thermes, Transcription-coupled and splicing-coupled strand asymmetries in eukaryotic genomes. *Nucl. Acids Res.* **32**, 4969–4978 (2004).

38. M. Touchon, S. Nicolay, B. Audit, E.-B. Brodie of Brodie, Y. d'Aubenton-Carafa, A. Arneodo, and C. Thermes, Replication-associated strand asymmetries in mammalian genomes: Towards detection of replication origins. *Proc. Natl. Acad. Sci. USA* **102**, 9836–9841 (2005).

39. E.-B. Brodie of Brodie, S. Nicolay, M. Touchon, B. Audit, Y. d'Aubenton-Carafa, C. Thermes, and A. Arneodo, From DNA sequence analysis to modelling replication in the human genome. *Phys. Rev. Lett.* **94**, 248103 (2005).

40. A. F. A. Smit, R. Hubley, and P. Green, RepeatMasker Open 3.0, http://www.repeatmasker.org. (1996–2004).

41. E.-M. Ladenburger, C. Keller, and R. Knippers, Identification of a binding region for human origin recognition complex proteins 1 and 2 that coincides with an origin of DNA replication. *Mol. Cell. Biol.* **22**, 1036–1048 (2002).

42. T. Taira, S. M. Iguchi-Ariga, and H. Ariga, A novel DNA replication origin identified in the human heat shock protein 70 gene promoter. *Mol. Cell. Biol.* **14**, 6386–6397 (1994).

43. C. Keller, E.-M. Ladenburger, M. Kremer, and R. Knippers, The origin recognition complex marks a replication origin in the human TOP1 gene promoter. *J. Biol. Chem.* **277**, 31430–31440 (2002).

44. L. Vassilev, and E. M. Johnson, An initiation zone of chromosomal DNA replication located upstream of the c-myc gene in proliferating HeLa cells. *Mol. Cell. Biol.* **10**, 4899–4904 (1990).

45. T. Nenguke, M. I. Aladjem, J. F. Gusella, N. S. Wexler, and N. Arnheim, Candidate DNA replication initiation regions at human trinucleotide repeat disease loci. *Hum. Mol. Genet.* **12**, 1021–1028 (2003).

46. F. D. Araujo, J. D. Knox, S. Ramchandani, R. Pelletier, P. Bigey, G. Price, M. Szyf, and M. Zannis-Hadjopoulos, Identification of initiation sites for DNA replication in the human dnmt1 (DNA-methyltransferase) locus. *J. Biol. Chem.* **274**, 9335–9341 (1999).

47. M. Giacca, L. Zentilin, P. Norio, S. Diviacco, D. Dimitrova, G. Contreas, G. Biamonti, G. Perini, F. Weighardt, et al. Fine mapping of a replication origin of human DNA. *Proc. Natl. Acad. Sci. USA* **91**, 7119–7123 (1994).

48. D. Kitsberg, S. Selig, I. Keshet, and H. Cedar, Replication structure of the human beta-globin gene domain. *Nature* **366**, 588–590 (1993).

49. S. Schwartz, Z. Zhang, K. A. Frazer, A. Smit, C. Riemer, J. Bouck, R. Gibbs, R. Hardison, and W. Miller, PipMaker—a web server for aligning two genomic DNA sequences. *Genome Res.* **10**, 577–586 (2000).

50. A. Arneodo, F. Argoul, E. Bacry, J. Elezgaray, and J.-F. Muzy, *Ondelettes Multifractales et Turbulences: de l'ADN aux Croissances Cristallines*, Diderot Editeur, Arts et Sciences, Paris, (1995).

51. E. S. Lander, L. M. Linton, B. Birren, C. Nusbaum, M. C. Zody, J. Baldwin, K. Devon, K. Dewar, M. Doyle, et al. Initial sequencing and analysis of the human genome. *Nature* **409**, 860–921 (2001).

52. G. Bernardi, The isochore organization of the human genome. *Annu. Rev. Genet.* **23**, 637–661 (1989).

53. G. Bernardi, The human genome: Organization and evolutionary history. *Annu. Rev. Genet.* **29**, 445–476 (1995).

54. A. Nekrutenko, and W. H. Li, Assessment of compositional heterogeneity within and between eukaryotic genomes. *Genome Res.* **10**, 1986–1995 (2000).

55. D. Häring, and J. Kypr, No isochores in the human chromosomes 21 and 22? *Biochem. Biophys. Res. Commun.* **280**, 567–573 (2001).

56. G. Bernardi, Misunderstandings about isochores. Part 1. *Gene* **276**, 3–13 (2001).

57. A. Eyre-Walker, and L. D. Hurst, The evolution of isochores. *Nat. Rev. Genet.* **2**, 549–555 (2001).

58. W. Li, Are isochore sequences homogeneous? *Gene* **300**, 129–139 (2002).

59. N. Cohen, T. Dagan, L. Stone, and D. Graur, GC composition of the human genome: in search of isochores. *Mol. Biol. Evol.* **22**, 1260–1272 (2005).

60. A. Pavlícek, J. Paces, O. Clay, and G. Bernardi, A compact view of isochores in the draft human genome sequence. *FEBS Lett.* **511**, 165–169 (2002).

61. H. Hori, and S. Osawa, Evolutionary change in 5S rRNA secondary structure and a phylogenic tree of 352 5S rRNA species. *Biosystems* **19**, 163–172 (1986).

62. G. D'Onofrio, D. Mouchiroud, B. Aïssani, C. Gautier, and G. Bernardi, Correlations between the compositional properties of human genes, codon usage, and amino acid composition of proteins. *J. Mol. Evol.* **32**, 504–510 (1991).

63. D. Graur, and W. H. Li, *Fundamentals of Molecular Evolution.* Sinauer Associates, Sunderland, MA, 1999.

64. J. R. Paulson, and U. K. Laemmli, The structure of histone-depleted metaphase chromosomes. *Cell* **12**, 817–828 (1977).

65. S. M. Gasser, and U. K. Laemmli, A glimpse at chromosomal order. *Trends Genet.* **3**, 16–22 (1987).

66. J. B. Rattner, and C. C. Lin, Radial loops and helical coils coexist in metaphase chromosomes. *Cell* **42**, 291–296 (1985).

67. E. Boy de la Tour, and U. K. Laemmli, The metaphase scaffold is helically folded: Sister chromatids have predominantly opposite helical handedness. *Cell* **55**, 937–944 (1988).

68. M. G. Poirier, A. Nemani, P. Gupta, S. Eroglu, and J. F. Marko, Probing chromosome structure with dynamic force relaxation. *Phys. Rev. Lett.* **86**, 360–363 (2001).

69. J. M. Bridger, and W. A. Bickmore, Putting the genome on the map. *Trends Genet.* **14**, 403–409 (1998).

70. T. Cremer, and C. Cremer, Chromosome territories, nuclear architecture and gene regulation in mammalian cells. *Nat. Rev. Genet.* **2**, 292–301 (2001).

71. A. S. Belmont, Visualizing chromosome dynamics with GFP. *Trends Cell Biol.* **11**, 250–257 (2001).

72. S. M. Gasser, Visualizing chromatin dynamics in interphase nuclei. *Science* **296**, 1412–1416 (2002).

73. D. Zink, T. Cremer, R. Saffrich, R. Fischer, M. F. Trendelenburg, W. Ansorge, and E. H. Stelzer, Structure and dynamics of human interphase chromosome territories *in vivo. Hum. Genet.* **102**, 241–251 (1998).

74. C. Münkel, and J. Langowski, Chromosome structure predicted by a polymer model. *Phys. Rev. E* **57**, 5888–5896 (1998).

75. L. Manuelidis, A view of interphase chromosomes. *Science* **250**, 1533–1540 (1990).

76. L. Manuelidis, and T. L. Chen, A unified model of eukaryotic chromosomes. *Cytometry* **11**, 8–25 (1990).

77. C. L. Woodcock, H. Woodcock, and R. A. Horowitz, Ultrastructure of chromatin. I. Negative staining of isolated fibers. *J. Cell Sci.* **99**, 99–106 (1991).

78. C. L. Woodcock, Chromatin fibers observed *in situ* in frozen hydrated sections. Native fiber diameter is not correlated with nucleosome repeat length. *J. Cell Biol.* **125**, 11–19 (1994).

79. A. S. Belmont, Large-scale chromatin organization, in *Genome Structure and Function*, Kluwer Academic Publishers, Dordrecht, 1997, p. 261.

80. W. G. Müller, D. Rieder, G. Kreth, C. Cremer, Z. Trajanoski, and J. G. McNally, Generic features of tertiary chromatin structure as detected in natural chromosomes. *Mol. Cell. Biol.* **24**, 9359–9370 (2004).

81. D. A. Jackson, and A. Pombo, Replicon clusters are stable units of chromosome structure: evidence that nuclear organization contributes to the efficient activation and propagation of S phase in human cells. *J. Cell Biol.* **140**, 1285–1295 (1998).

82. H. Ma, J. Samarabandu, R. S. Devdhar, R. Acharya, P. C. Cheng, C. Meng, and R. Berezney, Spatial and temporal dynamics of DNA replication sites in mammalian cells. *J. Cell Biol.* **143**, 1415–1425 (1998).

83. R. Berezney, D. D. Dubey, and J. A. Huberman, Heterogeneity of eukaryotic replicons, replicon clusters, and replication foci. *Chromosoma* **108**, 471–484 (2000).

84. H. J. Edenberg, and J. A. Huberman, Eukaryotic chromosome replication. *Annu. Rev. Genet.* **9**, 245–284 (1975).

85. R. Hand, Eucaryotic DNA: Organization of the genome for replication. *Cell* **15**, 317–325 (1978).

86. Y. B. Yurov, and N. A. Liapunova, The units of DNA replication in the mammalian chromosomes: evidence for a large size of replication units. *Chromosoma* **60**, 253–267 (1977).

87. N. A. Liapunova, Organization of replication units and DNA replication in mammalian cells as studied by DNA fiber radioautography. *Int. Rev. Cytol.* **154**, 261–308 (1994).

88. E. Chargaff, Structure and function of nucleic acids as cell constituents. *Fed. Proc.* **10**, 654–659 (1951).

89. R. Rudner, J. D. Karkas, and E. Chargaff, Separation of B. subtilis DNA into complementary strands. 3. Direct analysis. *Proc. Natl. Acad. Sci. USA* **60**, 921–922 (1968).

90. J. W. Fickett, D. C. Torney, and D. R. Wolf, Base compositional structure of genomes. *Genomics* **13**, 1056–1064 (1992).

91. J. R. Lobry, Properties of a general model of DNA evolution under no-strand-bias conditions. *J. Mol. Evol.* **40**, 326–330 (1995).

92. J. Mrázek, and S. Karlin, Strand compositional asymmetry in bacterial and large viral genomes. *Proc. Natl. Acad. Sci. USA* **95**, 3720–3725 (1998).

93. A. C. Frank, and J. R. Lobry, Asymmetric substitution patterns: A review of possible underlying mutational or selective mechanisms. *Gene* **238**, 65–77 (1999).

94. E. P. Rocha, A. Danchin, and A. Viari, Universal replication biases in bacteria. *Mol. Microbiol.* **32**, 11–16 (1999).

95. E. R. M. Tillier, and R. A. Collins, The contributions of replication orientation, gene direction, and signal sequences to base-composition asymmetries in bacterial genomes. *J. Mol. Evol.* **50**, 249–257 (2000).

96. E.-B. Brodie of Brodie, *De l'analyse des séquences d'ADN à la modélisation de la réplication chez les mammifères.* Ph.D. thesis, ENS de Lyon, France, 2005.

97. S. Nicolay, *Analyse des séquences d'ADN par la transformée en ondelettes: Extraction d'informations structurelles, dynamiques et fonctionnelles.* Ph.D. thesis, University of Liège, Belgium, 2006.

98. T. Gojobori, W. H. Li, and D. Graur, Patterns of nucleotide substitution in pseudogenes and functional genes. *J. Mol. Evol.* **18**, 360–369 (1982).

99. W. H. Li, C. I. Wu, and C. C. Luo, Nonrandomness of point mutation as reflected in nucleotide substitutions in pseudogenes and its evolutionary implications. *J. Mol. Evol.* **21**, 58–71 (1984).

100. D. A. Petrov, and D. L. Hartl, Patterns of nucleotide substitution in *Drosophila* and mammalian genomes. *Proc. Natl. Acad. Sci. USA* **96**, 1475–1479 (1999).

101. Z. Zhang, and M. Gerstein, Patterns of nucleotide substitution, insertion and deletion in the human genome inferred from pseudogenes. *Nucleic Acids Res.* **31**, 5338–5348 (2003).

102. J. M. Freeman, T. N. Plasterer, T. F. Smith, and S. C. Mohr, Patterns of genome organization in bacteria. *Science* **279**, 1827 (1998).

103. A. Beletskii, A. Grigoriev, S. Joyce, and A. S. Bhagwat, Mutations induced by bacteriophage T7 RNA polymerase and their effects on the composition of the T7 genome. *J. Mol. Biol.* **300**, 1057–1065 (2000).

104. M. P. Francino, and H. Ochman, Deamination as the basis of strand-asymmetric evolution in transcribed Escherichia coli sequences. *Mol. Biol. Evol.* **18**, 1147–1150 (2001).

105. L. Duret, Evolution of synonymous codon usage in metazoans. *Curr. Opin. Genet. Dev.* **12**, 640–649 (2002).

106. C. Shioiri, and N. Takahata, Skew of mononucleotide frequencies, relative abundance of dinucleotides, and DNA strand asymmetry. *J. Mol. Evol.* **53**, 364–376 (2001).

107. P. Green, B. Ewing, W. Miller, P. J. Thomas, and E. D. Green, Transcription-associated mutational asymmetry in mammalian evolution. *Nat. Genet.* **33**, 514–517 (2003).

108. J. Q. Svejstrup, Mechanisms of transcription-coupled DNA repair. *Nat. Rev. Mol. Cell Biol.* **3**, 21–29 (2002).

109. M. Touchon, *Biais de composition chez les mammifères.: Rôle de la transcription et de la réplication.* Ph.D. thesis, University Denis Diderot, Paris VII, France, 2005.

110. F. Jacob, S. Brenner, and F. Cuzin, On the regulation of DNA replication in bacteria. *Cold Spring Harb. Symp. Quant. Biol.* **28**, 329–342 (1963).

111. S. P. Bell, and A. Dutta, DNA replication in eukaryotic cells. *Annu. Rev. Biochem.* **71**, 333–374 (2002).

112. O. Hyrien, and M. Méchali, Chromosomal replication initiates and terminates at random sequences but at regular intervals in the ribosomal DNA of Xenopus early embryos. *EMBO J.* **12**, 4511–4520 (1993).

113. S. A. Gerbi, and A. K. Bielinsky, DNA replication and chromatin. *Curr. Opin. Genet. Dev.* **12**, 243–248 (2002).

114. D. Schübeler, D. Scalzo, C. Kooperberg, B. van Steensel, J. Delrow, and M. Groudine, Genome-wide DNA replication profile for *Drosophila melanogaster*: A link between transcription and replication timing. *Nat. Genet.* **32**, 438–442 (2002).

115. D. Fisher, and M. Méchali, Vertebrate HoxB gene expression requires DNA replication. *EMBO J.* **22**, 3737–3748 (2003).

116. M. Anglana, F. Apiou, A. Bensimon, and M. Debatisse, Dynamics of DNA replication in mammalian somatic cells: Nucleotide pool modulates origin choice and interorigin spacing. *Cell* **114**, 385–394 (2003).

117. D. M. Gilbert, Making sense of eukaryotic DNA replication origins. *Science* **294**, 96–100 (2001).

118. D. Coverley, and R. A. Laskey, Regulation of eukaryotic DNA replication. *Annu. Rev. Biochem.* **63**, 745–776 (1994).

119. T. Sasaki, T. Sawado, M. Yamaguchi, and T. Shinomiya, Specification of regions of DNA replication initiation during embryogenesis in the 65-kilobase DNApolalpha-dE2F locus of *Drosophila melanogaster.* *Mol. Cell. Biol.* **19**, 547–555 (1999).

120. J. A. Bogan, D. A. Natale, and M. L. Depamphilis, Initiation of eukaryotic DNA replication: conservative or liberal? *J. Cell. Physiol.* **184**, 139–150 (2000).

121. D. M. Gilbert, In search of the holy replicator. *Nat. Rev. Mol. Cell Biol.* **5**, 848–855 (2004).

122. M. Méchali, DNA replication origins: from sequence specificity to epigenetics. *Nat. Rev. Genet.* **2**, 640–645 (2001).

123. C. Demeret, Y. Vassetzky, and M. Méchali, Chromatin remodelling and DNA replication: From nucleosomes to loop domains. *Oncogene* **20**, 3086–3093 (2001).

124. A. J. McNairn, and D. M. Gilbert, Epigenomic replication: Linking epigenetics to DNA replication. *Bioessays* **25**, 647–656 (2003).

125. B. J. Brewer, When polymerases collide: Replication and the transcriptional organization of the *E. coli* chromosome. *Cell* **53**, 679–686 (1988).

126. E. P. C. Rocha, P. Guerdoux-Jamet, I. Moszer, A. Viari, and A. Danchin, Implication of gene distribution in the bacterial chromosome for the bacterial cell factory. *J. Biotech.* **78**, 209–219 (2000).

127. P. Lopez, and H. Philippe, Composition strand asymmetries in prokaryotic genomes: muta-tional bias and biased gene orientation. *C. R. Acad. Sci. III* **324**, 201–208 (2001).

128. E. P. C. Rocha, Is there a role for replication fork asymmetry in the distribution of genes in bacterial genomes. *Trends Microbiol.* **10**, 393–395 (2002).

129. M. Bulmer, Strand symmetry of mutation rates in the beta-globin region. *J. Mol. Evol.* **33**, 305–310 (1991).

130. M. P. Francino, and H. Ochman, Strand symmetry around the beta-globin origin of replication in primates. *Mol. Biol. Evol.* **17**, 416–422 (2000).

131. A. Gierlik, M. Kowalczuk, P. Mackiewicz, M. R. Dudek, and S. Cebrat, Is there replication-associated mutational pressure in the Saccharomyces cerevisiae genome? *J. Theor. Biol.* **202**, 305–314 (2000).

132. E. Louie, J. Ott, and J. Majewski, Nucleotide frequency variation across human genes. *Genome Res.* **13**, 2594–2601 (2003).

133. P. Kapranov, S. E. Cawley, J. Drenkow, S. Bekiranov, R. L. Strausberg, S. P. A. Fodor, and T. R. Gingeras, Large-scale transcriptional activity in chromosomes 21 and 22. *Science* **296**, 916–919 (2002).

134. J. Chen, M. Sun, S. Lee, G. Zhou, J. D. Rowley, and S. M. Wang, Identifying novel transcripts and novel genes in the human genome by using novel SAGE tags. *Proc. Natl. Acad. Sci. USA* **99**, 12257–12262 (2002).

135. J. L. Rinn, G. Euskirchen, P. Bertone, R. Martone, N. M. Luscombe, S. Hartman, P. M. Harrison, F. K. Nelson, P. Miller, et al. The transcriptional activity of human Chromosome 22. *Genes Dev.* **17**, 529–540 (2003).

136. D. Kampa, J. Cheng, P. Kapranov, M. Yamanaka, S. Brubaker, S. Cawley, J. Drenkow, A. Piccolboni, S. Bekiranov, et al. Novel RNAs identified from an in-depth analysis of the transcriptome of human chromosomes 21 and 22. *Genome Res.* **14**, 331–342 (2004).

137. T. Imanishi, T. Itoh, Y. Suzuki, C.O' Donovan, S. Fukuchi, K. O. Koyanagi, R. A. Barrero, T. Tamura, Y. Yamaguchi-Kabata, et al. Integrative annotation of 21,037 human genes validated by full-length cDNA clones. *PLoS Biol.* **2**, e162 (2004).

138. C. Girard-Reydet, D. Grégoire, Y. Vassetzky, and M. Méchali, DNA replication initiates at domains overlapping with nuclear matrix attachment regions in the Xenopus and mouse c-myc promoter. *Gene* **332**, 129–138 (2004).

139. L. T. Vassilev, W. C. Burhans, and M. L. DePamphilis, Mapping an origin of DNA replication at a single-copy locus in exponentially proliferating mammalian cells. *Mol. Cell. Biol.* **10**, 4685–4689 (1990).

140. S. Codlin, and J. Z. Dalgaard, Complex mechanism of site-specific DNA replication termination in fission yeast. *EMBO J.* **22**, 3431–3440 (2003).

141. D. Santamaria, E. Viguera, M. L. Martinez-Robles, O. Hyrien, P. Hernandez, D. B. Krimer, and J. B. Schvartzman, Bi-directional replication and random termination. *Nucleic Acids Res.* **28**, 2099–2107 (2000).

142. R. D. Little, T. H. Platt, and C. L. Schildkraut, Initiation and termination of DNA replication in human rRNA genes. *Mol. Cell. Biol.* **13**, 6600–6613 (1993).

143. E. J. White, O. Emanuelsson, D. Scalzo, T. Royce, S. Kosak, E. J. Oakeley, S. Weissman, M. Gerstein, M. Groudine, et al. DNA replication-timing analysis of human chromosome 22 at high resolution and different developmental states. *Proc. Natl. Acad. Sci. USA* **101**, 17771–17776 (2004).

144. H. G. Callan, Replication of DNA in the chromosomes of eukaryotes. *Proc. R. Soc. Lond. B Biol. Sci.* **181**, 19–41 (1972).

145. S. Asakura, and F. Oosawa, On interaction between two bodies immersed in a solution of macromolecules. *J. Chem. Phys.* **22**, 1255–1256 (1954).

146. Y. Snir, and R. D. Kamien, Entropically driven helix formation. *Science* **307**, 1067 (2005).

147. K. van Holde, and J. Zlatanova, What determines the folding of the chromatin fiber? *Proc. Natl. Acad. Sci. USA* **93**, 10548–10555 (1996).

148. B. Mergell, R. Everaers, and H. Schiessel, Nucleosome interactions in chromatin: Fiber stiffening and hairpin formation. *Phys. Rev. E* **70**, 011915 (2004).

149. A. Lesne, and J. M. Victor, Chromatin fiber functional organization: Some plausible models. *Eur. Phys. J. E*, **19**, 279–290 (2005).

150. C. L. Woodcock, S. A. Grigoryev, R. A. Horowitz, and N. Whitaker, A chromatin folding model that incorporates linker variability generates fibers resembling the native structures. *Proc. Natl. Acad. Sci. USA* **90**, 9021–9025 (1993).

151. A. Y. Grossberg, and A. R. Khoklov, in *Statistical Physics of Macromolecules, AIP Series in Polymers and Complex Materials*, R. Larson and P. A. Pincus, eds., AIP Press, Woodbury, 1994.

152. S. Jun, J. Bechhoefer, and B.-Y. Ha, Diffusion-limited loop formation of semiflexible polymers: Kramers theory and the intertwined time scales of chain relaxation and closing. *Europhys. Lett.* **64**, 420–426 (2003).

153. H. Yamakawa, and W. H. Stockmayer, Statistical mechanics of wormlike chains. II. Excluded volume effects. *J. Chem. Phys.* **57**, 2843–2854 (1972).

154. St. P. Jean, C. Vaillant, B. Audit, and A. Arneodo, Spontaneous emergence of rosette like folding of chromatin: A keystone to replication and transcription regulation. Preprint, 2006.

155. H. Reiss, H. L. Frisch, and J. L. Lebowitz, Statistical mechanics of rigid spheres. *J. Chem. Phys.* **31**, 369–380 (1959).

156. A. D. Dinsmore, A. G. Yodh, and D. J. Pine, Phase diagrams of nearly-hard-sphere binary colloids. *Phys. Rev. E* **52**, 4045–4057 (1995).

157. N. F. Carnahan, and K. E. Starling, Equation of state for nonattracting rigid spheres. *J. Chem. Phys.* **51**, 635–636 (1969).

158. A. P. Minton, The influence of macromolecular crowding and macromolecular confinement on biochemical reactions in physiological media. *J. Biol. Chem.* **276**, 10577–10580 (2001).

159. J. Herrick, P. Stanislawski, O. Hyrien, and A. Bensimon, Replication fork density increases during DNA synthesis in *X. laevis* egg extracts. *J. Mol. Biol.* **300**, 1133–1142 (2000).

160. J. Zlatanova, and S. H. Leuba, Chromatin fibers, one-at-a-time. *J. Mol. Biol.* **331**, 1–19 (2003).

161. C. Tassius, C. Moskalenko, P. Minard, M. Desmadril, J. Elezgaray, and F. Argoul, Probing the dynamics of a confined enzyme by surface plasmon resonance. *Physica A* **342**, 402–409 (2004).

BIOLOGICAL RHYTHMS AS TEMPORAL DISSIPATIVE STRUCTURES

ALBERT GOLDBETER

Service de Chimie Physique et Biologie Théorique, Faculté des Sciences, Université Libre de Bruxelles, B-1050 Brussels, Belgium

CONTENTS

Special Volume in Memory of Ilya Prigogine: Advances in Chemical Physics, Volume 135, edited by Stuart A. Rice

I. INTRODUCTION

From the very beginning of his scientific path, which spanned more than six decades, Ilya Prigogine was attracted by the question of how order in time and space spontaneously arises in chemical and biological systems. The title of an article he published in 1969, "Structure, Dissipation and Life," reflects this theme, which long remained a central preoccupation in his research. In this chapter I will show how the views of Ilya Prigogine on nonequilibrium self-organization found multifarious applications in the life sciences. I will focus on temporal self-organization in the form of oscillatory behavior, which is ubiquitous in biological systems. One question that naturally arises is, Why are there so many biological rhythms?

Until the 1950s, the rare periodic phenomena known in chemistry, such as the reaction of Bray [1], represented laboratory curiosities. Some oscillatory reactions were also known in electrochemistry. The link was made between the cardiac rhythm and electrical oscillators [2]. New examples of oscillatory chemical reactions were later discovered [3, 4]. From a theoretical point of view, the first kinetic model for oscillatory reactions was analyzed by Lotka [5], while similar equations were proposed soon after by Volterra [6] to account for oscillations in predator–prey systems in ecology. The next important advance on biological oscillations came from the experimental and theoretical studies of Hodgkin and Huxley [7], which clarified the physicochemical bases of the action potential in electrically excitable cells. The theory that they developed was later applied [8] to account for sustained oscillations of the membrane potential in these cells. Remarkably, the classic study by Hodgkin and Huxley appeared in the same year as Turing's pioneering analysis of spatial patterns in chemical systems [9].

The approach of periodic phenomena in physicochemical terms made further progress when Prigogine and Balescu [10] showed that sustained oscillations can occur far from thermodynamic equilibrium in open chemical systems governed by appropriate, nonlinear kinetic laws. The model analyzed in that study was a chemical analogue of the Lotka–Volterra system for predator–prey oscillations in ecology. The results were later extended by the analysis of abstract models of oscillatory reactions, such as the *Brusselator*, whose name was given by Tyson [11] to a theoretical model studied in detail in Brussels by Lefever, Nicolis, and Prigogine [12]. As shown by these studies, chemical oscillations can occur at a critical distance from equilibrium, around a steady state that has become unstable owing to the presence of autocatalytic steps in the reaction kinetics [12–16].

In the phase space formed by the concentrations of the chemical variables involved in the reaction, sustained oscillations correspond to the evolution towards a closed curve called a limit cycle [17]. The time taken to travel once along the closed curve represents the period of the oscillations. When a single limit cycle exists, the system always evolves towards the same closed curve characterized by a fixed amplitude and period, for a given set of parameter values, regardless of the initial conditions. It is in this sense that oscillations of the limit cycle type differ from Lotka–Volterra oscillations, for which an infinity of closed curves, corresponding to oscillations of different periods and amplitudes, surround the steady state in the phase space. Then the choice of any one of the closed trajectories depends on the initial conditions [15, 17].

The developments of the Thermodynamics of Irreversible Processes in the nonlinear domain permitted Prigogine to place periodic phenomena within the field of nonequilibrium processes of self-organization [13, 14, 18]. Much as spatial structures arise in chemical systems beyond a critical point of instability with respect to diffusion [9], rhythms correspond to a temporal organization that appears beyond a critical point of instability of a nonequilibrium steady state. These two types of nonequilibrium self-organization represent dissipative structures [13, 14] that can be maintained only by the energy dissipation associated with the exchange of matter between the chemical system and its environment. Sustained oscillations of the limit cycle type can thus be viewed as temporal dissipative structures [13, 14, 15, 18]. When it occurs in constant environmental conditions, periodic behavior provides the clearest sign that a chemical or biological system operates beyond a point of nonequilibrium instability. Endogenous rhythms, produced by a system and not by its environment, are indeed the signature of an instability.

From a mathematical point of view, the onset of sustained oscillations generally corresponds to the passage through a Hopf bifurcation point [19]: For a critical value of a control parameter, the steady state becomes unstable as a focus. Before the bifurcation point, the system displays damped oscillations and eventually reaches the steady state, which is a stable focus. Beyond the bifurcation point, a stable solution arises in the form of a small-amplitude limit cycle surrounding the unstable steady state [15, 17]. By reason of their stability or regularity, most biological rhythms correspond to oscillations of the limit cycle type rather than to Lotka–Volterra oscillations. Such is the case for the periodic phenomena in biochemical and cellular systems discussed in this chapter. The phase plane analysis of two-variable models indicates that the oscillatory dynamics of neurons also corresponds to the evolution toward a limit cycle [20]. A similar evolution is predicted [21] by models for predator–prey interactions in ecology.

The 1970s saw an explosion of theoretical and experimental studies devoted to oscillating reactions. This domain continues to expand as more and more complex phenomena are observed in the experiments or predicted theoretically. The initial impetus for the study of oscillations owes much to the concomitance of several factors. The discovery of temporal and spatiotemporal organization in the Belousov–Zhabotinsky reaction [22], which has remained the most important example of a chemical reaction giving rise to oscillations and waves, and the elucidation of its reaction mechanism [23] occurred at a time when thermodynamic advances were establishing the theoretical bases of temporal and spatial self-organization in chemical systems under nonequilibrium conditions [10, 13–15, 18].

At the same time as the Belousov–Zhabotinsky reaction provided a chemical prototype for oscillatory behavior, the first experimental studies on the reaction catalyzed by peroxidase [24] and on the glycolytic system in yeast (to be discussed in Section III) demonstrated the occurrence of biochemical oscillations *in vitro*. These advances opened the way to the study of the molecular bases of oscillations in biological systems.

Oscillations represent one of the most striking manifestations of dynamic behavior in biological systems. In 1936, Fessard [25] published a book entitled *Rhythmic Properties of Living Matter*. This book was solely devoted to the oscillatory properties of nerve cells. It has now become clear that rhythms are encountered at all levels of biological organization, with periods ranging from a fraction of a second to years, spanning more than 10 orders of magnitude. The main types of biological rhythms are listed in Table I, where they are ordered according to their period, from the fastest rhythms in nerve and muscle cells to the rhythms of longest period observed in ecology and for the flowering of some plant species.

New examples of cellular rhythms have recently been uncovered (Table II). These include periodic changes in the intracellular concentration of the transcription factor NF-KB and of the tumor suppressors p53, stress-induced oscillations in the transport of the transcription factor Msn2 between cytoplasm and nucleus in yeast, the segmentation clock that is responsible for the

TABLE I
Main Biological Rhythms

Biological Rhythm	Period
Neural rhythms[a]	0.001 s to 10 s
Cardiac rhythm[a]	1 s
Calcium oscillations[a]	sec to min
Biochemical oscillations[a]	30 s to 20 min
Mitotic oscillator[a]	10 min to 24 h
Hormonal rhythms[a]	10 min to 3–5 h (24 h)
Circadian rhythms[a]	24 h
Ovarian cycle	28 days (human)
Annual rhythms	1 year
Rhythms in ecology and epidemiology	years

[a]These rhythms can already occur at the cellular level.

Source: Goldbeter [31].

TABLE II
Some Recently Discovered Cellular Rhythms[a]

Cellular Rhythm	Period
Segmentation clock	90 min
NFκB	3 h
P53	3 h
Msn2 in yeast	6 min
Yeast transcriptome	40 min

[a]See section VIII for details.

formation of somites in vertebrates, and whole genome oscillations in yeast. Some synthetic oscillatory gene circuits were recently constructed, as exemplified by the *Repressilator* [26]. Given the rapidly rising interest in the dynamic behavior of genetic circuits, it is likely that additional examples of cellular rhythms will be found in a near future.

II.   DISSIPATIVE STRUCTURES IN TIME AND SPACE

In the course of time open systems that exchange matter and energy with their environment generally reach a stable steady state. However, as shown by Glansdorff and Prigogine, once the system operates sufficiently far from equilibrium and when its kinetics acquire a nonlinear nature, the steady state may become unstable [15, 18]. Feedback regulatory processes and cooperativity are two major sources of nonlinearity that favor the occurrence of instabilities in biological systems.

Some of the main types of cellular regulation associated with rhythmic behavior are listed in Table III. Regulation of ion channels gives rise to the periodic variation of the membrane potential in nerve and cardiac cells [27, 28; for a recent review of neural rhythms see, for example, Ref. 29]. Regulation of enzyme activity is associated with metabolic oscillations, such as those that occur in glycolysis in yeast and muscle cells. Calcium oscillations originate

TABLE III
Biological Regulations and Examples of Associated Cellular Rhythms

Regulation of	Examples of Associated Rhythms
Ion channel	Neural and cardiac rhythms
Enzyme	Glycolytic oscillations in yeast
Receptor	cAMP oscillations in *Dictyostelium*
Transport	Ca^{2+} oscillations
Gene expression	Circadian rhythms, segmentation clock

from the control of transport processes within the cell. Regulation of receptors, coupled to the regulation of enzyme activity, can give rise to periodic behavior, as exemplified by oscillations of cyclic AMP (cAMP) in *Dictyostelium* cells. Regulation of gene expression represents a key type of cellular regulation involved in the mechanism of circadian rhythms and of the segmentation clock.

When the steady state becomes unstable, the system moves away from it and often undergoes sustained oscillations around the unstable steady state. In the phase space defined by the system's variables, sustained oscillations generally correspond to the evolution toward a limit cycle (Fig. 1). Evolution toward a limit cycle is not the only possible behavior when a steady state becomes unstable in a spatially homogeneous system. The system may evolve toward another stable steady state—when such a state exists. The most common case of multiple steady states, referred to as bistability, is of two stable steady states separated by an unstable one. This phenomenon is thought to play a role in differentiation [30]. When spatial inhomogeneities develop, instabilities may lead to the emergence of spatial or spatiotemporal dissipative structures [15]. These can take the form of propagating concentration waves, which are closely related to oscillations.

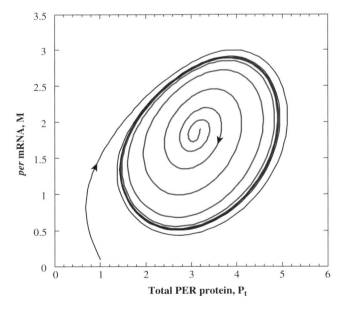

Figure 1. In most examples of biological rhythms, sustained oscillations correspond to the evolution toward a limit cycle. The limit cycle shown here was obtained in a model for circadian oscillations of the PER protein and *per* mRNA in *Drosophila* [107].

Elucidating the molecular mechanism of a biological rhythm largely reduces to identifying the feedback processes that lie at the core of the oscillations. The latter may originate from positive or negative feedback, or from a mixture of both. The interplay between a large number of variables coupled through multiple regulatory interactions makes it difficult, if not impossible, to fully grasp the dynamics of oscillatory behavior without resorting to modeling and computer simulations [31, 32].

As indicated above, theoretical models for biological rhythms were first used in ecology to study the oscillations resulting from interactions between populations of predators and preys [6]. Neural rhythms represent another field where such models were used at an early stage: The formalism developed by Hodgkin and Huxley [7] still forms the core of most models for oscillations of the membrane potential in nerve and cardiac cells [33–35]. Models were subsequently proposed for oscillations that arise at the cellular level from regulation of enzyme, receptor, or gene activity (see Ref. 31 for a detailed list of references).

Some of the main examples of biological rhythms of nonelectrical nature are discussed below, among which are glycolytic oscillations (Section III), oscillations and waves of cytosolic Ca^{2+} (Section IV), cAMP oscillations that underlie pulsatile intercellular communication in *Dictyostelium* amoebae (Section V), circadian rhythms (Section VI), and the cell cycle clock (Section VII). Section VIII is devoted to some recently discovered cellular rhythms. The transition from simple periodic behavior to complex oscillations including bursting and chaos is briefly dealt with in Section IX. Concluding remarks are presented in Section X.

III. GLYCOLYTIC OSCILLATIONS

Glycolytic oscillations in yeast cells provided one of the first examples of oscillatory behavior in a biochemical system. They continue to serve as a prototype for cellular rhythms. This oscillatory phenomenon, discovered some 40 years ago [36, 37] and still vigorously investigated today [38], was important in several respects: First, it illustrated the occurrence of periodic behavior in a key metabolic pathway. Second, because they were soon observed in cell extracts, glycolytic oscillations provided an instance of a biochemical clock amenable to *in vitro* studies. Initially observed in yeast cells and extracts, glycolytic oscillations were later observed in muscle cells and evidence exists for their occurrence in pancreatic β-cells in which they could underlie the pulsatile secretion of insulin [39].

The molecular mechanism of glycolytic oscillations has been discussed for long [31, 38, 40–42]. Because glycolysis represents a system of enzymatic reactions coupled through different intermediates such as ATP and NADH, which impinge on multiple steps in the pathway, it is difficult to isolate a single enzymatic step that would be responsible for oscillatory behavior. However, there is a large, if not unanimous, consensus in attributing to the enzyme phosphofructokinase (PFK) a prominent role in the instability-generating mechanism that leads to glycolytic oscillations. This role, recognized since the early experimental studies on the phenomenon, is due to the peculiar regulation of PFK, which is activated by a reaction product, ADP. Such product activation means that the PFK reaction is autocatalytic, a feature long shown to be associated with nonequilibrium instabilities [15, 18]. Self-amplification of PFK due to product activation of the enzyme was at the core of early models proposed for glycolytic oscillations [43–45].

A two-variable model taking into account the allosteric (i.e. cooperative) nature of the enzyme and the autocatalytic regulation exerted by the product shows the occurrence of sustained oscillations. Beyond a critical parameter value, the steady state admitted by the system becomes unstable and the system evolves toward a stable limit cycle corresponding to periodic behavior. The model accounts for most experimental data, particularly the existence of a domain of substrate injection rates producing sustained oscillations, bounded by two critical values of this control parameter, and the decrease in period observed when the substrate input rate increases [31, 45, 46].

Whereas two bifurcation values for the glucose input rate define the domain of oscillations in yeast extracts [40], only a single bifurcation value below which oscillations occur is found in intact yeast cells [47]. This does not necessarily imply a difference in oscillatory mechanism but merely indicates that in intact cells the glucose transporter becomes saturated before the intracellular glucose input has reached the upper bifurcation value above which oscillations disappear in yeast extracts [38].

If the primary role of PFK in generating glycolytic oscillations has long been stressed and substantiated by models based on its regulatory properties, other reactions of the glycolytic pathways are coupled to PFK and may thus influence its dynamic behavior. More complex models incorporating a large number of enzymatic reactions and of glycolytic intermediates have been proposed. This alternative approach to modeling was pioneered more than four decades ago by Garfinkel and Hess [48], who early on presented a comprehensive computer model for the glycolytic pathway. This work represents one of the first studies in a field currently known as (computational) systems biology. Other full-scale models of the yeast glycolytic system were subsequently proposed [49, 50].

The question of how glycolytic oscillations synchronize in a population of yeast cells is of great current interest [51]. It has long been known that the oscillations disappear in a yeast suspension when the cell density decreases below a critical value. Acetaldehyde appears to act as synchronizing factor in such suspensions [52], and the way it allows cells to synchronize is being studied in both an experimental and theoretical manner. The link between glycolytic oscillations and the pulsatile secretion of insulin in pancreatic β cells [53] is another topic of current concern. Models for the latter phenomenon rely on the coupling between intracellular metabolic oscillations and an ionic mechanism generating action potentials. Such coupling results in bursting oscillations of the membrane potential, which are known to accompany insulin secretion in these cells [54, 55].

IV. CALCIUM OSCILLATIONS

The three best-known examples of biochemical oscillations were found during the decade 1965–1975 [40, 41]. These include the peroxidase reaction, glycolytic oscillations in yeast and muscle, and the pulsatile release of cAMP signals in *Dictyostelium* amoebae (see Section V). Another decade passed before the development of Ca^{2+} fluorescent probes led to the discovery of oscillations in intracellular Ca^{2+}. Oscillations in cytosolic Ca^{2+} have since been found in a variety of cells where they can arise spontaneously, or after stimulation by hormones or neurotransmitters. Their period can range from seconds to minutes, depending on the cell type [56]. The oscillations are often accompanied by propagation of intracellular or intercellular Ca^{2+} waves. The importance of Ca^{2+} oscillations and waves stems from the major role played by this ion in the control of many key cellular processes—for example, gene expression or neurotransmitter secretion.

In cells that use Ca^{2+} as second messenger, binding of an external signal to a cell membrane receptor activates phospholipase C (PLC), which, in turn, synthesizes inositol 1,4,5-trisphosphate (InsP$_3$). This metabolite binds to an InsP$_3$ receptor located on the membrane of internal Ca^{2+} stores (endoplasmic or sarcoplasmic reticulum) and thereby triggers the release of Ca^{2+} into the cytoplasm of the cell [56]. A conspicuous feature of Ca^{2+} release is that it is *self-amplified*: Cytosolic Ca^{2+} triggers the release of Ca^{2+} from intracellular stores into the cytosol, a process known as Ca^{2+}-induced Ca^{2+} release (CICR) [57, 58].

A first model for cytosolic Ca^{2+} oscillations was based on the activation of PLC by Ca^{2+} [59]. Although this positive feedback has been observed in some cell types, CICR represents a more general self-amplifying process underlying the oscillations. Several processes limit the explosive nature of self-amplification. A simple two-variable model for signal-induced Ca^{2+} oscillations based on CICR accounts for oscillations of cytosolic Ca^{2+} [60]. Sustained oscillations occur between two critical values of the stimulus intensity—for example, two critical levels of an extracellular hormonal signal (see Fig. 2). Below the lower critical value, a low steady-state level of cytosolic Ca^{2+} is established; above the larger critical value, the system evolves toward a higher, stable steady-state level

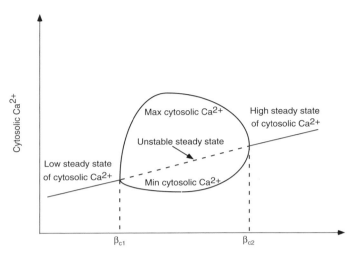

Figure 2. Schematic bifurcation diagram showing the domain and amplitude of intracellular Ca^{2+} oscillations as a function of the degree of external stimulation, β, which is used as control parameter. Sustained Ca^{2+} oscillations occur in a range of stimulation between the two critical β values denoted β_{c1} and β_{c2}. The maximum and minimum of cytosolic Ca^{2+} oscillations are plotted as a function of β in this range, in which the dashed line refers to the unstable steady state. On the left and right sides of the oscillatory domain, the system evolves to a stable steady state (solid line) corresponding to a low and high level of cytosolic Ca^{2+}, respectively. The bifurcation diagram is obtained in a two-variable model for Ca^{2+} oscillations based on CICR (see Goldbeter et al. [60] for a nonschematic version of the diagram). A similar bifurcation diagram with a domain of sustained oscillations bounded by two bifurcation values of the control parameter is obtained for glycolytic oscillations as a function of the substrate injection rate in yeast extracts [46, 193]. In intact yeast cells, however, the upper bifurcation point cannot be reached, likely because, as described in Section III, the glucose transporter is saturated before the bifurcation value for the substrate input is reached inside the cell.

of cytosolic Ca^{2+}. The model predicts that the frequency of Ca^{2+} oscillations rises with the degree of stimulation, as observed experimentally. In this minimal model the level of intracellular $InsP_3$ is treated as a control parameter reflecting the degree of external stimulation. More complex models for Ca^{2+} oscillations are based on more detailed descriptions of $InsP_3$ receptor kinetics [61; for a recent review see Ref. 62] but still attribute to CICR a primary role in the origin of repetitive Ca^{2+} spiking.

Mathematical models for Ca^{2+} signaling were subsequently developed in two additional directions. First, waves of intra- or intercellular Ca^{2+} can be modeled by incorporating the diffusion of cytosolic Ca^{2+} or the passage of Ca^{2+} or $InsP_3$ from cell to cell through gap junctions [62–65]. While most models for

Ca^{2+} waves are deterministic, stochastic simulations were used to clarify the nature of local increases of cytosolic Ca^{2+} known as blips or puffs which are thought to trigger the onset of waves [56, 66]. Second, models are used to probe mechanisms for encoding Ca^{2+} spikes in terms of their frequency. A variety of physiological responses are controlled by the frequency and waveform of Ca^{2+} oscillations, such as gene expression during development [67]. Among the processes that could underlie such frequency encoding are protein (de)phosphorylation by a Ca^{2+}-dependent kinase (phosphatase) [60], or the Ca^{2+}-dependence of calmodulin-kinase II [68, 69]. A study combining experimental and modeling approaches showed the possibility of frequency encoding of Ca^{2+} spikes by interplay with cyclic AMP signaling [70].

The characteristics of cytosolic Ca^{2+} oscillations vary from one cell type to another. One source for this variability is the existence of three isoforms of the inositol 1,4,5-trisphosphate receptor, $InsP_3R$, whose proportions vary in different cells. Upon binding of $InsP_3$, the $InsP_3R$ functions as a Ca^{2+} channel on the endoplasmic reticulum, allowing passage of Ca^{2+} into the cytoplasm of the cell. The relative amounts of each isoform of the $InsP_3$ receptor affect the time course of Ca^{2+} changes after agonist stimulation, because the effect of Ca^{2+} on the three isoforms are different. The different modes of Ca^{2+} oscillatory behavior have recently been modeled as a function of the relative proportions of the three $InsP_3R$ isoforms [71]. Based on a comparative study of various models, Sneyd et al. [72] recently proposed a method for determining the dependence of Ca^{2+} oscillations on $InsP_3$ oscillations, to determine whether $InsP_3$ plays an active role in the oscillatory mechanism or passively follows the periodic spikes in Ca^{2+}.

The role of Ca^{2+} oscillations in some physiological disorders begins to be characterized. Two recent studies provide examples of how changes in Ca^{2+} oscillatory signaling possess profound implications for developmental processes. First, Beltramello et al. [73] showed that a mutation associated with hereditary deafness reduces metabolic coupling mediated by $InsP_3$ and impairs the propagation of intercellular Ca^{2+} waves. Second, Uhlen et al. [74] recently demonstrated that the Noonan syndrome, a human developmental disorder often accompanied by congenital heart abnormalities, is caused by alterations in the Ca^{2+} oscillatory control of the transcription factor NFAT. These examples illustrate the impact of changes in the normal patterns of oscillations and waves of Ca^{2+} on the pathogenesis of some inherited human diseases.

V. PULSATILE INTERCELLULAR COMMUNICATION IN *DICTYOSTELIUM*

While intracellular information can be encoded in the frequency of signal-induced Ca^{2+} spikes, some extracellular signals can themselves be produced in a periodic, pulsatile manner. Examples of pulsatile intercellular communication include episodic hormone secretion and pulsatile signals of cAMP in the slime mold *Dictyostelium discoideum*. The latter phenomenon represents a prototype both for spatiotemporal self-organization and for pulsatile signaling in intercellular communication [31].

A. Oscillations of cAMP

After starvation, *Dictyostelium* amoebae undergo a transition from a unicellular to a multicellular phase of their life cycle. By a chemotactic response to cAMP signals, up to 10^5 amoebae collect around cells behaving as aggregation centers. These centers release cAMP with a period of about 5 min; surrounding cells relay the chemotactic signal toward the periphery of the aggregation field. Relay and oscillations of cAMP result in the formation of concentric or spiral waves of aggregating cells [75].

Models help to clarify the mechanism of cAMP oscillations in *Dictyostelium* [76, 77]. The mechanism involves both positive and negative feedback. Binding of extracellular cAMP to a cell surface receptor leads to the activation of adenylate cyclase, which catalyzes the synthesis of intracellular cAMP. Transport of cAMP into the extracellular medium creates a positive feedback loop, which elicits a rapid rise in cAMP synthesis. For sustained oscillations to occur, this rise in cAMP must be self-limiting, so that cAMP first levels off before decreasing to its minimum level. Models confirm that negative feedback due to cAMP-induced receptor desensitization through reversible phosphorylation can play such a role in limiting self-amplification [76]. Once the levels of intra- and extracellular cAMP are sufficiently low, dephosphorylation can resensitize the receptor. The ensuing buildup of extracellular cAMP progressively brings it to the threshold above which self-amplification triggers a new pulse.

Numerical simulations indicate that relay of cAMP pulses represents a different mode of dynamic behavior, closely related to oscillations. Just before autonomous oscillations break out, cells in a stable steady state can amplify suprathreshold variations in extracellular cAMP in a pulsatory manner. Thus, relay and oscillations of cAMP are produced by a unique mechanism in adjacent domains in parameter space. The two types of dynamic behavior are analogous to the excitable or pacemaker behavior of nerve cells.

Theoretical models shed light on additional aspects of pulsatile cAMP signaling in *Dictyostelium*. First, like Ca^{2+} spikes, cAMP pulses are frequency encoded. Only pulses delivered at 5-min intervals are capable of accelerating slime mold development after starvation. Simulations indicate that frequency encoding is based on reversible receptor desensitization [76]. The kinetics of receptor resensitization dictates the interval between successive pulses required for a maximum relay response [78]. Second, cAMP oscillations in

Dictyostelium provide a prototype for the ontogenesis of biological rhythms. The amoebae become capable of relaying extracellular cAMP pulses only a few hours after the beginning of starvation, before acquiring the property of autonomous oscillations. Models show that these developmental transitions can be brought about by the continuous increase in certain biochemical parameters such as the activities of adenylate cyclase or phosphodiesterase, the enzyme that degrades cAMP. In parameter space, these biochemical changes define a *developmental path* that successively crosses domains corresponding to different types of dynamic behavior, from no relay to relay, and finally to oscillations [31, 79].

Models are also being used to probe the mechanisms underlying the formation of concentric or spiral waves of cAMP responsible for the spatiotemporal patterns observed during aggregation [80]. Among the factors shown to play a role in the transition between the two types of waves are the activity of extracellular phosphodiesterase [81] and desynchronization of the cells that follow the developmental path after starvation [82]. The model based on the positive feedback mechanism coupled to receptor desensitization also accounts for the propagation of planar and scroll waves within the multicellular slug formed by the amoebae after aggregation [83].

In recent years, work by Loomis and co-workers has raised the possibility that cAMP oscillations in *D. discoideum* may originate from an intracellular regulatory network rather than from the mixed positive and negative feedback exerted by extracellular cAMP [84, 85]. These authors obtained evidence for an intracellular feedback loop involving MAP kinase and the cAMP-dependent protein kinase, PKA. The later enzyme would inactivate adenylate cyclase after a cAMP pulse. Numerical simulations of a model based on this intracellular negative feedback loop confirm that it can produce sustained oscillations of cAMP.

To establish which of the two feedback loops plays a prominent role in the origin of cAMP oscillations, Cox and co-workers recently examined the patterns of wavelike aggregation in a variety of mutants lacking components of the intracellular and extracellular regulatory loops. They reached the conclusion that the primary (but not necessarily sole) source of the oscillations resides in the regulation exerted by extracellular cAMP upon binding to its membrane receptor [86]. Interestingly, the possibility of cAMP oscillations due to intracellular regulation of adenylate cyclase by PKA seems to exist not only in *Dictyostelium* but also in yeast. In this organism, Jacquet et al. [87] recently observed a stress-induced oscillatory shuttling of the transcription factor Msn2 between cytosol and nucleus. They since obtained evidence suggesting that this periodic phenomenon is caused by intracellular cAMP oscillations, via the control of adenylate cyclase by PKA (see Section VIII). In this view, periodic activation of PKA by cAMP oscillations in yeast would underlie the repetitive, coherent shuttling of the transcription factor Msn2 into and out of the nucleus.

B. Link with Pulsatile Hormone Secretion

Pulsatile cAMP signaling in *Dictyostelium* is closely related with pulsatile hormone secretion in higher organisms. It is now clear that most hormones are secreted in a pulsatile rather than continuous manner [88] and that the temporal pattern of a hormone is often as important as its concentration in the blood [89]. The best examples of pulsatile hormone secretion are the gonadotropin-releasing hormone (GnRH) released by the hypothalamus with a periodicity of 1 h in man and rhesus monkey [90], the growth hormone (GH) secreted by the hypothalamus with a period of 3–5 h [91], and insulin secreted by pancreatic β cells with a period close to 13 min in man [53]. In the cases of GnRH and GH—the effect is less clear-cut for insulin—the frequency of the pulses governs the physiological efficacy of hormone stimulation [90, 91].

A general model for a two-state receptor subjected to periodic ligand variations shows that frequency encoding of hormone pulses may rely on reversible desensitization in target cells, as in the case of cAMP pulses in *Dictyostelium* [78, 92]. The mechanism of the hypothalamic GnRH pulse generator is still unknown and provides an important challenge for both experiments and theory. The basis of pulsatile GH secretion has been studied by a modeling approach [93]. In β cells, pulsatile insulin release could originate from insulin feedback on glucose transport into the cells [94] or from oscillatory membrane activity driven by glycolytic oscillations [53–55]. Together with such metabolic oscillations, membrane potential bursting and Ca^{2+} oscillations in β cells illustrate the multiplicity of rhythms that can be encountered in a given cell type.

VI. CIRCADIAN RHYTHMS

The most ubiquitous biological rhythms are those that occur with a period close to 24 h in all eukaryotes and in some prokaryotes such as cyanobacteria. These circadian rhythms allow organisms to adapt to the natural periodicity of the terrestrial environment, which is characterized by the alternation of day and night due to rotation of the earth on its axis. Circadian clocks provide cells with an endogenous mechanism, allowing them to anticipate the time of day.

Experimental advances during the last decade have clarified the molecular bases of circadian rhythms, first in *Drosophila* and *Neurospora*, and more recently in cyanobacteria, plants, and mammals [95–99]. In nearly all cases investigated so far, it appears that circadian rhythms originate from the negative feedback exerted by a protein on the expression of its gene [100]. Circadian rhythms in cyanobacteria appear to be based on a different molecular mechanism, which can be uncoupled from transcriptional control. Thus the circadian oscillation in the phosphorylation of the cyanobacterial KaiC clock protein has recently been reconstituted *in vitro* [101].

Before details on the molecular mechanism of circadian rhythms began to be uncovered, theoretical models borrowed from physics were used to investigate the dynamic properties of circadian clocks. The relative simplicity of these models explains why their use continues to this day. Thus, the Van der Pol equations, derived for an electrical oscillator, served for modeling the response of human circadian oscillations to light [102] and to account for experimental observations on increased fitness due to resonance of the circadian clock with the external light–dark (LD) cycle in cyanobacteria [103, 104]. The earliest model predicting oscillations due to negative feedback was proposed by Goodwin [105], at a time when the role played by such a regulatory mechanism in the origin of circadian rhythms was not yet known. Models based on Goodwin's equations are still being used in studies of circadian oscillations— for example, in *Neurospora* [106].

A. Circadian Rhythms in *Drosophila*

Molecular models for circadian rhythms were initially proposed [107] for circadian oscillations of the PER protein and its mRNA in *Drosophila*, the first organism for which detailed information on the oscillatory mechanism became available [100]. The case of circadian rhythms in *Drosophila* illustrates how the need to incorporate experimental advances leads to a progressive increase in the complexity of theoretical models. A first model governed by a set of five kinetic equations is shown in Fig. 3A; it is based on the negative control exerted by the PER protein on the expression of the *per* gene [107]. Numerical simulations show that for appropriate parameter values, the steady state becomes unstable and limit cycle oscillations appear (Fig. 1).

The early model based on PER alone did not account for the effect of light on the circadian system. Experiments subsequently showed that a second protein, TIM, forms a complex with PER and that light acts by inducing TIM degradation [96]. An extended, 10-variable model was then proposed [108], in which the negative regulation is exerted by the PER–TIM complex (Fig. 3B). This model produces essentially the same result, sustained oscillations in continuous darkness. In addition, it accounts for the behavior of mutants and explicitly incorporates the effect of light on the TIM degradation rate. Thereby the model can account for the entrainment of the oscillations by light-dark (LD) cycles and for the phase shifts induced by light pulses.

Subsequent experimental studies have shown that the mechanism of circadian rhythms in *Drosophila* is more complex, since the negative

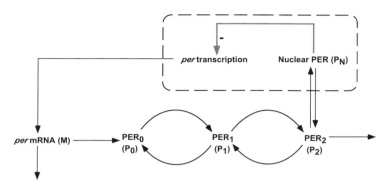

Figure 3. Molecular models of increasing complexity considered for circadian oscillations. (A) Model for circadian oscillations in *Drosophila* based on negative autoregulation of the *per* gene by its protein product PER [31, 107]. The model incorporates gene transcription into *per* mRNA, transport of *per* mRNA into the cytosol as well as mRNA degradation, synthesis of the PER protein at a rate proportional to the *per* mRNA level, reversible phosphorylation and degradation of PER, and transport of PER into the nucleus where it represses the transcription of the *per* gene. The model is described by a set of five kinetic equations. (B) Model for circadian oscillations in *Drosophila* incorporating the formation of a complex between the PER and TIM proteins [108]. The model is described by a set of 10 kinetic equations. (C) Model for circadian oscillations in mammals incorporating indirect, negative autoregulation of the *Per* and *Cry* genes through binding of the PER-CRY dimer to the complex formed between the two activating proteins CLOCK and BMAL1. Also considered is the negative feedback exerted by the latter proteins on the expression of their genes. Synthesis, reversible phosphorylation, and degradation of the various proteins are taken into account. The model is described by a set of 16 kinetic equations, or 19 when the Rev-Erbα gene is incorporated into the model [114]. For appropriate parameter values, all three models admit sustained circadian oscillations in conditions corresponding to continuous darkness. The effect of light is taken into account in models (B) and (C) by incorporating light-induced TIM degradation or light-induced *Per* expression, respectively.

autoregulation exerted by the PER–TIM complex on gene expression is indirect (see below).

B.   The Mammalian Circadian Clock

The pacemaker generating circadian rhythms in mammals is located in the suprachiasmatic nuclei of the hypothalamus. Recent studies have shown, however, that a number of peripheral circadian oscillators operate in tissues such as liver and heart [109]. Theoretical models for circadian rhythms in *Drosophila* bear on the mechanism of circadian oscillations in mammals, where homologues of the *per* gene exist and negative autoregulation of gene expression is also found [96]. However, in mammals, the role of TIM as a partner for PER is played by the CRY protein, and light acts by inducing gene expression rather than protein degradation as in *Drosophila*. A further analogy between

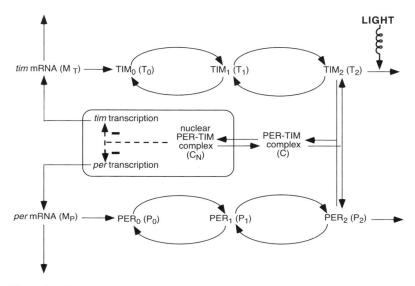

Figure 3. (*Continued*)

Drosophila and mammals is that the negative feedback on gene expression is indirect: The PER-TIM or PER-CRY complexes exert their repressive effect by binding to a complex of two proteins, CLOCK-CYC or CLOCK-BMAL1 in the fly [110] and in mammals [111], respectively. These proteins activate *per* and *tim* (or *cry*) gene expression. Thus negative feedback occurs by counteracting the effect of gene activators. Additional feedback loops are present, such as the negative feedback exerted by CLOCK or BMAL1 on the expression of their genes. These controls are removed upon formation of the complex with the PER-TIM or PER-CRY dimers.

Further extensions of the model are required to address the dynamical consequences of these additional regulatory loops and of the indirect nature of the negative feedback on gene expression. Such extended models have been proposed for *Drosophila* [112, 113] and mammals [113]. The model for the circadian clock mechanism in mammals is schematized in Fig. 3C. The presence of additional mRNA and protein species, as well as of multiple complexes formed between the various clock proteins, complicates the model, which is now governed by a system of 16 or 19 kinetic equations. Sustained or damped oscillations can occur in this model for parameter values corresponding to continuous darkness. As observed in the experiments on the mammalian clock, *Bmal1* mRNA oscillates in opposite phase with respect to *Per* and *Cry* mRNAs [97]. The model displays the property of entrainment by the LD cycle

Figure 3. (*Continued*)

when incorporating the light-induced increase in the rate of *Per* expression. A more detailed model containing a much larger number of variables has been proposed for the mammalian circadian clock [115].

Knowledge of the detailed mechanism underlying circadian rhythms continues to be refined as new experiments reveal novel facets of the oscillatory machinery. Thus, a link has recently been established between chromatin structure and the circadian oscillatory mechanism. The CLOCK protein indeed functions as a histone acetyltransferase [116]. This enzyme activity is required for oscillations so that histone modification and the associated chromatin remodeling are implicated in the origin of circadian rhythmicity.

C.   Link with Disorders of the Sleep–Wake Cycle

The results obtained with the model for the mammalian circadian clock provide cues for circadian-rhythm-related sleep disorders in humans [117]. Thus permanent phase shifts in LD conditions could account for (a) the familial advanced sleep phase syndrome (FASPS) associated with PER hypophosphorylation [118, 119] and (b) the delayed sleep phase syndrome, which is also related to PER [120]. People affected by FASPS fall asleep around 7:30 P.M. and awake around 4:30 A.M. The duration of sleep is thus normal, but the phase is advanced by several hours. Moreover, the autonomous period measured for circadian rhythms in constant conditions is shorter [121]. The model shows that a decrease in the activity of the kinase responsible for PER phosphorylation is indeed accompanied by a reduction of the circadian period in continuous darkness and by a phase advance upon entrainment of the rhythm by the LD cycle [114].

For some parameter values the model for the mammalian clock fails to allow entrainment by 24-h LD cycles, regardless of the amplitude of the light-induced change in *Per* expression. The question arises whether there exists a syndrome corresponding to this mode of dynamic behavior predicted by the model. Indeed there exists such a syndrome, known as the non-24-h sleep–wake syndrome, in which the phase of the sleep–wake pattern continuously varies with respect to the LD cycle; that is, the patient free-runs in LD conditions [117]. Disorders of the sleep–wake cycle associated with alterations in the dynamics of the circadian clock belong to the broad class of "dynamical diseases" [122, 123], although the term "syndrome" seems more appropriate for some of these conditions.

Another common perturbation of the circadian clock is the jet lag, which results from an abrupt shift in the phase of the LD cycle to which the rhythm is naturally entrained. The molecular bases of the jet lag are currently being investigated [124]. The model for the circadian clock is being used to probe the various ways by which the clock returns to the limit cycle trajectory after a sudden shift in the phase of the LD cycle.

D. Long-Term Suppression of Circadian Rhythms by a Single Light Pulse

Circadian rhythms illustrate how theoretical models can provide surprising, counterintuitive insights. A case in point is the puzzling observation that in some organisms, circadian rhythms in continuous darkness can be suppressed by a single pulse of light and restored by a second such pulse. A first theoretical explanation for this long-term suppression, proposed by Winfree [125], assumes that the limit cycle in each oscillating cell surrounds an unstable steady state. The light pulse would act as a critical perturbation that would bring the clock to the singularity—that is, the steady state. Because the steady state is unstable, each cell would eventually return to the limit cycle, but with a random phase. The population of oscillating cells would then be spread out over the entire cycle so that the cells would be desynchronized and no global rhythm would be established.

An alternative explanation is based on the *coexistence* of sustained oscillations with a stable steady state. Such coexistence has been observed [126] in the model for circadian rhythms in *Drosophila* based on negative autoregulation by the PER-TIM complex (Fig. 3B). In such a situation, the effect of the light pulse is to bring the clock mechanism into the basin of attraction of the stable steady state in each oscillating cell, so that the rhythm is suppressed. A second light pulse then brings the system back to the basin of attraction of the limit cycle corresponding to circadian oscillations. Without a model it is impossible to predict the coexistence between a stable steady state and a stable rhythm. The question remains open as to which one of the two explanations accounts for long-term suppression of circadian rhythms by a single light pulse.

E. Stochastic Versus Deterministic Models for Circadian Rhythms

Only deterministic models for cellular rhythms have been discussed so far. Do such models remain valid when the numbers of molecules involved are small, as may occur in cellular conditions? Barkai and Leibler [127] stressed that in the presence of small amounts of mRNA or protein molecules, the effect of molecular noise on circadian rhythms may become significant and may compromise the emergence of coherent periodic oscillations. The way to assess the influence of molecular noise on circadian rhythms is to resort to stochastic simulations [127–129]. Stochastic simulations of the models schematized in Fig. 3A,B show that the dynamic behavior predicted by the corresponding deterministic equations remains valid as long as the maximum numbers of mRNA and protein molecules involved in the circadian clock mechanism are of the order of a few tens and hundreds, respectively [128]. In the presence of molecular noise, the trajectory in the phase space transforms into a cloud of points surrounding the deterministic limit cycle.

Stochastic simulations confirm the existence of bifurcation values of the control parameters bounding a domain in which sustained oscillations occur. The effect of noise diminishes as the number of molecules increases. Only when the maximum numbers of molecules of mRNA and protein become smaller than a few tens does noise begin to obliterate the circadian rhythm. The robustness of circadian rhythms with respect to molecular noise is enhanced when the rate of binding of the repressor molecule to the gene promoter increases [128]. Conditions that enhance the resistance of genetic oscillators to random fluctuations have been investigated [130].

VII.   THE CELL-CYCLE CLOCK

The cell cycle is a key process that recurs in a periodic manner. Early cell cycles in amphibian embryos are driven by a mitotic oscillator. This oscillator produces the repetitive activation of the cyclin-dependent kinase cdk1, also known as cdc2 [131]. Cyclin synthesis is sufficient to drive repetitive cell division cycles in amphibian embryonic cells [132]. The period of these relatively simple cell cycles is of the order of 30 min. In somatic cells the cell cycle becomes longer, with durations of up to 24 h or more, owing to the presence of checkpoints that ensure that a cell cycle phase is properly completed before the cell progresses to the next phase. The cell cycle goes successively through the phases G1, S (DNA replication), G2, and M (mitosis) before a new cycle starts in G1. After mitosis cells can also enter a quiescent phase G0, from which they enter G1 under mitogenic stimulation.

Models of reduced complexity have first been proposed for the early cell cycles in amphibian embryos. These models are based on the activation of the kinase cdc2 upon binding of cyclin. One of these models predicts that limit cycle oscillations in cdc2 activity may arise from the activation by cdc2 of cyclin degradation. Indeed, cdc2 activates the anaphase-promoting complex (APC), which leads to cyclin destruction and subsequently to cdc2 inactivation. Such a negative feedback regulation is capable of producing sustained oscillatory behavior in the presence of thresholds and delays, both of which are linked and naturally arise in the control of cdc2 by phosphorylation-dephosphorylation [31, 133].

Positive feedback is also involved in the control of cdc2 by reversible phosphorylation. Thus, cdc2 activates the phosphatase cdc25, which catalyzes the dephosphorylation and concomitant activation of the kinase cdc2. The model based on negative feedback in cyclin-cdc2 interactions can be extended to take this positive feedback into account. Sustained oscillations can be obtained in these conditions [31, 134], but the waveform of cdc2 in the course of oscillations now displays a plateau. This plateau is due to the occurrence of a phenomenon of bistability, which is accompanied by hysteresis, as shown theoretically and experimentally in *Xenopus* egg extracts by Pomerening et al. [135] and Sha et al. [136]. The effect of suppressing the positive feedback loop on the occurrence of cdc2 oscillations was investigated in recent experiments [137].

The interplay between oscillations and bistability has been addressed in detailed molecular models for the cell cycles of amphibian embryos, yeast and somatic cells [138–141]. The predictions of a detailed model for the cell cycle in yeast were successfully compared with observations of more than a hundred mutants [142]. Other theoretical studies focus on the dynamical properties of particular modules of the cell cycle machinery such as that controlling the G1/S transition [143].

If the cell cycle in amphibian embryonic cells appears to be driven by a limit cycle oscillator, the question arises as to the precise dynamical nature of more complex cell cycles in yeast and somatic cells. Novak et al. [144] constructed a detailed bifurcation diagram for the yeast cell cycle, piecing together the diagrams obtained as a function of increasing cell mass for the transitions between the successive phases of the cell cycle. In these studies, cell mass plays the role of control parameter; a critical mass has to be reached for cell division to occur, provided that it coincides with a surge in cdk1 activity which triggers the G2/M transition.

The periodic recurrence of cell division suggests that globally the cell cycle functions like an autonomous oscillator. An extended model incorporating the sequential activation of the various cyclin-dependent kinases, followed by their inactivation, shows that even in the absence of control by cell mass, this sequence of biochemical events can operate as a limit cycle oscillator [145]. This supports the union of the two views of the cell cycle as dominoes and clock [146]. Because of the existence of checkpoints, however, the cell cycle stops at the end of certain phases before engaging in the next one. Thus the cell cycle looks more like an oscillator that slows down and makes occasional stops. A metaphor for such behavior is provided by the movement of the round plate on the table in a Chinese restaurant, which would rotate continuously under the movement imparted by the participants, were it not for frequent stops.

An alternative approach for modeling the cell cycle considers the sequential transitions between the G1, S, G2, and M phases without taking into account the underlying molecular mechanism. Based on a previous study of the dynamics of hair cycles [147], this phenomenological approach represents the cell cycle as a stochastic automaton capable of switching between the successive phases, with a probability related to their duration. The automaton model can reproduce the distributions between the various phases of the cell cycle at steady state [148]. This phenomenological model is being used to investigate the effect of periodic administration of anticancer drugs that interfere with the cell division cycle.

Recent experimental studies have uncovered a direct link between the cell cycle and circadian rhythms. Thus, the circadian clock protein BMAL1 induces the expression of the gene *Wee1*, which codes for the protein kinase that inactivates through phosphorylation the kinase cdk1 that controls the G2/M transition [149]. This link allows the coupling of cell division to the circadian clock and explains how the latter may entrain the cell cycle clock in a variety of cell types.

VIII.   NEWLY DISCOVERED CELLULAR RHYTHMS

In the last decade, and particularly in the last five years, several new examples of cellular rhythms have been uncovered (see Table II). These include oscillations in the tumor suppressor p53 and in the transcription factor NF-KB, the segmentation clock that controls the formation of somites in vertebrates, and the oscillatory nucleocytoplasmic shuttling of the transcription factor Msn2 in yeast. Other examples are the genome-wide periodicity of about 80 min observed for gene expression in yeast [150, 151] and the "transcriptional clock" based on the estrogen-receptor-mediated ordered, cyclical recruitment of protein cofactors involved in target gene transcription [152].

A.   Oscillations of p53 and NF-KB

The tumor suppressor p53 plays an important role in the control of the cell cycle, and it is inactivated in many types of human tumors. In response to genomic stress, p53 activation may elicit cell-cycle arrest or apoptotic cell death, as well as contribute to DNA repair processes. Regulation of p53 is mediated by its interactions with the protein Mdm2. The binding of Mdm2 to p53 inhibits the transcriptional functions of p53 and also leads to its degradation. At the same time, p53 stimulates the transcription of the *mdm2* gene. These interactions define a negative feedback loop ensuring that the p53 response is brought to an end once a p53-activating stress signal has been effectively dealt with.

The dynamics of the p53-Mdm2 feedback loop was analyzed mathematically by Lev Bar-Or et al. [153], who pointed out that this negative feedback regulation can give rise to oscillatory behavior if there is a sufficient delay in the induction of Mdm2 by p53. They verified experimentally that oscillations of both p53 and Mdm2 indeed occur on exposure of various cell types to ionizing radiation. Lahav et al. [154] pursued this study by investigating the dynamics of p53 and Mdm2 in individual cells. They showed that p53 was expressed in a series of discrete pulses after DNA damage. The number of p53 pulses, but not their amplitude, varied in different cells and increased with DNA damage. Ma et al. [155] recently proposed a model for this "digital" response of individual cells to DNA damage. The model is based on the coupling of DNA damage to the p53-Mdm2 oscillator. An alternative model for oscillations in the p53/Mdm2 module has also been proposed [156].

A negative feedback loop likewise controls the activity of the transcription factor NF-KB, which is rapidly turned off by a protein inhibitor, I-KB, which exists under three isoforms denoted α, β, and ε. Only the I-KBα isoform participates in the negative feedback. When NF-KB dissociates from I-KB in the cytosol, it enters the nucleus, where it induces the transcription of a number of target genes, one of which codes for the inhibitor I-KBα. Hoffmann et al. [157] analyzed a computational model based on the interactions of NF-KB and its inhibitor I-KBα. They predicted and verified experimentally that these interactions can give rise to oscillatory changes in NF-KB activity characterized by a period of the order of several hours. Using single-cell time-lapse imaging and computational modeling, Nelson et al. [158] showed that NF-KB localization oscillates between the nucleus and the cytosol following cell stimulation. The frequency of these oscillations decreased with increased I-KBα transcription. The question of whether the pattern of NF-KB oscillations can selectively control the expression of certain genes remains open [157–159].

B. Segmentation Clock

The formation of somites in the course of vertebrate development is associated with body segmentation. This phenomenon represents a striking example of spatial pattern in morphogenesis. It has long been suggested that a temporal periodic process is also at work in somitogenesis, since pairs of somites form progressively, one at a time, along the presomitic mesoderm (PSM). Thus Cooke and Zeeman [160] proposed a "clock and wavefront" model, which postulated the existence of (a) a wavefront moving from the anterior to the posterior end of the PSM and (b) a clock that would periodically induce the formation of a pair of somites at the location of the wavefront. Neither the nature of the clock nor the wavefront-defining process was characterized in this abstract model, which nevertheless proved highly seminal. Experimental evidence for an oscillator involved in somitogenesis was later obtained [161, 162]. This oscillator is based on periodic gene expression and is known as the vertebrate "segmentation clock" [163–165]. Its period is of the order of 90 min to 2 h, depending on the organism considered. A unique feature of the segmentation clock is that is links a temporal oscillation with the formation of a stable spatial pattern, in an important developmental process [163].

The mechanism of the segmentation clock relies on the negative feedback exerted on the expression of genes that participate in the signaling pathway controlled by Notch and other transcription factors [163-165]. Thus, periodic Notch inhibition by the product of the gene *lunatic fringe* (*Lfng*) underlies the chick segmentation clock [166]. The product of *Lfng* establishes a negative feedback loop that results in the periodic inhibition of Notch, which, in turn, controls the rhythmic expression of cyclic genes in the chick PSM. This feedback loop provides a molecular basis for the oscillator underlying the avian segmentation clock. A model based on such negative autoregulatory feedback on gene expression (Fig. 4) confirms that it can produce sustained oscillations [167].

Subsequent experimental studies have shown that another signal pathway controlled by WNT may drive oscillations in the Notch pathway [168]. The Wnt pathway is also regulated by negative feedback and could thus give rise to oscillatory behavior (Fig. 5), as shown by a modeling study [167]. Oscillations could be transduced from one pathway to the other by a common intermediate such as the protein kinase GSK3, which is involved both in Wnt and Notch signaling.

There is thus a multiplicity of negative feedback processes that could, in principle, give rise to oscillations in Notch signaling, and on which the

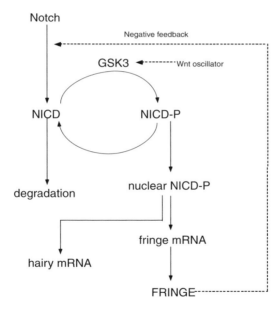

Figure 4. A negative feedback mechanism in the Notch signaling pathway can give rise to periodic expression of genes such as *Lunatic fringe* (*Lnfg*). This negative feedback on transcription is thought to play a key role in the segmentation clock controlling somatogenesis in vertebrates (see Section VIII.B). Upon cleavage, the Notch ligand produces the form NICD, which, after phosphorylation, possibly by the kinase GSK3, migrates to the nucleus where it induces the expression of genes like *Hairy* and *Lfng*. The FRINGE protein inhibits the cleavage of Notch into NICD. The model based on this negative feedback regulation shows that it can give rise to sustained oscillations.

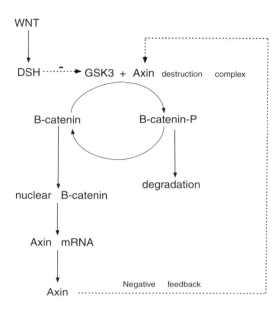

Figure 5. A negative feedback mechanism in the Wnt signaling pathway can also give rise to periodic gene expression. This negative feedback on transcription could form the core of the segmentation clock mechanism by driving oscillations in the Notch pathway. Wnt signaling activates the ligand DSH which inhibits the β-catenin destruction complex formed by the protein Axin and the kinase GSK3. This destruction complex phosphorylates β-catenin and thereby marks it for degradation. Negative feedback originates from the fact that β-catenin induces the expression of the Axin gene; translation of Axin mRNA results in the accumulation of the Axin protein, which leads to β-catenin degradation. Here again a model based on this negative feedback regulation shows that it can produce sustained oscillations. Coupling of the Notch oscillator to the Wnt oscillator could be mediated by oscillations in the activity of the kinase GSK3.

segmentation clock could be based [165]. The possibility of sustained oscillations resulting from such negative feedback loops has been investigated in theoretical models [169, 170] that emphasize the role of time delays in the appearance of sustained oscillations. A role for intercellular signaling through the Notch-Delta pathway has also been pointed out in the model proposed for the zebra fish segmentation clock [169].

One intriguing aspect of the models for the segmentation clock is their link with mechanisms proposed for circadian rhythms. Both types of oscillations are based on negative autoregulation of gene expression. The question arises as to how similar mechanisms produce oscillations with a period in the range 30 min to 2 h for the segmentation clock and around 24 h for circadian rhythms. It would be interesting to further characterize the differences that lead to a 10-fold change in period in the two situations. Such differences may pertain, for example, to the half-life of proteins or mRNAs or to the post-transcriptional regulation of proteins involved in the oscillatory mechanism.

Oscillations of the segmentation clock with a period of 2 h have also been observed in fibroblast cell cultures following serum shock. There also, oscillations in the expression of the gene *Hes1* related to the Notch pathway have been attributed to negative feedback on transcription [171]. The periodic operation of the segmentation clock was recently demonstrated in cells of the PSM, where intercellular coupling is needed to prevent damping of the oscillations [172].

Progress has also been made on the experimental characterization of the biochemical process that mediates the anterior to posterior progression of the determination wavefront along the PSM during somitogenesis. Fibroblast growth factor (FGF) signaling is involed in the coupling between the segmentation clock and the formation of somites [173]. Moreover, a gradient in *fgf* mRNA, starting at the posterior end of the PSM, extends in the direction of the anterior end where its progressive degradation results in the anterior to posterior movement of the wavefront [174]. The study of a theoretical model has recently shown [175] that the bistability assumed in the clock and wavefront model [160] could originate from the antagonistic gradients of mutually inhibiting FGF and retinoic acid [176] along the PSM. Mutual inhibition is known to give rise to bistability, as demonstrated, for example, in a synthetic genetic network [177].

C. Nucleocytoplasmic Oscillations of the Transcription Factor Msn2 in Yeast

In the yeast *Saccharomyces cerevisiae*, two related transcriptional activators Msn2 and Msn4 are activated in various stress conditions. These proteins are located in the cytoplasm, but they migrate to the nucleus upon activation. Translocation to the nucleus is inhibited by high activity of the cAMP-PKA pathway. Using time-lapse video microscopy on single cells, Jacquet et al. [87] followed the kinetics of translocation of Msn2 fused to the Green fluorescence protein (GFP). They showed that light emission of the microscope is sufficient to induced migration of Msn2 to the nucleus and therefore is sensed as a stress by the cell. Unexpectedly, the population of Msn2 molecules displayed an oscillatory behavior, shuttling repetitively between nucleus and cytoplasm upon light stress, with a periodicity of the order of several minutes. The phenomenon presents a large variability between individual cells. Upon additional stress the oscillatory behavior is maintained but the average time spent in the nucleus is increased. A plausible theoretical model was proposed to account for such oscillations, based on the hypothesis that this transcriptional regulator is involved in an autoregulatory loop controlling its nuclear localization [87].

An alternative possibility is that a biochemical oscillator controls the periodic shuttling of Msn2 between cytosol and nucleus. So far the existence of such a putative biochemical oscillator driving Msn2 oscillatory shuttling has not been substantiated by experimental observations. Among possible biochemical oscillators with periods in the range of minutes, glycolysis could be a natural candidate (see Section IV), but glycolytic oscillations are controlled by glucose rather than stress. Calcium oscillations have not been observed in yeast, and their occurrence may be precluded by the fact that this organism lacks InsP$_3$ receptors. Recent observations [178] suggest that oscillatory shuttling of Msn2 may well be controlled by oscillations of cAMP, via the protein kinase PKA, which is activated by cAMP. It appears indeed that the cellular localization of Msn2 is governed through phosphorylation by PKA.

As shown by a theoretical model [178], the periodic variation of cAMP could originate from the negative feedback exerted via PKA on the synthesis of cAMP (Fig. 6). Thus, PKA could exert its negative control by inactivating through phosphorylation the GAP protein, which is involved in the activation of adenylate cyclase, or by activating the enzyme phosphodiesterase, which degrades cAMP. Such a mechanism producing intracellular oscillations of cAMP could operate in other cell types. Thus it is closely related to the intracellular mechanism proposed for cAMP oscillations in *Dictyostelium* [84, 85]. Oscillations of cAMP were also proposed to underlie pulsatile hormone release in GnRH secreting cells [179]. The evidence for cAMP oscillations in yeast so far remains indirect, since the variations of this metabolite cannot be followed continuously. The oscillatory nucleocytoplasmic shuttling of Msn2 nevertheless provides an indirect sign that cAMP might oscillate in yeast.

IX. COMPLEX OSCILLATORY BEHAVIOR

The transition from simple to complex oscillatory phenomena is often observed in biochemical and cellular systems. Thus, bursting represents one type of complex oscillations that is particularly common in neurobiology [29]. An active phase of spike generation is followed by a quiescent phase, after which a new active phase begins. Mathematical models throw light on the conditions that generate such complex periodic oscillations [180]. Chaos is another common mode of complex oscillatory behavior that has been studied intensively in physical, chemical, and biological systems [31, 122, 181]. These irregular oscillations are characterized by their sensitivity to initial conditions, which accounts for the unpredictable nature of chaotic dynamics.

Yet another type of complex oscillatory behavior involves the coexistence of multiple attractors. Hard excitation refers to the coexistence of a stable steady state and a stable limit cycle—a situation that might occur in the case of circadian rhythm suppression discussed in Section VI. Two stable limit

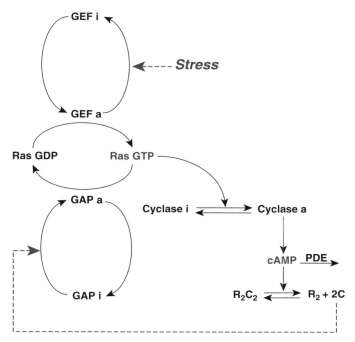

Figure 6. Regulation of the cAMP-PKA pathway in yeast. Synthesis of cAMP by adenylate cyclase is enhanced when this enzyme is activated by Ras-GTP, the active form of the Ras protein. The transformation of the inactive form Ras-GDP into Ras-GTP is triggered by the active form GEFa of the GEF protein, while the reverse, inactivating step is elicited by the active form GAPa of the GAP protein. Protein kinase A is activated when cAMP binds to the holoenzyme form R_2C_2 and frees the active, catalytic subunit C. Negative feedback originates from the fact that the catalytic subunit C of PKA phosphorylates and thereby activates the inactive form GAPi into GAPa. This leads to a decrease in Ras-GTP and, subsequently, to a drop in cAMP. A model based on this negative feedback regulation shows the possibility of sustained oscillations in cAMP and PKA activity [178]. These oscillations could be triggered by the stress-induced inactivation of GEFa. Stress-induced, nucleocytoplasmic oscillations of the transcription factor Msn2 in yeast would originate from the control of its subcellular localization through phosphorylation by the catalytic subunit of PKA (see Section VIII.C).

cycles may also coexist, separated by an unstable limit cycle. This phenomenon, referred to as birhythmicity [182], is the oscillatory counterpart of bistability in which two stable steady states, separated by an unstable state, coexist. Birhythmicity was predicted theoretically before being observed experimentally.

The study of models indicates the existence of two main routes to complex oscillatory phenomena. The first relies on forcing a system that displays simple periodic oscillations by a periodic input [122]. In an appropriate range of input frequency and amplitude, one can often observe the transition from simple to complex oscillatory behavior such as bursting and chaos. For other frequencies and amplitudes of the forcing, entrainment or quasiperiodic oscillations occur. Because circadian rhythms are naturally subjected to periodic forcing by LD cycles, the possibility arises that such forcing might lead to chaos. Such a situation would be detrimental to the organism, since entrainment by the LD cycle is precisely the most conspicuous function of circadian rhythms. Numerical simulations of a model for the circadian clock indicate that entrainment, quasiperiodic oscillations, and chaos may indeed occur, depending on the magnitude of the periodic changes induced by the LD cycle in the light-sensitive parameter. The waveform of the forcing, however, is also important since the domain of entrainment enlarges at the expense of chaos when the periodic variation in the light-sensitive parameter changes from square wave to sinusoidal [183]. Given that the real LD cycle differs from a square wave, the possibility that chaotic dynamics in the circadian control system seems unlikely in natural conditions.

Complex oscillations can also occur in autonomous systems that operate in a constant environment. The study of models for a variety of cellular oscillations shows that complex oscillatory phenomena may arise through the interplay between several instability-generating mechanisms, each of which is capable of producing sustained oscillations [31, 182]. The case of Ca^{2+} signaling is particularly revealing because of the multiplicity of feedback mechanisms that could potentially be involved in the onset of oscillations. Thus, among the many nonlinear processes that could take part in an instability-generating loop are (1) Ca^{2+}-induced Ca^{2+} release, (2) desensitization of the InsP$_3$ receptor, (3) bell-shaped dependence of the InsP$_3$ receptor on Ca^{2+} that reflects its activation and inhibition at different Ca^{2+} levels, (4) capacitative Ca^{2+} entry, (5) PLC or/ and InsP$_3$ 3-kinase activation by Ca^{2+}, (6) control of Ca^{2+} by mitochondria, (7) G-protein regulation by Ca^{2+}, and (8) coupling of the membrane potential to cytosolic Ca^{2+}. Several models in which at least two of these regulatory processes are coupled were shown to admit birhythmicity, bursting, or chaotic oscillations [62, 184–187]. Chaotic dynamics has been observed experimentally in glycolysis in autonomous conditions [188], presumably as a result of the interplay between two endogenous instability-generating mechanisms.

X. CONCLUDING REMARKS

Since the onset of studies on dissipative structures at the end of the 1960s [12, 14, 15, 18, 189], the field devoted to the investigation of nonequilibrium structures in chemistry, physics, and the life sciences has grown tremendously. In this chapter I focused on temporal dissipative structures in biology, a field where they abound. My aim was to provide an overview of oscillatory processes in biological systems, including recent developments and new examples. Rhythmic phenomena are common at all levels of biological organization because the thermodynamic and kinetic conditions for the occurrence of rhythmic behavior are particularly well-satisfied [15, 18]. Indeed, biological systems are open, since they exchange matter and energy with their surroundings; they often function far from equilibrium; and their evolution is governed by nonlinear kinetic laws.

The sources of nonlinearity are manifold in biological systems. Besides cooperativity, a major source of nonlinearity is provided by the variety of regulatory processes encountered at the cellular level (see Table III). The existence of regulatory feedback is thus the main reason why rhythmic behavior is among the most conspicuous properties of living organisms. Positive feedback is generally associated with multiple steady states [30], while negative feedback is capable of generating oscillations provided a minimum delay in the negative feedback loop exists. Inhibition by the product is indeed at the core of many rhythmic phenomena in biological systems. To name but a few among those discussed in this chapter, circadian rhythms, oscillations in NF-KB or p53, and the segmentation clock all appear to be based on direct or indirect, negative autoregulation. Negative feedback is also involved in the *Repressilator*, which is a synthetic oscillatory gene circuit consisting of three repressors coupled in a cyclical manner [26, 190]. Beyond the particular nature of the molecules involved, mathematical models help to establish links between various oscillatory mechanisms based on inhibitory feedback [170].

In oscillatory mechanisms based on negative feedback, a positive effect is generally needed to sustain periodic behavior, but it can just be an induction process—as is the case for the action of CLOCK/BMAL1 in the circadian clock—and does not need to take the form of positive feedback. Positive feedback is self-amplifying and amounts to activation by the product. Such mode of regulation is less common than negative feedback but plays a key role in the origin of several biological rhythms. Besides oscillations of the membrane potential in nerve and muscle cells, which were not considered in detail in this chapter devoted to cellular rhythms of nonelectrical nature, examples of rhythms based on positive feedback include glycolytic oscillations, Ca^{2+} spiking, and cAMP oscillations in *Dictyostelium* cells.

To give rise to oscillatory behavior instead of a biochemical explosion, self-amplification must, however, be coupled to a limiting process. Such a limiting process can be viewed as a form of negative feedback because it occurs as a consequence of the positive feedback that precedes it. Thus, in the case of glycolytic oscillations, the activation of phosphofructokinase by a reaction product is followed by a counteracting fall in the rate of the enzymatic reaction, due to the enhanced substrate consumption associated with enzyme activation. In Ca^{2+} pulsatile signaling, the explosive rise in cytosolic Ca^{2+} due to Ca^{2+}-induced Ca^{2+} release from the endoplasmic reticulum is limited by the emptying of the store and by the inhibition of the Ca^{2+} channel at high levels of cytosolic Ca^{2+}. Likewise, in *Dictyostelium*, the rapid rise in cAMP due to positive feedback necessarily leads to its subsequent decrease through cAMP-induced desensitization of the cAMP receptor on the plasma membrane.

Sometimes a mixture of positive and negative feedback can produce relaxation oscillations based on bistability. This situation is illustrated by the oscillations in cdk1 activity, which drive the early cell division cycles in amphibian embryos (see Section VII). While oscillations may originate from the negative feedback exerted by the cyclin-dependent kinase cdk1 (cdc2) through activation of the cyclin degradation pathway [133], bistability has been shown to arise from positive feedback of cdk1 through activation of phosphatase cdc25 and inhibition of kinase wee1, which respectively activate and inhibit cdk1. This bistability, coupled to the negative feedback loop, results in repetitive cycles of hysteresis, which correspond to robust, sustained oscillations of cdk1 activity [135–137].

A striking property that is well illustrated by recently discovered examples of rhythms in genetic networks is that multiple oscillatory mechanisms can coexist in a given biological system. This is exemplified by the case of circadian rhythms discussed in Section VI and by the vertebrate segmentation clock considered in Section VIII. In the latter system, several negative feedback loops are present in the Notch or Wnt signaling pathways, providing for multiple potential sources of oscillatory behavior [165]. The reasons for such a multiplicity of oscillatory mechanisms may be manifold. One may be to provide redundancy so that back-up mechanisms are available in case of failure of one of the regulatory loops. That a feedback loop exists does not guarantee, however, that it operates in the parameter domain producing sustained oscillations. Even when multiple feedback loops are present within a system, some of them may mot be capable of producing oscillations by themselves because the values of the parameters that characterize the feedback module correspond to the evolution toward a stable steady state. Another reason for the multiplicity of instability-generating mechanisms may simply be to strengthen the oscillations by enlarging the domain of rhythmic behavior in parameter space.

As recalled in Section IX, models indicate that the interplay between several endogenous mechanisms may give rise to complex oscillations in the form of bursting or chaos. The multiplicity of oscillatory mechanisms within a given metabolic or genetic network may therefore bear, positively—as in the case of bursting oscillations or, adversely, as perhaps in the case of chaos—on the physiological functions of cellular rhythms. Even in the presence of interactions between multiple instability-generating mechanisms, however, simple periodic behavior or oscillations of the bursting type remain more common than chaos in parameter space.

In several instances, intracellular oscillations are linked to the formation of spatial or spatiotemporal patterns, as shown by the role of the vertebrate segmentation clock in somitogenesis, by the propagation of intra- or intercellular waves in Ca^{2+} signaling in many cell types, and by cAMP signaling in *Dictyostelium* amoebae. Recent observations on the occurrence of intracellular waves in activated leukocytes [191] provide yet another example of close link between temporal and spatial organization at the cellular level. The close intertwining of spatial and temporal patterns led Duboule [192] to write that "animal development is, in fact, nothing but time."

Since the initial impetus given by Ilya Prigogine to the study of oscillatory phenomena in chemical and biological systems, the number of examples of periodic behavior has grown immensely. While some examples were known for long, new ones were added to the list, which will undoubtedly continue to expand at an increasing pace. The views of Ilya Prigogine on nonequilibrium self-organization in the form of dissipative structures provide a conceptual framework that allows us to unify the multifarious rhythms that occur in biological systems with periods spanning more than 10 orders of magnitude. This global perspective underlines the links between rhythmic phenomena occurring in widely different biological settings, from genetic to metabolic and neural networks and from cell to animal populations.

Acknowledgments

I wish to dedicate this chapter to the memory of Ilya Prigogine, who triggered my interest in the modeling of rhythmic phenomena in biological systems and deeply influenced this research by the originality of his views and his unwavering enthusiasm and generosity. To this homage I wish to associate the name of Benno Hess, who did much to extend the study of nonequilibrium self-organization phenomena to the field of biological sciences. Thanks are due to members of my research group for fruitful discussions. This work was supported by grant #3.4636.04 from the *Fonds de la Recherche Scientifique Médicale* (F.R.S.M., Belgium) and by the European Union through the Network of Excellence BioSim, Contract No. LSHB-CT-2004-005137. The chapter was prepared while the author held a *Chaire Internationale de Recherche Blaise Pascal de l'Etat et de la Région Ile-de-France, gérée par la Fondation de l'Ecole Normale Supérieure* in the Institute of Genetics and Microbiology at the University of Paris-Sud 11 (Orsay, France).

References

1. W. C. Bray, A periodic chemical reaction and its mechanism. *J. Am. Chem. Soc.* **43**, 1262 (1921).

2. B. van der Pol and J. van der Markt, The heart beat considered as a relaxation oscillation, and an electrical model of the heart. *Philos. Mag.* **6**, 763–775 (1928).

3. G. Nicolis and J. Portnow, Chemical oscillations. *Chem. Rev.* **73**, 365–384 (1973).

4. R. M. Noyes and R. J. Field, Oscillatory chemical reactions. *Annu. Rev. Phys. Chem.* **25**, 95–119 (1974).

5. A. J. Lotka, Undamped oscillations derived from the law of mass action. *J. Am. Chem. Soc.* **42**, 1595 (1920).

6. V. Volterra, Fluctuations in the abundance of a species considered mathematically. *Nature* **118**, 558–560 (1926).

7. A. L. Hodgkin and A. F. Huxley, A quantitative description of membrane currents and its application to conduction and excitation in nerve. *J. Physiol. (Lond.)* **117**, 500–544 (1952).

8. A. H. Huxley, Ion movements during nerve activity. *Ann. N.Y. Acad. Sci.* **81**, 221–246 (1959).

9. A. M. Turing, The chemical basis of morphogenesis. *Philos. Trans. R. Soc. Lond. B* **237**, 37–72 (1952).

10. I. Prigogine and R. Balescu, Phénomènes cycliques dans la thermodynamique des processus irréversibles. *Bull. Cl. Sci. Acad. R. Belg.* **XLII**, 256–265 (1956).

11. J. J. Tyson, Some further studies of nonlinear oscillations in chemical systems. *J. Chem. Phys.* **58**, 3919–3930 (1973).

12. R. Lefever, G. Nicolis, and I. Prigogine, On the occurrence of oscillations around the steady state in systems of chemical reactions far from equilibrium. *J. Chem. Phys.* **47**, 1045–1047 (1967).

13. I. Prigogine, *Introduction to Thermodynamics of Irreversible Processes*, John Wiley & Sons, New York, 1967.

14. I. Prigogine, Structure, dissipation and life, in *Theoretical Physics and Biology*. M. Marois, ed., North-Holland, Amsterdam, pp. 23–52, 1969.

15. G. Nicolis and I. Prigogine, *Self-Organization in Nonequilibrium Systems. From Dissipative Structures to Order through Fluctuations*, John Wiley & Sons, New York, 1977.

16. R. Lefever, G. Nicolis, and P. Borckmans, The Brusselator: It does oscillate all the same. *J. Chem. Soc. Faraday Trans. 1* **84**, 1013–1023 (1988).

17. N. Minorsky, *Nonlinear Oscillations*, Van Nostrand, Princeton, NJ, 1962.

18. P. Glansdorff and I. Prigogine, *Thermodynamic Theory of Structure, Stability and Fluctuations*, John Wiley & Sons, New York.

19. J. Guckenheimer and P. Holmes, *Nonlinear Oscillations, Dynamical Systems, and Bifurcations of Vector Fields*, Springer, New York, 1983.

20. R. Fitzhugh, Impulses and physiological states in theoretical models of nerve membranes, *Biophys. J.* **1**, 445–466 (1961).

21. R. M. May, Limit cycles in predator–prey communities. *Science* **177**, 900–902 (1972).

22. A. M. Zhabotinsky, Periodic process of the oxidation of malonic acid in solution. Study of the kinetics of Belousov's reaction. *Biofizika* **9**, 1306 (1964).

23. R. J. Field, E. Köros, and R. M. Noyes, Oscillations in chemical systems. II. Thorough analysis of temporal oscillation in the bromate–cerium–malonic acid system. *J. Am. Chem. Soc.* **94**, 8649–8664 (1972).

24. S. Nakamura, K. Yokota, and I. Yamazaki, Sustained oscillations in a lactoperoxidase, NADPH and O_2 system. *Nature* **222** 794 (1969).

25. A. Fessard, *Propriétés Rythmiques de la Matière Vivante*, Hermann, Paris, 1936.

26. M. B. Elowitz and S. Leibler, A synthetic oscillatory network of transcriptional regulators. *Nature* **403**, 335–338 (2000).

27. R. Llinas, The intrinsic electrophysiological properties of mammalian neurons: a new insight into CNS function, *Science* **242**, 1654–1664 (1988).

28. D. DiFrancesco, Pacemaker mechanisms in cardiac tissue. *Annu. Rev. Physiol.* **55**, 455–472 (1993).

29. A. Destexhe and T. J. Sejnowski, Interactions between membrane conductances underlying thalamocortical slow wave oscillations. *Physiol. Rev* **83**, 1401–1453 (2003).

30. R. Thomas and R. d'Ari, *Biological Feedback*, CRC Press, Boca Raton, FL, 1990.

31. A. Goldbeter, *Biochemical Oscillations and Cellular Rhythms. The Molecular Bases of Periodic and Chaotic Behaviour*, Cambridge University Press, Cambridge, UK, 1996.

32. A. Goldbeter, Computational approaches to cellular rhythms. *Nature* **420**, 238–245 (2002).

33. C. Koch and I. Segev, eds., *Methods in Neuronal Modeling. From Synapses to Networks*, 2nd ed., MIT Press, Cambridge, MA, 1998.

34. J. P. Keener and J. Sneyd, *Mathematical Physiology*, Springer, New York, 1998.

35. D. Noble, Modeling the heart—from genes to cells to the whole organ. *Science* **295**, 1678–1682 (2002).

36. B. Chance, B. Schoener, and S. Elsaesser, Control of the waveform of oscillations of the reduced pyridine nucleotide level in a cell-free extract. *Proc. Natl. Acad. Sci. USA* **52**, 337–341 (1964).

37. B. Hess, K. Brand, and K. Pye, Continuous oscillations in a cell-free extract of *S. carlsbergensis*. *Biochem. Biophys. Res. Commun.* **23**, 102–108 (1966).

38. M. F. Madsen, S. Dano, and P. G. Sorensen, On the mechanisms of glycolytic oscillations in yeast. *FEBS J.* **272**, 2648–2660 (2005).

39. H. F. Chou, N. Berman, and E. Ipp, Oscillations of lactate released from islets of Langerhans: Evidence for oscillatory glycolysis in β-cells. *Am. J. Physiol.* **262**, E800–E805 (1992).

40. B. Hess and A. Boiteux, Oscillatory phenomena in biochemistry. *Annu. Rev. Biochem.* **40**, 237–258 (1971).

41. A. Goldbeter and S. R. Caplan, Oscillatory enzymes. *Annu. Rev. Biophys. Bioeng.* **5**, 449–476 (1976).

42. K. A. Reijenga, H. V. Westerhoff, B. N. Kholodenko, and J. L. Snoep, Control analysis for autonomously oscillating biochemical networks. *Biophys. J.* **82**, 99–108 (2002).

43. J. Higgins, A chemical mechanism for oscillation of glycolytic intermediates in yeast cells. *Proc. Natl. Acad. Sci. USA* **51**, 989–994 (1964).

44. E. E. Sel'kov, Self-oscillations in glycolysis. 1. A simple kinetic model. *Eur. J. Biochem.* **4**, 79–86 (1968).

45. A. Goldbeter and R. Lefever, Dissipative structures for an allosteric model. Application to glycolytic oscillations. *Biophys. J.* **12**, 1302–1315 (1972).

46. A. Boiteux, A. Goldbeter, and B. Hess, Control of oscillating glycolysis of yeast by stochastic, periodic, and steady source of substrate: a model and experimental study. *Proc. Natl. Acad. Sci. USA* **72**, 3829–3833 (1975).

47. S. Dano, P. G. Sorensen, and F. Hynne, Sustained oscillations in living cells. *Nature* **402**, 320–322 (1999).

48. D. Garfinkel and B. Hess, Metabolic control mechanisms. VII. A detailed computer model of the glycolytic pathway in ascites cells. *J. Biol. Chem.* **239**, 971–983 (1964).

49. Y. Termonia and J. Ross, Oscillations and control features in glycolysis: Numerical analysis of a comprehensive model. *Proc. Natl. Acad. Sci. USA* **78**, 2952–2956 (1981).

50. F. Hynne, S. Dano, and P. G. Sorensen, Full-scale model of glycolysis in *Saccharomyces cerevisiae*. *Biophys. Chem.* **94**, 121–163 (2001).

51. J. Wolf, J. Passarge, O. J. Somsen, J. L. Snoep, R. Heinrich, and H. V. Westerhoff, Transduction of intracellular and intercellular dynamics in yeast glycolytic oscillations. *Biophys. J.* **78**, 1145–1153 (2000).

52. P. Richard, B. M. Bakker, B. Teusink, K. Van Dam, H. V. Westerhoff, Acetaldehyde mediates the synchronization of sustained glycolytic oscillations in populations of yeast cells. *Eur. J. Biochem.* **235**, 238–241 (1996).

53. K. Tornheim, Are metabolic oscillations responsible for normal oscillatory insulin secretion? *Diabetes* **46**, 1375–1380 (1997).

54. M. G. Pedersen, R. Bertram, and A. Sherman, Intra- and inter-islet synchronization of metabolically driven insulin secretion. *Biophys. J.* **89**, 107–119 (2005).

55. K. Wierschem and R. Bertram, Complex bursting in pancreatic islets: a potential glycolytic mechanism. *J. Theor. Biol.* **228**, 513–521 (2004).

56. M. J. Berridge, Elementary and global aspects of calcium signalling. *J. Physiol. (London)* **499**, 291–306 (1997).

57. A. Fabiato, Calcium-induced release of calcium from the cardiac sarcoplasmic reticulum. *Am. J. Physiol.* **245**, C1–C14 (1983).

58. M. Wakui, Y. V. Osipchuk, and O. H. Petersen, Receptor-activated cytoplasmic Ca^{2+} spiking mediated by inositol trisphosphate is due to Ca^{2+}-induced Ca^{2+} release. *Cell* **63**, 1025–1032 (1990).

59. T. Meyer and L. Stryer, Molecular model for receptor-stimulated calcium spiking. *Proc. Natl. Acad. Sci. USA* **85**, 5051–5055 (1988).

60. A. Goldbeter, G. Dupont, and M. J. Berridge, Minimal model for signal-induced Ca^{2+} oscillations and for their frequency encoding through protein phosphorylation. *Proc. Natl. Acad. Sci. USA* **87**, 1461–1465 (1990).

61. G. W. De Young and J. Keizer, A single-pool inositol 1,4,5-trisphosphate-receptor-based model for agonist-stimulated oscillations in Ca^{2+} concentration. *Proc. Natl. Acad. Sci. USA* **89**, 9895–9899 (1992).

62. S. Schuster, M. Marhl, and T. Höfer, Modelling of simple and complex calcium oscillations. From single-cell responses to intercellular signalling. *Eur. J. Biochem.* **269**, 1333–1355 (2002).

63. G. Dupont and A. Goldbeter, Properties of intracellular Ca^{2+} waves generated by a model based on Ca^{2+}-induced Ca^{2+} release. *Biophys. J.* **67**, 2191–2204 (1994).

64. J. Sneyd, A. C. Charles, and M. J. Sanderson, A model for the propagation of intercellular calcium waves. *Am. J. Physiol.* **266**, C293–C302 (1994).

65. G. Dupont, T. Tordjmann, C. Clair, S. Swillens, M. Claret, and L. Combettes, Mechanism of receptor-oriented intercellular calcium wave propagation in hepatocytes. *FASEB J.* **14**, 279–289 (2000).

66. S. Swillens, G. Dupont, L. Combettes, and P. Champeil, From calcium blips to calcium puffs: Theoretical analysis of the requirements for interchannel communication. *Proc. Natl. Acad. Sci. USA* **96**, 13750–13755 (1999).

67. N. C. Spitzer, N. J. Lautermilch, R. D. Smith, and T. M. Gomez, Coding of neuronal differentiation by calcium transients. *BioEssays* **22**, 811–817 (2000).

68. P. De Koninck and H. Schulman, Sensitivity of CaM kinase II to the frequency of Ca^{2+} oscillations. *Science* **279**, 227–230 (1998).

69. G. Dupont, G. Houart, and P. De Koninck, Sensitivity of CaM kinase II to the frequency of Ca^{2+} oscillations: A simple model. *Cell Calcium* **34**, 485–497 (2003).

70. Y. V. Gorbunova and N. C. Spitzer, Dynamic interactions of cyclic AMP transients and spontaneous Ca^{2+} spikes. *Nature* **418**, 93–96 (2002).

71. G. Dupont and L. Combettes, Modelling the effect of specific inositol 1,4,5-trisphosphate receptor isoforms on cellular Ca^{2+} signals. *Biol. Cell* **98**, 171–182 (2006).

72. J. Sneyd, K. Tsaneva-Atanasova, V. Reznikov, Y. Bai, M. J. Sanderson, and D. I. Yule, A method for determining the dependence of calcium oscillations on inositol trisphosphate oscillations. *Proc. Natl. Acad. Sci. USA* **103**, 1675–1680 (2006).

73. M. Beltramello, V. Piazza, F. F. Bukauskas, T. Pozzan, and F. Mammano, Impaired permeability to Ins(1,4,5)P3 in a mutant connexin underlies recessive hereditary deafness. *Nat. Cell. Biol.* **7**, 63–69 (2005).

74. P. Uhlen, P. M. Burch, C. I. Zito, M. Estrada, B. E. Ehrlich, and A. M. Bennett, Gain-of-function/Noonan syndrome SHP-2/Ptpn11 mutants enhance calcium oscillations and impair NFAT signaling. *Proc. Natl. Acad. Sci. USA* **103**, 2160–2165 (2006).

75. D. Dormann, J. Y. Kim, P. Devreotes, and C. J. Weijer, cAMP receptor affinity controls wave dynamics, geometry and morphogenesis in *Dictyostelium*. *J. Cell. Sci.* **114**, 2513–2523 (2001).

76. J. L. Martiel and A. Goldbeter, A model based on receptor desensitization for cyclic AMP signaling in *Dictyostelium* cells. *Biophys. J.* **52**, 807–828 (1987).

77. Y. Tang and H. G. Othmer, Excitation, oscillations and wave propagation in a G-protein-based model of signal transduction in *Dictyostelium discoideum*. *Philos. Trans. R. Soc. Lond. B. Biol. Sci.* **349**, 179–195 (1995).

78. Y. X. Li and A. Goldbeter, Frequency encoding of pulsatile signals of cyclic AMP based on receptor desensitization in *Dictyostelium* cells. *J. Theor. Biol.* **146**, 355–367 (1990).

79. A. Goldbeter and L. A. Segel, Control of developmental transitions in the cyclic AMP signaling system of *Dictyostelium discoideum*. *Differentiation* **17**, 127–135 (1980).

80. H. Levine, I. Aranson, L. Tsimring, and T. V. Truong, Positive genetic feedback governs cAMP spiral wave formation in *Dictyostelium*. *Proc. Natl. Acad. Sci. USA* **93**, 6382–6386 (1996).

81. E. Palsson and E. C. Cox, Origin and evolution of circular waves and spirals in *Dictyostelium discoideum* territories. *Proc. Natl. Acad. Sci. USA* **93**, 1151–1155 (1996).

82. J. Lauzeral, J. Halloy, and A. Goldbeter, Desynchronization of cells on the developmental path triggers the formation of spiral waves of cAMP during *Dictyostelium* aggregation. *Proc. Natl. Acad. Sci. USA* **94**, 9153–9158 (1997).

83. T. Bretschneider, F. Siegert, and C. J. Weijer, Three-dimensional scroll waves of cAMP could direct cell movement and gene expression in *Dictyostelium* slugs. *Proc. Natl. Acad. Sci. USA* **92**, 4387–4391 (1995).

84. M. T. Laub and W. F. Loomis, A molecular network that produces spontaneous oscillations in excitable cells of *Dictyostelium*. *Mol. Biol. Cell* **9**, 3521–3532 (1998).

85. M. Maeda, S. Lu, G. Shaulsky, Y. Miyazaki, H. Kuwayama, Y. Tanaka, A. Kuspa, and W. F. Loomis, Periodic signaling controlled by an oscillatory circuit that includes protein kinases ERK2 and PKA. *Science* **304**, 875–878 (2004).

86. S. Sawai, P. A. Thomason, and E. C. Cox, An autoregulatory circuit for long-range self-organization in *Dictyostelium* cell populations. *Nature* **433**, 323–326 (2005).

87. M. Jacquet, G. Renault, S. Lallet, J. De Mey, A. Goldbeter, Oscillatory nucleocytoplasmic shuttling of the general stress response transcriptional activators Msn2 and Msn4 in *Saccharomyces cerevisiae*. *J. Cell Biol.* **161**, 497–505 (2003).

88. D. J. Chadwick and J. A. Goode, eds., *Mechanisms and Biological Significance of Pulsatile Hormone Secretion. Novartis Foundation Symposium* 227, John Wiley & Sons, Chichester, UK, 2000.

89. E. Knobil, Patterns of hormone signals and hormone action. *N. Engl. J. Med.* **305**, 1582–1583 (1981).

90. P. E. Belchetz, T. M. Plant, Y. Nakai, E. J. Keogh, and E. Knobil, Hypophysial responses to continuous and intermittent delivery of hypothalamic gonadotropin-releasing hormone. *Science* **202**, 631–633 (1978).

91. P. C. Hindmarsh, R. Stanhope, M. A. Preece, and C. G. D. Brook, Frequency of administration of growth hormone—An important factor in determining growth response to exogenous growth hormone. *Horm. Res.* **33**(Suppl. 4), 83–89 (1990).

92. Y. X. Li and A. Goldbeter, Frequency specificity in intercellular communication: The influence of patterns of periodic signalling on target cell responsiveness. *Biophys. J.* **55**, 125–145 (1989).

93. C. Wagner, S. R. Caplan, and G. S. Tannenbaum, Genesis of the ultradian rhythm of GH secretion: A new model unifying experimental observations in rats. *Am. J. Physiol.* **275**, E1046–1054 (1998).

94. L. W. Maki and J. Keizer, Mathematical analysis of a proposed mechanism for oscillatory insulin secretion in perifused HIT-15 cells. *Bull. Math. Biol.* **57**, 569–591 (1995).

95. J. C. Dunlap, Molecular bases for circadian clocks. *Cell* **96**, 271–290 (1999).

96. M. W. Young and S. A. Kay, Time zones: A comparative genetics of circadian clocks. *Nat. Rev. Genet.* **2**, 702–715 (2001).

97. S. M. Reppert and D. R. Weaver, Coordination of circadian timing in mammals. *Nature* **418**, 935–941 (2002).

98. P. E. Hardin, The circadian timekeeping system of *Drosophila. Curr. Biol.* **15**, R714–R722 (2005).

99. D. Bell-Pedersen, V. M. Cassone, D. J. Earnest, S. S. Golden, P. E. Hardin, T. L. Thomas, and M. J. Zoran, Circadian rhythms from multiple oscillators: Lessons from diverse organisms. *Nat. Rev. Genet.* **6**, 544–556 (2005).

100. P. E. Hardin, J. C. Hall, and M. Rosbash, Feedback of the *Drosophila* period gene product on circadian cycling of its messenger RNA levels. *Nature* **343**, 536–540 (1990).

101. M. Nakajima, K. Imai, H. Ito, T. Nishiwaki, Y. Murayama, H. Iwasaki, T. Oyama, and T. Kondo, Reconstitution of circadian oscillation of cyanobacterial KaiC phosphorylation *in vitro. Science* **308**, 414–415 (2005).

102. R. E. Kronauer, D. B. Forger, and M. E. Jewett, Quantifying human circadian pacemaker response to brief, extended, and repeated light stimuli over the phototopic range. *J. Biol. Rhythms* **14**, 500–515 (1999).

103. Y., Ouyang, C. R. Andersson, T. Kondo, S. S. Golden, and C. H. Johnson. Resonating circadian clocks enhance fitness in cyanobacteria. *Proc. Natl. Acad. Sci. USA* **95**, 8660–8664 (1998).

104. D. Gonze, M. Roussel, and A. Goldbeter, A model for the enhancement of fitness in cyanobacteria based on resonance of a circadian oscillator with the external light–dark cycle. *J. Theor. Biol.* **214**, 577–597 (2002).

105. B. C. Goodwin, Oscillatory behavior in enzymatic control processes. *Adv. Enzyme Regul.* **3**, 425–438 (1965).

106. P. Ruoff, M. Vinsjevik, C. Monnerjahn, and L. Rensing, The Goodwin model: Simulating the effect of light pulses on the circadian sporulation rhythm of *Neurospora crassa. J. Theor. Biol.* **209**, 29–42 (2001).

107. A. Goldbeter, A model for circadian oscillations in the *Drosophila* period protein (PER). *Proc. R. Soc. Lond. B Biol. Sci.* **261**, 319–324 (1995).

108. J. C. Leloup and A. Goldbeter, A model for circadian rhythms in *Drosophila* incorporating the formation of a complex between the PER and TIM proteins. *J. Biol. Rhythms* **13**, 70–87 (1998).

109. U. Schibler and P. Sassone-Corsi, A web of circadian pacemakers. *Cell* **111**, 919–922 (2002).

110. N. R. Glossop, L. C. Lyons, and P. E. Hardin, Interlocked feedback loops within the *Drosophila* circadian oscillator. *Science* **286**, 766–768 (1999).

111. L. P. Shearman, S. Sriram, D. R. Weaver, E. S. Maywood, I. Chaves, B. Zheng, K. Kume, C. C. Lee, G. T. van der Horst, M. H. Hastings, and S. M. Reppert, Interacting molecular loops in the mammalian circadian clock. *Science* **288**, 1013–1019 (2000).

112. H. R. Ueda, M. Hagiwara, and H. Kitano, Robust oscillations within the interlocked feedback model of *Drosophila* circadian rhythm. *J. Theor. Biol.* **210**, 401–406 (2001).

113. P. Smolen, D. A. Baxter, and J. H. Byrne, Modeling circadian oscillations with interlocking positive and negative feedback loops. *J. Neurosci.* **21**, 6644–6656 (2001).

114. J. C. Leloup and A. Goldbeter, Toward a detailed computational model for the mammalian circadian clock. *Proc. Natl. Acad. Sci. USA* **100**, 7051–7056 (2003).

115. D. B. Forger and C. S. Peskin, A detailed predictive model of the mammalian circadian clock. *Proc. Natl. Acad. Sci. USA* **100**, 14806–14811 (2003).

116. M. Doi, J. Hirayama, and P. Sassone-Corsi, Circadian regulator CLOCK is a histone acetyltransferase. *Cell* **125**, 497–508 (2006).

117. G. S. Richardson and H. V. Malin, Circadian rhythm sleep disorders: Pathophysiology and treatment. *J. Clin. Neurophysiol.* **13**, 17–31 (1996).

118. K. L. Toh, C. R. Jones, Y. He, E. J. Eide, W. A. Hinz, D. M. Virshup, L. J. Ptacek, and Y. H. Fu, An hPer2 phosphorylation site mutation in familial advanced sleep phase syndrome. *Science* **291**, 1040–1043 (2001).

119. Y. Xu, Q. S. Padiath, R. E. Shapiro, C. R. Jones, S. C. Wu, N. Saigoh, K. Saigoh, L. J. Ptacek, and Y. H. Fu, Functional consequences of a CKIdelta mutation causing familial advanced sleep phase syndrome. *Nature* **434**, 640–644 (2005).

120. T. Ebisawa, M. Uchiyama, N. Kajimura, K. Mishima, Y. Kamei, M. Katoh, T. Watanabe, M. Sekimoto, K. Shibui, K. Kim, Y. Kudo, Y. Ozeki, M. Sugishita, R. Toyoshima, Y. Inoue, N. Yamada, T. Nagase, N. Ozaki, O. Ohara, N. Ishida, M. Okawa, K. Takahashi, and T. Yamauchi, Association of structural polymorphisms in the human *period3* gene with delayed sleep phase syndrome. *EMBO Rep.* **2**, 342–346 (2001).

121. C. R. Jones, S. S. Campbell, S. E. Zone, F. Cooper, A. DeSano, P. J. Murphy, B. Jones, L. Czajkowski, and L. J. Ptacek, Familial advanced sleep-phase syndrome: A short-period circadian rhythm variant in humans. *Nat. Med.* **5**, 1062–1065 (1999).

122. L. Glass and M. C. Mackey, *From Clocks to Chaos: The Rhythms of Life*, Princeton University Press, Princeton, NJ, 1988.

123. M. C. Mackey and J. G. Milton, Dynamical diseases. *Ann NY Acad. Sci.* **504**, 16–32 (1987).

124. M. Nagano, A. Adachi, K. Nakahama, T. Nakamura, M. Tamada, E. Meyer-Bernstein, A. Sehgal, and Y. Shigeyoshi, An abrupt shift in the day/night cycle causes desynchrony in the mammalian circadian center. *J. Neurosci.* **23**, 6141–6151 (2003).

125. A. T. Winfree, *The Geometry of Biological Time*, 2nd ed., Springer, New York, 2001.

126. J. C. Leloup and A. Goldbeter, A molecular explanation for the long-term suppression of circadian rhythms by a single light pulse. *Am. J. Physiol. Regul. Integr. Comp. Physiol.* **280**, R1206–R1212 (2001).

127. N. Barkai and S. Leibler, Circadian clocks limited by noise. *Nature* **403**, 267–268 (2000).

128. D. Gonze, J. Halloy, and A. Goldbeter, Robustness of circadian rhythms with respect to molecular noise. *Proc. Natl. Acad. Sci. USA* **99**, 673–678 (2002).

129. D. B. Forger and C. S. Peskin, Stochastic simulation of the mammalian circadian clock. *Proc. Natl. Acad. Sci. USA* **102**, 321–324 (2005).

130. J. M. Vilar, H. Y. Kueh, N. Barkai, and S. Leibler, Mechanisms of noise-resistance in genetic oscillators. *Proc. Natl. Acad. Sci. USA* **99**, 5988–5992 (2002).

131. A. W. Murray and T. Hunt, *The Cell Cycle: An Introduction*, Oxford University Press, Oxford, 1993.

132. A. W. Murray and M. W. Kirschner, Cyclin synthesis drives the early embryonic cell cycle. *Nature* **339**, 275–280 (1989).

133. A. Goldbeter, A minimal cascade model for the mitotic oscillator involving cyclin and cdc2 kinase. *Proc. Natl. Acad. Sci. USA* **88**, 9107–9111 (1991).

134. A. Goldbeter, Modeling the mitotic oscillator driving the cell division cycle. *Comments Theor. Biol.* **3**, 75–107 (1993).

135. J. R. Pomerening, E. D. Sontag, and J. E. Ferrell, Jr. Building a cell cycle oscillator: Hysteresis and bistability in the activation of Cdc2. *Nat. Cell Biol.* **5**, 346–351 (2003).

136. W. Sha, J. Moore, K. Chen, A. D. Lassaletta, C. S. Yi, J. J. Tyson, and J. C. Sible, Hysteresis drives cell-cycle transitions in *Xenopus laevis* egg extracts. *Proc. Natl. Acad. Sci. USA* **100**, 975–980 (2003).

137. J. R. Pomerening, S. Y. Kim, and J. E. Ferrell, Jr. Systems-level dissection of the cell-cycle oscillator: Bypassing positive feedback produces damped oscillations. *Cell* **122**, 565–578 (2005).

138. B. Novak and J. J. Tyson, Numerical analysis of a comprehensive model of M-phase control in *Xenopus* oocyte extracts and intact embryos. *J. Cell Sci.* **106**, 1153–1168 (1993).

139. B. Novak and J. J. Tyson, A model for restriction point control of the mammalian cell cycle. *J. Theor. Biol.* **230**, 563–579 (2004).

140. J. J. Tyson and B. Novak, Regulation of the eukaryotic cell cycle: Molecular antagonism, hysteresis, and irreversible transitions. *J. Theor. Biol.* **210**, 249–263 (2001).

141. J. J. Tyson, K. Chen, and B. Novak, Network dynamics and cell physiology. *Nat. Rev. Mol. Cell Biol.* **2**, 908–916 (2001).

142. K. C. Chen, L. Calzone, A. Csikasz-Nagy, F. R. Cross, B. Novak, and J. J. Tyson, Integrative analysis of cell cycle control in budding yeast. *Mol. Biol. Cell* **15**, 3841–3862 (2004).

143. M. Swat, A. Kel, and H. Herzel, Bifurcation analysis of the regulatory modules of the mammalian G1/S transition. *Bioinformatics* **20**, 1506–1511 (2004).

144. B. Novak, Z. Pataki, A. Ciliberto, and J. J. Tyson, Mathematical model of the cell division cycle of fission yeast. *Chaos* **11**, 277–286 (2001).

145. C. Gérard and A. Goldbeter, In preparation.

146. A. W. Murray and M. W. Kirschner, Dominoes and clocks: The union of two views of the cell cycle. *Science* **246**, 614–621 (1989).

147. J. Halloy, B. A. Bernard, G. Loussouarn, and A. Goldbeter, Modeling the dynamics of human hair cycles by a follicular automaton. *Proc. Natl. Acad. Sci. USA* **97**, 8328–8333 (2000).

148. A. Altinok and A. Goldbeter, In preparation.

149. T. Matsuo, S. Yamaguchi, S. Mitsui, A. Emi, F. Shimoda, and H. Okamura, Control mechanism of the circadian clock for timing of cell division *in vivo*. *Science* **302**, 255–259 (2003).

150. R. R. Klevecz, J. Bolen, G. Forrest, and D. B. Murray, A genomewide oscillation in transcription gates DNA replication and cell cycle. *Proc. Natl. Acad. Sci. USA* **101**, 1200–1205 (2004).

151. M. W. Young, An ultradian clock shapes genome expression in yeast. *Proc. Natl. Acad. Sci. USA* **101**, 1118–1119 (2004).

152. R. Metivier, G. Penot, M. R. Hubner, G. Reid, H. Brand, M. Kos, and F. Gannon, Estrogen receptor-alpha directs ordered, cyclical, and combinatorial recruitment of cofactors on a natural target promoter. *Cell* **115**, 751–763 (2003).

153. R. Lev Bar-Or, R. Maya, L. A. Segel, U. Alon, A. J. Levine, and M. Oren, Generation of oscillations by the p53-Mdm2 feedback loop: A theoretical and experimental study. *Proc. Natl. Acad. Sci. USA* **97**, 11250–11255 (2000).

154. G. Lahav, N. Rosenfeld, A. Sigal, N. Geva-Zatorsky, A. J. Levine, M. B. Elowitz, and U. Alon, Dynamics of the p53-Mdm2 feedback loop in individual cells. *Nat. Genet.* **36**, 147–150 (2004).

155. L. Ma, J. Wagner, J. J. Rice, W. Hu, A. J. Levine, and G. A. Stolovitzky, A plausible model for the digital response of p53 to DNA damage. *Proc. Natl. Acad. Sci. USA* **102**, 14266–14271 (2005).

156. A. Ciliberto, B. Novak, and J. J. Tyson, Steady states and oscillations in the p53/Mdm2 network. *Cell Cycle* **4**, 488–493 (2005).

157. A. Hoffmann, A. Levchenko, M. L. Scott, and D. Baltimore, The IkappaB-NF-kappaB signaling module: Temporal control and selective gene activation. *Science* **298**, 1241–1245 (2002).

158. D. E. Nelson, A. E. Ihekwaba, M. Elliott, J. R. Johnson, C. A. Gibney, B. E. Foreman, G. Nelson, V. See, C. A. Horton, D. G. Spiller, S. W. Edwards, H. P. McDowell, J. F. Unitt, E. Sullivan, R. Grimley, N. Benson, D. Broomhead, D. B. Kell, and M. R. White, Oscillations in NF-kappaB signaling control the dynamics of gene expression. *Science* **306**, 704–708 (2004).

159. D. Barken, C. J. Wang, J. Kearns, R. Cheong, A. Hoffmann, and A. Levchenko, Comment on "Oscillations in NF-kappaB signaling control the dynamics of gene expression." *Science* **308**, 52 (2005).

160. J. Cooke and E. C. Zeeman, A clock and wavefront model for control of the number of repeated structures during animal morphogenesis. *J. Theor. Biol.* **58**, 455–476 (1976).

161. I. Palmeirim, D. Henrique, D. Ish-Horowicz, and O. Pourquié, Avian *hairy* gene expression identifies a molecular clock linked to vertebrate segmentation and somitogenesis. *Cell* **91**, 639–648 (1997).

162. M. Maroto and O. Pourquié, A molecular clock involved in somite segmentation. *Curr. Top. Dev. Biol.* **51**, 221–248 (2001).

163. O. Pourquié, The segmentation clock: Converting embryonic time into spatial pattern. *Science* **301**, 328–330 (2003).

164. Y. Bessho and R. Kageyama, Oscillations, clocks and segmentation. *Curr. Opin. Genet. Dev.* **13**, 379–384 (2003).

165. F. Giudicelli and J. Lewis, The vertebrate segmentation clock. *Curr. Opin. Genet. Dev.* **14**, 407–414 (2004).

166. J. K. Dale, M. Maroto, M. L. Dequeant, P. Malapert, M. McGrew, and O. Pourquie, Periodic notch inhibition by lunatic fringe underlies the chick segmentation clock. *Nature* **421**, 275–278 (2003).

167. A. Goldbeter and O. Pourquié, Unpublished results.

168. A. Aulehla, C. Wehrle, B. Brand-Saberi, R. Kemler, A. Gossler, B. Kanzler, and B. G. Herrmann, Wnt3a plays a major role in the segmentation clock controlling somitogenesis. *Dev. Cell.* **4**, 395–406 (2003).

169. J. Lewis, Autoinhibition with transcriptional delay: A simple mechanism for the zebrafish somitogenesis oscillator. *Curr. Biol.* **13**, 1398–1408 (2003).

170. N. A. Monk, Oscillatory expression of Hes1, p53, and NF-kappaB driven by transcriptional time delays. *Curr. Biol.* **13**, 1409–1413 (2003).

171. H. Hirata, S. Yoshiura, T. Ohtsuka, Y. Bessho, T. Harada, K. Yoshikawa, and R. Kageyama, Oscillatory expression of the bHLH factor Hes1 regulated by a negative feedback loop. *Science* **298**, 840–843 (2002).

172. Y. Masamizu, T. Ohtsuka, Y. Takashima, H. Nagahara, Y. Takenaka, K. Yoshikawa, H. Okamura, and R. Kageyama, Real-time imaging of the somite segmentation clock: Revelation of unstable oscillators in the individual presomitic mesoderm cells. *Proc. Natl. Acad. Sci. USA* **103**, 1313–1318 (2006).

173. J. Dubrulle, M. J. McGrew, and O. Pourquié, FGF signaling controls somite boundary position and regulates segmentation clock control of spatiotemporal Hox gene activation. *Cell* **106**, 219–232 (2001).

174. J. Dubrulle and O. Pourquié, fgf8 mRNA decay establishes a gradient that couples axial elongation to patterning in the vertebrate embryo. *Nature* **427**, 419–422 (2004).

175. A. Goldbeter, D. Gonze, and O. Pourquié, Sharp developmental thresholds defined through bistability by antagonistic gradients of retinoic acid and FGF signaling. Submitted.

176. R. Diez del Corral and K. G. Storey, Opposing FGF and retinoid pathways: A signalling switch that controls differentiation and patterning onset in the extending vertebrate body axis. *BioEssays* **26**, 857–869 (2004).

177. T. S. Gardner, C. R. Cantor, and J. J. Collins, Construction of a genetic toggle switch in *Escherichia coli. Nature* **403**, 339–342 (2000).

178. C. Garmendia-Torres, A. Goldbeter, and M. Jacquet, Nucleocytoplasmic oscilliations of the transcription factor Msn2 in yeast are mediated by phosphorylation of its NLS by protein kinase A. Submitted.

179. E. A. Vitalis, J. L. Costantin, P. S. Tsai, H. Sakakibara, S. Paruthiyil, T. Iiri, J. F. Martini, M. Taga, A. L. H. Choi, A. C. Charles, and R. I. Weiner, Role of the cAMP signaling pathway in the regulation of gonadotropin-releasing hormone secretion in GT1 cells. *Proc. Natl. Acad. Sci. USA* **97**, 1861–1866 (2000).

180. J. Rinzel, A formal classification of bursting mechanisms in excitable systems. *Lect. Notes Biomath* **71**, 267–281 (1987).

181. L. F. Olsen and H. Degn, Chaos in biological systems. *Q. Rev. Biophys.* **18**, 165–225 (1985).

182. O. Decroly and A. Goldbeter, Birhythmicity, chaos, and other patterns of temporal self-organization in a multiply regulated biochemical system. *Proc. Natl. Acad. Sci. USA* **79**, 6917–6921 (1982).

183. D. Gonze and A. Goldbeter, Entrainment versus chaos in a model for a circadian oscillator driven by light–dark cycles. *J. Stat. Phys.* **101**, 649–663 (2000).

184. A. Goldbeter, D. Gonze, G. Houart, J. C. Leloup, J. Halloy, and G. Dupont, From simple to complex oscillatory behavior in metabolic and genetic control networks. *Chaos* **11**, 247–260 (2001).

185. P. Shen and R. Larter, Chaos in intracellular Ca^{2+} oscillations in a new model for nonexcitable cells. *Cell Calcium* **17**, 225–232 (1995).

186. G. Houart, G. Dupont, and A. Goldbeter, Bursting, chaos and birhythmicity originating from self-modulation of the inositol 1,4,5-triphosphate signal in a model for intracellular Ca^{2+} oscillations. *Bull. Math. Biol.* **61**, 507–530 (1999).

187. U. Kummer, L. F. Olsen, C. J. Dixon, A. K. Green, E. Bornberg-Bauer, and G. Baier, Switching from simple to complex oscillations in calcium signaling. *Biophys. J.* **79**, 1188–1195 (2000).

188. K. Nielsen, P. G. Sörensen, and F. Hynne, Chaos in glycolysis. *J. Theor. Biol.* **186**, 303–306 (1997).

189. I. Prigogine, R. Lefever, A. Goldbeter, and M. Herschkowitz-Kaufman, Symmetry-breaking instabilities in biological systems. *Nature* **223**, 913–916 (1969).

190. J. Garcia-Ojalvo, M. B. Elowitz, and S. H. Strogatz, Modeling a synthetic multicellular clock: Repressilators coupled by quorum sensing. *Proc. Natl. Acad. Sci. USA* **101**, 10955–10960 (2004).

191. H. R. Petty, Neutrophil oscillations: Temporal and spatiotemporal aspects of cell behavior. *Immunol. Res.* **23**, 85–94 (2001).

192. D. Duboule, Time for chronomics? *Science* **301**, 277 (2003).

AUTHOR INDEX

Numbers in parentheses are reference numbers and indicate that the author's work is referred to although his name is not mentioned in the text. Numbers in *italic* show the pages on which the complete references are listed.

Special Volume in Memory of Ilya Prigogine: Advances in Chemical Physics, Volume 135, edited by Stuart A. Rice
Copyright © 2007 John Wiley & Sons, Inc.

SUBJECT INDEX

Special Volume in Memory of Ilya Prigogine: Advances in Chemical Physics, Volume 135, edited by Stuart A. Rice
Copyright © 2007 John Wiley & Sons, Inc.